Turfgrass Installation
Management and Maintenance

Turfgrass Installation
Management and Maintenance

Rodney Johns

McGraw-Hill

New York Chicago San Francisco Lisbon London Madrid
Mexico City Milan New Delhi San Juan Seoul
Singapore Sydney Toronto

The McGraw·Hill Companies

Copyright © 2004 by The McGraw-Hill Companies, Inc. All rights reserved. Printed in the United States of America. Except as permitted under the United States Copyright Act of 1976, no part of this publication may be reproduced or distributed in any form or by any means, or stored in a data base or retrieval system, without the prior written permission of the publisher.

1 2 3 4 5 6 7 8 9 0 DOC/DOC 0 9 8 7 6 5 4 3

ISBN 0-07-141008-2

The sponsoring editor for this book was Larry S. Hager and the production supervisor was Pamela A. Pelton. It was set in Stone Sans by Lone Wolf Enterprises, Ltd.

Printed and bound by RR Donnelley.

McGraw-Hill books are available at special quantity discounts to use as premiums and sales promotions, or for use in corporate training programs. For more information, please write to the Director of Special Sales, McGraw-Hill Professional, Two Penn Plaza, New York, NY 10121-2298. Or contact your local bookstore.

 This book is printed on recycled, acid-free paper containing a minimum of 50% recycled, de-inked fiber.

Information contained in this work has been obtained by The McGraw-Hill Companies, Inc. ("McGraw-Hill") from sources believed to be reliable. However, neither McGraw-Hill nor its authors guarantee the accuracy or completeness of any information published herein and neither McGraw-Hill nor its authors shall be responsible for any errors, omissions, or damages arising out of use of this information. This work is published with the understanding that McGraw-Hill and its authors are supplying information but are not attempting to render engineering or other professional services. If such services are required, the assistance of an appropriate professional should be sought.

DEDICATION

To Taylor Kaye Johns, my daughter.

ABOUT THE AUTHOR

Rodney Johns has a bachelor's degree in Plant Science from the University of Missouri at Columbia, as well as an AS in Business and Turf and a greenhouse management diploma. He has nearly 10 years of experience in the golf course industry, and was a superintendent of a golf course for 4 years.

Mr. Johns also has spent nearly 15 years in the lawn and landscaping industry. He currently owns Arki-Tec Landscaping and Sales LLC, a company that offers complete landscaping and lawn care services. This business also involves a retail garden center full of equipment and supplies. *www.arki-tec.com* is one of the top websites in the world for used lawn and garden equipment.

CONTENTS

CHAPTER 1
Plant Biology 1
Introduction to Soils 1
Soil pH and Fertility 3
Soil Types 9

CHAPTER 2
Plant Biology 17
Plant Responses 17
Turfgrass Plant Anatomy 18
Seeding Structures 22
Plant Roles in the Ecosystem 27
Turfgrass Environment 28

CHAPTER 3
Turfgrass Species 33
Turfgrass Zones and Species 33
    Cool-Season Grasses 36
        Kentucky Bluegrass 36
        Rough-stalked Bluegrass 42
        Annual Bluegrass 42
        Perennial Ryegrass 43

    Annual Ryegrass *45*
    Tall Fescue. *46*
    Fine Fescues *48*
    Creeping Bentgrass. *49*
  Warm-Season Grasses 52
    Zoysia Grass *52*
    Bermuda Grass *55*
    St. Augustine Grass *57*
    Centipede Grass *58*
    Buffalo Grass. *59*

CHAPTER 4
Careers in Turfgrass 61
Golf Course Careers. 63
Lawn Care Professions. 68
Other Areas for Employment. 73

CHAPTER 5
Cultural Practices 81
Mowing 81
  Golf Course Mowing. 82
    Mowing Areas. *84*
Fertilizers. 88
  Fertilizer Types. 88
  Application Methods. 94
    Cool-Season Fertilization *95*
    Warm-Season Fertilization *100*
Irrigation. 101
Integrated Pest Management 110

CHAPTER 6
Turf Establishment.............................113
Seeding and Reestablishment of Turfgrass...................113
    Is Renovation the Answer?.............................114
    New Installations......................................125
Sodding 138
Establishment of Warm-Season Grasses.....................141

CHAPTER 7
Residential and Homeowner Lawn Care Business....145
Finding and Servicing the Customer 148
Maintenance Schedule 153
Billing and Invoicing 156
Charges and Finance....................................159
Appearance...161
Special Residential Cultural Practices 163

CHAPTER 8
Commercial Lawn Care 167
Bids and Contracts 171
On the Job 178
Safety 180
Licenses and Taxes 183
Pesticide Treatments 186

CHAPTER 9
Turf in the Landscape 191

CHAPTER 10
Irrigation 209
by Bruce Carleton

Basic System Hardware 209
    Controllers. 211
    Valves 213
    Anti-Siphon Protection 219
    Emission Devices 221
Maintaining a Commerical System 223
    A Well-Manicured System 223
    Drive-by Maintenance 224
    Get on the Conservation Bandwagon 226
Irrigation Management 229
    A Matter of Semantics. 229
    Managing a System Wisely 232
    Got Dirt?. 233
    Scheduling Irrigation. 236
    You Can't Manage What You Can't Measure. 237
    Time is on Your Side 240

CHAPTER 11
Golf Course Management 243
Course Maintenance 247
Budgeting and Finance 256

CHAPTER 12
The Putting Green 263
by Danny Quast and Wayne Otto

The Art and Science of Turfgrass Management 263
    Science 264

  Mowing. *264*
  Irrigation *265*
  Nutrition *266*
  Cultivation. *267*
  Pest Control *267*
  Topdressing. *268*
About Green Speed. 269
Cultural Practices and Circumstances That Influence Green Speed . 269
  Mowing. 269
  Rolling. 271
  Soil Moisture 273
  Nutrition 273
  Plant Growth Regulators 276
Winter Play? 277

CHAPTER 13
Fairways **279**
  by Danny Quast and Wayne Otto
The Importance of Fairways. 279
Fairway Conversion 280
  Reasons for Fairway Conversion 281
  Benefits of Fairway Conversion 281
  Equipment and Supplies 283
  Table Steps 284
Maintenance 288
  Bentgrass Fairway Care 289
  Bluegrass Fairways 290
  Maintenance of Poa Annua Fairways at $^1/_2$ inch 291
    Winter Damage. *291*
    Spring *293*

    Mowing. *293*
    Spring Aerification of Poa Annua Fairways. *296*
    Herbicide Use on Poa Annua Fairways. *296*
    Watering of Poa Annua Fairways. *297*
  Damage to Poa Annua *298*
    Anthracnose (Colletotrichum graminicola) *299*
    Dollar Spot (Sclerotinia homoeocarpa) *300*
    Pythium Blight. *300*
    Summer Patch (Magnaporthe poae) *301*
    Insect Problems. *302*

CHAPTER 14
The Teeing Ground. 305
    by Danny Quast and Wayne Otto
The Teeing Ground Surface. 305
Maintenance and Cultural Practices. 305
    Mowing. 306
    Irrigation 307
    Nutrition 307
    Cultivation and Topdressing. 308
    Pest Control 310
Practice Tee. 311

CHAPTER 15
Topdressing 315
    by Danny Quast and Wayne Otto
Introduction 315
Effective Sand Topdressing Programs 317
Benefits of Topdressing Putting Green Surfaces. 318
Incorporating Sand into the Turf Canopy 321

CHAPTER 16
Soil and Water Testing.........................325
by Danny Quast and Wayne Otto

Soil 325
   Soil as a Source of Plant Food 326
   Soil Sampling 327
Irrigation Water Quality.. 328
   Salt-Related Problems 331
      Characteristics of Salt-Affected Soils.................. 332
      High pH, Carbonates, and Bicarbonates 334
      Sodium Absorption Ratio (SAR)....................... 335
      Electrical Conductivity (EC) 335
   Ion Toxicity in Water... 336
Long-Term Effects of Irrigation Water on Soils................ 336

CHAPTER 17
Pesticide Application.........................339

Testing and Licensing 342
Safety 344
Types of Pesticides.. 356
Application Equipment 360
Disposal of Chemicals 364

CHAPTER 18
Pest Management for Turfgrass 371

Weed Control 373
Disease Control 378
Insect Control 383
Other Pest Problems in Turfgrass................................ 388

CHAPTER 19
Weeds Found in Turf and Landscapes 395
by Matt Fagerness

Weed Life Cycles: Annuals, Biennials, and Perennials............ 398
Weed Biology: Broadleaf Weeds, Grasses, and Sedges........... 418
Broadleaf Weeds Commonly Found in Turf and Landscapes 421
Grassy Weeds and Sedges Commonly Found in Turf
and Landscapes.. 443
Weed Ecology and Cultural Ways of Controlling Weeds 467

CHAPTER 20
Basic Chemical and Pesticide Principles............. 475
by Matt Fagerness

Pesticide Formulations...................................... 478
Chemical and Pesticide Safety Information—Using an MSDS Form . 504
Understanding the Chemical/Pesticide Label.................. 516
Environmental Impact of Chemicals and Pesticides 521
    Alternative Fates of a Pesticide........................... 524
    Pesticide Mobility 526
    Economic and Aesthetic Pest Thresholds 528

CHAPTER 21
Disease Problems and Fungicides Used in Turfgrass. . 537
by Matt Fagerness

Disease Problems in Turf.................................... 538
Fungicide Families and Modes of Action 562
Control Strategies for Turfgrass Diseases 569

Index.. 579

Chapter 1

Soils

INTRODUCTION TO SOILS

Since soil is the basis for almost all living plant material, it is important to know and understand how soil directly affects the plants that are grown within it. If you have an understanding of soil and how it can be managed, the task of maintaining turf on that soil is much easier. All plants have basic needs for survival, and most of these are derived from some kind of growing medium. When I refer to a growing medium, I am talking primarily about soil.

Soil has many purposes and functions. It will be looked at differently by an engineer who is building a highway and by a homeowner who is installing a lawn. (Figure 1.1) The most common mistake is to refer to soil as dirt. Dirt is simply soil that is out of place. Plants are not planted in dirt but rather in a rich medium that enables them to grow.

Soils take on a wide range of characteristics and are different in almost every part of the world. There are clay soils, loam soils, basin soils, desert soils, and numerous other types. Since soils are so different, the types of plants that grow in each soil type are quite dif-

TURFTIP

A common mistake is to refer to soil as dirt. Dirt is simply soil that is out of place. Plants are not planted in dirt but rather in a rich medium that enables them to grow.

FIGURE 1.1 In a highway construction situation, topsoil may not be desirable.

ferent as well. Each soil has a unique structure, making it very different from one plot of ground to the next. Soils can even vary from one side of a field to another. Many of the differences between soils come from their individual biological components. Some areas, for instance, that are high in limestone content may have a higher pH. A soil that is high in sand or quartz content may drain more quickly than another soil composed of clay.

The reason soil content is important is because all plants have specific requirements. These requirements include:

- *Air*
- *Water*
- *Nutrients*
- *Light*
- *Heat*
- *Organic matter*
- *Mechanical support*

This varies among plant and turfgrass species so that only certain plants will grow in certain soils. If any one of these factors is not present, a plant will not have optimum growing conditions.

SOIL PH AND FERTILITY

If you are a good manager of your soil or are involved in growing turf of any kind, the first step to healthy turf is proper preparation of the soil. The two basic elements for plant growth are pH and fertility. Although soils vary in every region, from a turfgrass standpoint, pH must be in a certain range and there must also be a certain level of nutrition available in the soil.

TURFTIP

The two basic elements for plant growth are pH and fertility.

> ## TURFTIP
>
> The optimum range for turfgrass is usually 6.2- 7.0 on a pH table. Turfgrass is probably the pickiest of all plants when it comes to the requirements of soil pH and the amount of each nutrient it requires.

The major factor overlooked in the lawn care industry today is probably pH. When was the last time your yard care company asked to put down a lime application? A soil that does not have the correct pH balance will not have the correct balance of beneficial nutrients or organisms. The optimum range for turfgrass is usually 6.2- 7.0 on a pH table. (Figure 1.2) Turfgrass is probably the pickiest of all plants when it comes to the requirements of soil pH and the amount of each nutrient it requires. Studies have shown that there is an optimum range of pH in which nutrients are available to plants. You could put on pounds and pounds of fertilizer and the grass would not be able to take up these nutrients if the pH is not in the recommended range. I have heard so often, "Well, I put on fertilizer every year but my yard never seems to improve." For starters, the reason can usually be tracked down to a pH problem in the soil.

There are also a wide range of microorganisms that live in the soil and that help to improve soil structure and provide an optimum growing environment. Once again, these organisms will occur in much smaller numbers in a soil that is not in that optimum pH range of 6.2-7.0. So you can see that if a lawn care professional simply starts out with a pH test, a lot of problems can be diagnosed right away. Not only is this test relatively easy to perform , but pH balance is easy to correct should the test reveal that it is not in the correct range.

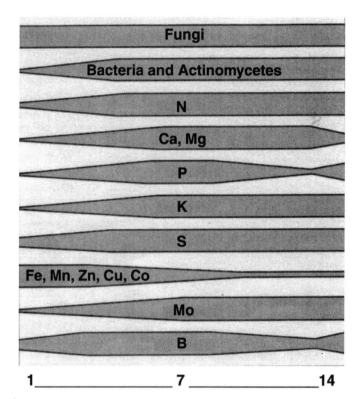

FIGURE 1.2 Ph directly affects nutrient availability in turfgrass but also can affect microbial activity and other soil organisms. For best results, the optimum range for ph levels is in the 6.2-7.0 range for most turf areas.

TurfTip

Of all the problems that can exist in a lawn, a pH adjustment is probably the easiest to diagnose and correct. A simple soil test will reveal the pH of a lawn and the nutrient deficiencies that are present.

Of all the problems that can exist in a lawn, a pH adjustment is probably the easiest to diagnose and correct. A simple soil test will reveal the pH of a lawn and the nutrient deficiencies that are present. This is a very inexpensive test and can usually be performed by a local co-op or state extension office. If the test indicates a low pH, then a simple lime application will correct the problem. Lime, which is easy to apply and readily available in most regions, is an inexpensive product that will fix a low pH. It may be in different forms or particle sizes depending on the availability in a particular area. It is possible that a soil can test high on a pH chart, but it is very uncommon. This would indicate a need for some sort of soil acidifier. Soil acidifiers are seldom needed in most situations.

So now that we have established the importance of pH and how to correct a pH problem, what next? Even though a lime application corrects a low pH, it will take several months before the actual pH of the soil will change. This is a reason why soil pH and fertility should be addressed in the fall. Now that the soil is adjusting, the fertilizers and treatments on turf will become much more effective.

The reason that fertilizer is applied is to provide nutrients to the soil that will be available for plants to take up through their roots. These nutrients are required for turfgrass plants to grow, maintain a green color, and reproduce. The primary nutrients required by turfgrass are nitrogen, phosphorus, and potassium (potash). These three nutrients are called macronutrients, and they need to be

---

TURFTIP

Lime, which is easy to apply and readily available in most regions, is an inexpensive product that will fix a low pH.

> # TURFTIP
>
> The primary nutrients required by turfgrass are nitrogen, phosphorus, and potassium (potash). These three nutrients are called macronutrients, and they need to be applied to turfgrass almost yearly in the form of some sort of fertilizer.

applied to turfgrass almost yearly in the form of some sort of fertilizer. These are the most common elements found in most fertilizers and will be shown on a fertilizer bag as N-P-K. (Sometimes, instead of N, P, K there will be three numbers in a ratio. The ratio of the three numbers represents the percentage of each element in the fertilizer. This is referred to as the analysis.)

There are other nutrients that are found in some of the higher-quality fertilizers and that turfgrass requires in smaller amounts. These elements are called secondary nutrients and include calcium, magnesium, and sulfur. Still other elements required by turfgrass in relatively low amounts are called micronutrients or trace elements. These include:

- *Manganese*
- *Iron*
- *Boron*
- *Copper*
- *Zinc*
- *Molybdenum*
- *Chlorine*

These micronutrients are not present in most fertilizers and are the least important in the overall health of turfgrass plants. I say they are the least important, but a soil that is totally lacking in any element may reveal an odd characteristic of turf color or growth. This total arrangement of elements makes up the balance of what a healthy turf plant will need to grow. Of course, as mentioned before, all these elements are only readily available to the plant within a certain range of soil pH. There can be a situation where there is plenty of an element present in the soil yet little to none of this element is plant-available.

A soil test will provide a better insight into which nutrients need to be applied to the soil and in what amounts. It is important to recognize that too much or too little of any fertilizer element can provide an undesirable result. When using fertilizers, "if a little is good then more is better" is not a good rule to follow.

Nitrogen is the most required element for healthy grass and needs to be applied in fairly large quantities to the turf. This element is an essential component of chlorophyll and other plant substances. Since chlorophyll provides the green coloration that is seen in leaves, a plant low in nitrogen will not have a bright green color. It is possible, however, that if too much nitrogen is applied to a lawn, a burning of the turf can occur much like that of a herbicide mishap.

The atmosphere is composed primarily of nitrogen yet plants are not able to use this nitrogen for their benefit. Of course all grass is

TURFTIP

When using fertilizers, "if a little is good then more is better" is not a good rule to follow.

> **TURFTIP**
>
> Nitrogen is the most required element for healthy grass and needs to be applied in fairly large quantities to the turf.

green, and plants are able to get some nitrogen from the soil. Much of the nitrogen that is in the soil is actually fixed to the soil particles by lightning. There is often not enough nitrogen in the soil to keep plants lush and thick as desired by most homeowners and lawn care professionals. Supplemental nitrogen is one of the primary reasons why fertilizers are applied to lawns, but the other elements in fertilizers are critical to growth as well.

All fertilizers have a rate of N-P-K in their analysis, as I mentioned before. The ideal ratio of N-P-K should be 3-1-2 for most turf species. This is not always the case, and some fertilizers may not have one of the elements at all. These fertilizers are needed for special circumstances or when a soil test might reveal a specific need.

SOIL TYPES

Hopefully you are now realizing that a healthy soil is the building block for healthy turf. One cannot exist without the other, so a good turf manager must understand that soil content is very crucial. Every turf manager is confronted with many different soil types and situations.

Soils are all composed of minerals and materials that in many cases have been present for thousand of years. A mineral soil is composed of four major components: inorganic or mineral materials, organic matter, water, and air. All soils have these components to some level or another. It is their variation that creates different soil types. (Figure 1.3).

All soils have a different content of organic and inorganic materials. These materials are arranged in different proportions of variably sized particles. This property is referred to as soil texture. The terms sand, silt, and loam refer to various categories of soil texture. They describe a specific soil structure, texture, and composition.

The three major types of inorganic soil particles are sand, silt, and clay (Figure 1.4). These particles compose the majority of any soil's composition. Not all three types will necessarily be found in one soil. Soils will have different quantities of those materials along with water, air, and the organic particles. The organic particles found in soils are in much smaller quantities. Usually organic content in a soil is as low as 1% to 6% of the soil's total composition. Organic matter is very beneficial to soil and can be added to almost any soil type to improve

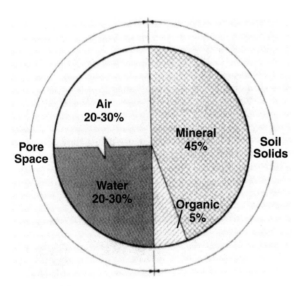

FIGURE 1.3 All soils have a ratio of components as shown in this soil composition circle. It is possible in poor soils for any of the factors to not be present. Also, the percentages can vary from environmental factors.

Three Major Organic Soil Particles General Properties

Property	Range in diameter of particles in millimeters (mm)		
	Sand (0.05-2 mm)	Silt (0.002-0.05 mm)	Clay (<0.002 mm)
Means of observation	Naked Eye	Microscope	Electron Microscope
Dominant minerals	Primary	Primary & Secondary	Secondary
Attraction of particles to each other	Low	Medium	High
Attraction of particles to water	Low	Medium	High
Ability to hold chemical nutrients & supply them to plants	Very Low	Low	High
Consistency properties when wet	Loose, gritty	Smooth	Sticky, plastic
Consistency properties when dry	Very loose, gritty	Powdery, some clods	Hard Clods

FIGURE 1.4 Organic soil properties.

TURFTIP

The three major types of inorganic soil particles are sand, silt, and clay. These particles compose the majority of any soil's composition.

the soil structure and texture. All of the components of soil are connected in the way they affect one another. A very organic loamy soil, for instance, will have more air spaces and better water retention compared to that of a sandy soil, which will have good air space but very poor water retention. You can see that the composition will directly affect how much water and fertilizer would need to be applied to a soil.

A very helpful tool in analyzing soil composition is the soil textural triangle (Figure 1.5). The soil textural triangle is a descriptive tool created to help title various soil types according to their composition. The soil types are based on the quantities of sand, silt, and

TURFTIP

A very helpful tool in analyzing soil composition is the soil textural triangle. The soil textural triangle is a descriptive tool created to help title various soil types according to their composition.

clay present. These descriptions are used by all soil scientists to describe soils in every region of world. Every soil that exists will fall somewhere within the soil triangle diagram.

Recognizing the type and composition of soil will help to develop a better understanding of how to maintain the plants that are growing in the soil. A turf manager who realizes a sandy soil will need more water than a clay soil can eliminate turf stress during dry periods. These differences in soil structure require that certain cultural practices be followed in order to maximize efficiency and minimize expense.

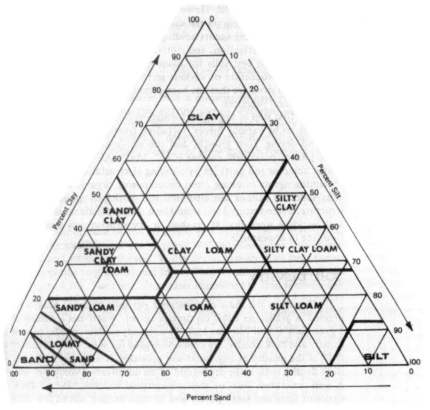

FIGURE 1.5 Textural triangle with soil textural classes reflecting the relative percentages of sand, silt, and clay.

> **TURFTIP**
>
> The presence of air in the soil is critical to a plant's survival by allowing the plant to draw oxygen, water, and nutrients into the roots.

The presence of air in the soil is critical to a plant's survival by allowing the plant to draw oxygen, water, and nutrients into the roots. Soil that is compacted will be very low in porosity, as is common in clay soil types. There are, of course, ways to aerate a soil. Any improvements made to increase soil porosity will help to reduce turf stress and damage. This in turn will provide a more healthy growing environment for the plants.

Aeration of the soil is one way to achieve this air exchange. Ground that is being prepared for turfgrass can be improved much more easily than already established turf. Organic matter can be added to the soil, and rototilling can be performed to "fluff" it up. Rototilling

> **TURFTIP**
>
> Aerification is any mechanical means of improving the aeration in the soil. This allows better porosity and ventilation within the soil. The idea is to increase a plant's ability to take up nutrients and water that are needed for survival.

is just one form of aerification that provides better porosity in the soil. Aerification is any mechanical means of improving the aeration in the soil. This allows better porosity and ventilation within the soil. The idea is to increase a plant's ability to take up nutrients and water that are needed for survival. Soils can become "sealed" up to the point that there is little penetration of moisture or fertilizer content. Some sort of aeration will improve the growing environment.

Aeration does not have to be man-made or done mechanically. There are also many natural forms of aeration. Certainly the most recognized form of natural aeration is earthworm activity. Although a homeowner may often complain about the bumps and lumps that earthworms cause in the yard, this is nature's way of allowing air exchange with the soil. This is why I often discourage customers or friends from continuously rolling their lawns. Lawn rolling can lead to compaction over time and discourages earthworm activity.

As I pointed out, the main idea behind aeration is to improve moisture uptake by grass plants. The reason that moisture is so critical to healthy turf is because water is the largest component of grass. Grass is mostly composed of water, and a lack of water for a period of time will cause death. The "point of no return" where a plant becomes so deficient from water that it will not survive is called the permanent wilting point. Many plants will wilt from lack of water but are able to quickly recover once water is provided to them. This

TURFTIP

The "point of no return" where a plant becomes so deficient from water that it will not survive is called the permanent wilting point.

depends on the type of plant and a combination of other environmental factors such as temperature and even wind.

Wilting is simply a lack of moisture in plant cells that causes a drooping appearance. Once the plants receive moisture, they become turgid again giving the plant its normal appearance, assuming that the plant has not reached the permanent wilting point. Turf plants have an unusually high tolerance for lack of moisture. Bluegrass, for example, has the ability to go dormant during hot, dry periods and then almost "magically" regenerate once rains come again. This is typical of most grass species, but certain types adapt better than others to dry periods. If the soil is more susceptible to moisture and drainage, then it is better able to benefit from rain than a compacted soil where much of the moisture might run right off the top. The importance of soil porosity is the direct relationship it has with plant-available water.

As you can tell, soil is the basis for everything that a plant needs to grow, develop, and establish itself. This is why the maintenance of the soil itself can often be as important as the maintenance of the plants that grow in the soil. Plants get all of the moisture and nutrients they need for survival right out of the ground. Without moisture or nutrition in the ground, soil does little more than hold the plant up and support it. It is the combination of cultural practices that will make turfgrass establishment a true success.

Chapter 2
Plant Biology

PLANT RESPONSES

Plants have a certain response to soil, water, temperature, and many other environmental factors. The reason this occurs is that plants are living material and are composed of cells. Turfgrass managers often forget that in reality we are dealing with something that is very much alive and composed of living cells and tissues. In many instances a good turf manager or maintenance provider is very much like a plant doctor. When plants show distress, problems, or disease, it is up to us to fix these problems.

Plants have certain responses to injury or disease just as a human body does to a virus or "bug." If a person is ill or suffers from a lack of some vitamin or nutrient, there is a obvious physical reaction. Plants respond in very similar ways. For instance, a plant's reaction to a lack of water is to wilt. If a plant is deficient in a mineral, it will react with an odd color or growth pattern. This is why it is so critical to look at plants as living material and to "treat" plants in a manner to provide vigor and health.

A grass plant has the unique capability to withstand frequent mowing and is able to regenerate after this occurs. Not only is it harm-

> **TURFTIP**
>
> When mower blades come into contact with the crown of the plant, which is where all the leaves are attached and grow from, it often can result in turf injury or death. Many plants will simply never recover from such mowing practices and will die off.

less to the grass plant, but it is in fact necessary for the plant to spread or establish itself. The reason that the plant is not damaged in any way is because the crown of the plant is protected at ground level. This is just one of the reasons that "scalping" a lawn can be so detrimental to the overall health of the lawn. When mower blades come into contact with the crown of the plant, which is where all the leaves are attached and grow from, it often can result in turf injury or death. Many plants will simply never recover from such mowing practices and will die off.

TURFGRASS PLANT ANATOMY

The practice of mowing stimulates the grass in such a way that new leaves are constantly forming at the crown of the plant as the older leaves and materials are cut off. If these leaves are not cut off, eventually they will simply fall to the ground and die off as the new ones are formed. Grass plants form seed heads at the tips if they are allowed to mature to a certain level. Usually these seed heads are not desirable or attractive and serve no purpose on lawns. The only time seed heads would be desirable is for seed production or in the case of ornamental grass plants. Immature grass plants do not have the capa-

> **TURFTIP**
>
> A rhizome is an underground "sucker" that creeps out and eventually will root an entirely new plant.

bility to produce a seed head. Maturation time varies in each individual species of grass.

Another result of mowing is the stimulation of either rhizomes or stolons. Rhizomes and stolons both possess the capacity to form totally new grass plants and in turn create a much thicker lawn. A rhizome is an underground "sucker" that creeps out and eventually will root an entirely new plant. A stolon is a surface "sucker" that will grow on top of the ground but still ultimately create a new plant. Some grasses have the ability to spread and be quite vigorous, while other types cannot spread out at all. This is why the placement and location of turf species can directly affect the turf success.

Most grasses also have tillers, which are new shoots that form at the base of the crown. These tillers allow the grass to spread out from the crown. Tillers are quite different from the rhizome and stolon

> **TURFTIP**
>
> A stolon is a surface "sucker" that will grow on top of the ground but still ultimately create a new plant.

> **TURFTIP**
>
> Some plants are limited to growth in a clump form, which is simply an enlarged crown with many leaves that may or may not have tillers and cannot spread out on their own unless they are able to seed out.

structures and do not allow turfgrass plants to spread out in the same fashion. Not all grass plants possess the ability to produce rhizomes, stolons, and tillers. Some plants are limited to growth in a clump form, which is simply an enlarged crown with many leaves that may or may not have tillers and cannot spread out on their own unless they are able to seed out. These grass species require numerous plants to create a lush look because of the nature of growth. Since these grasses do not have the ability to spread, many individual plants must be used to achieve the desired effect. Fescue is a good example of this type of plant. This is also why plants that produce rhizomes and stolons are seeded at lower rates than plants that cannot produce these plant structures. Since fescue does not spread and cannot fill in by itself, you can imagine that there would be numerous bare spots in the lawn if not all of the seeds were to germinate. Bluegrass, on the other hand, has the ability to spread very easily through rhizome structures. What that means essentially is that if there are bare spots in a bluegrass lawn, potentially the spots would fill in on their own after repeated mowing.

The largest and most noticeable part of all grasses is of course the leaves. As with a tree, the leaves are what people see and recognize when they think about plants. I have already described what the functions of some root structures are, but there are also other root structures you should be aware of. There are two types of roots present on all grass plants: the seminal roots, which grow deep in

TURFTIP

There are two types of roots present on all grass plants: the seminal roots, which grow deep in the ground, and the adventitious roots, which spread out and are shallow in the ground.

the ground, and the adventitious roots, which spread out and are shallow in the ground.

The seminal or primary roots, as they are more commonly called, live only a short period of time. These roots are created during germination and extend deeper in the soil searching for nutrients and moisture to help the new plant survive.

Adventitious roots, or secondary roots, arise from stem nodes and are only found on mature plants. These roots help to support the plant and eventually will develop into the entire root system of the grass plant.

TURFTIP

Turfgrass professionals who are able to identify the grass structures are at a huge advantage in actually identifying specific grass species. One of the structures used for identification is the ligule. The ligule is a membranous hairy structure that is found at the junction of the blade and sheath. The sheath is simply the lower part of the leaf.

The crown is found at the base of the plant and is almost like the trunk of a tree in the way it supports the entire plant structure. There are a couple of other grass structures that are very important from an identification standpoint. Turfgrass professionals who are able to identify the grass structures are at a huge advantage in actually identifying specific grass species. One of the structures used for identification is the ligule. The ligule is a membranous hairy structure that is found at the junction of the blade and sheath. The sheath is simply the lower part of the leaf.

Opposite the ligule on the outside of the plant sheath is the collar. This structure will appear to be a whitish band that wraps around the plant. On certain grass species, auricles may also be present. Auricles are clawlike appendages on the edges of the collar. All of these structures are important features of grass plants, and they are especially helpful in identifying the many types of grassy weeds that can be found in turf. All grass species have many similar characteristics, and knowing the different parts of the grass will certainly help you identify both grasses and weeds much more easily. Refer to Figure 2.1.

SEEDING STRUCTURES

The inflorescence or seed head is always located at the top of the plant. This is where a turfgrass plant will produce seeds if allowed

TURFTIP

Only mature plants produce seed structures; however, there is no predetermined height for a plant to seed out or produce a seed head.

to grow to a height that will allow seeding out. Since most grass plants are mowed off long before they go to seed, most turf professionals do not know the seed structures or are unable to identify plants based on the seed head type.

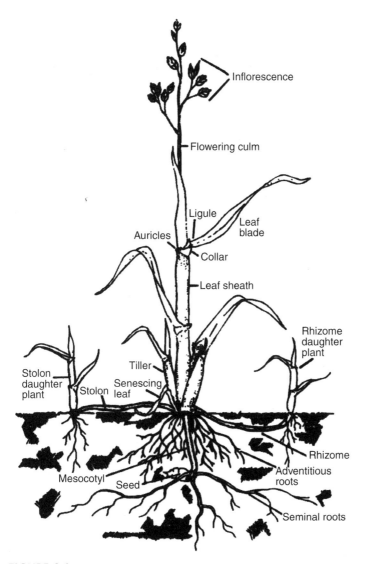

FIGURE 2.1   Diagram of the grass plant

Only mature plants produce seed structures; however, there is no predetermined height for a plant to seed out or produce a seed head. Annual ryegrass, for instance, can produce a seed head even after being mowed to $5/32$ inch in height. Fescue is a taller grass type; it could never survive at that mowing height and will seed out at over $1 1/2$ feet in height. Weeds often seed out and then adapt as mowing and cultivation occur. A dandelion plant might seed out at three inches the first time it blooms, but as mowing occurs, the plant will reduce the seeding height each time to maximize seed production. I have seen dandelions on rare occasions form a seed head at putting green height.

The flowering portion of a grass plant is referred to as the inflorescence (Figure 2.2). The inflorescence has no leaves and is actually composed of floral bracts within spikelets. Each spikelet contains one or more flowers within two floral bracts, called the lemma and the palea. Spikelets may be attached directly, which is called a sessile, or on short stalks, which is known as pediceled. The arrangement of the inflorescence has three primary types. In one type all spikelets are located on a sessile on a main axis. This is prevalent in grass types like Bermuda grasses, ryegrasses, and wheat grasses. A raceme has spikelets on individual flower stalks, called pedicels, which are attached

TURFTIP

The flowering portion of a grass plant is referred to as the inflorescence. The inflorescence has no leaves and is actually composed of floral bracts within spikelets. Each spikelet contains one or more flowers within two floral bracts called the lemma and the palea.

to the main axis. This pattern of flower can be found in St. Augustine grass and in carpetgrass. The last common type is panicle inflorescence. In a panicle arrangement, spikelets occur on flowering branches that look almost like limbs on a tree. This flowering pattern is common to bluegrass, bentgrass, and many other common turfgrass types.

The most important fact for a turf manager to remember is that all plants are composed of cells. Cells are alive and possess the ability to grow and reproduce while growing. Cells also have requirements for healthy growth and nutrition. These cells are much different than human cells but have the same capability to be infected by disease, reproduce, and die. There are also tissues, which are present in all vegetation, that allow the plants to perform a variety of tasks. A couple of the more important tasks are to take in nutrients and water.

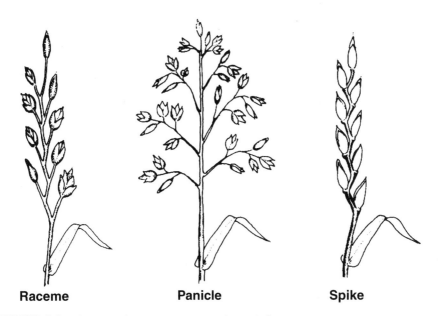

Raceme Panicle Spike

FIGURE 2.2 There are three primary types of grass inflorescence.

There are two types of plant tissue that are responsible for these two tasks. The first is xylem tissue, which helps to transport water up through the plant. It allows the plant to take water and often a combination of nutrients up through the plant, delivering moisture to all of its components. This is a very complex tissue, which uses the polarity of water to pull up the moisture almost like a drinking straw.

The phloem tissues are responsible for just the opposite. Phloem tissues are food-conducting elements that move materials down through a plant. A very easy way to remember this is "xylem up and phloem down." This expression may be valuable when thinking of plant anatomy and the way materials travel through plants.

All plants have a green color and contain chlorophyll in their leaves. The reason for the green color is the chlorophyll content in the leaves. More importantly, the chlorophyll is critical in the process of photosynthesis. Photosynthesis explains why plants are green, how they live, and even determines their place in the ecosystem. It is one of the most complex processes that occurs within a plant. Without photosynthesis, oxygen and life would not exist on the earth.

Photosynthesis is an energy-storing process that takes place in the green parts of plants in the presence of light. Refer to Figure 2.3. Light energy is stored in simple sugar molecules that are produced from carbon dioxide present in air and water and absorbed by the plant. The carbon dioxide and water combine to form a sugar molecule in the chloroplast. The by-product of this whole reaction is oxygen, which is released into the air.

TURFTIP

Photosynthesis is an energy-storing process that takes place in the green parts of plants in the presence of light.

$$6\ CO_2 + 6\ H_{22} + \text{Light Energy} \xrightarrow[\text{Enzymes}]{\text{Chlorophyll}} \alpha\ C_6H_{12}O_6 + 6\ O_2$$

FIGURE 2.3 Photosynthesis diagram

PLANT ROLES IN THE ECOSYSTEM

Plants have important roles in our ecosystem and environment. Not only do they produce oxygen, but they also help prevent erosion, reduce pollution, and have several other beneficial functions. Turfgrass in the landscape is important to the environment and helps to break up large areas of concrete parking lots and add beauty to areas as well. By adding turfgrass, heat from concrete areas can be reduced and runoff water can be slowed and purified. Turfgrass utilizes the large amount of water that runs off parking lots, sidewalks, and streets. Sometimes the runoff water can be contaminated with impurities and pollutants.

Although turf helps to purify the toxins and substances contained in runoff water, some of these substances can be harmful to the grass. Pollutants such as carbon monoxide, salts, and other chemicals can be absorbed by plants in some way or another. This is an important consideration for turf placement when there are possibilities for high pollution. These substances can create a "sick" appearance to the turf that can be confused with a nutrient deficiency or lack of a plant element. This is another case of the need for careful turf diagnosis.

There is also the possibility of an unhealthy plant response caused by an element such as dog urine or other animal activity. I have had customers accuse spouses of spraying Round Up™ on a lawn when actually the cause to the turf injury was due to the neighborhood dog. There are many other animals and pests that can cause damage to turfgrass.

> # TURFTIP
>
> The plant's primary environment consists of temperature, moisture, light, wind, location, and even human factors. Certainly the combination of these factors is a good indication of how the plant will do in a specific area.

TURFGRASS ENVIRONMENT

Since a grass plant is subject to so many responses to elements in the environment, we must consider every possibility. Understanding how the environment affects the plant is just as much a part of managing turf as solving problems associated with disease. The plant's primary environment consists of temperature, moisture, light, wind, location, and even human factors. Certainly the combination of these factors is a good indication of how the plant will do in a specific area.

Each of these factors can also be influenced by other variables, making the overall picture even more complicated. Take, for instance, the presence of light. Elements such as clouds, trees, and buildings can all change the amount of light that a turfgrass plant receives. Although it may appear to any of us that there is more than sufficient light to support a plant's development in a particular location, that may not be the case. The human eye is a very poor judge of light when it comes to the consideration of plant use. Many responses of plants are triggered by the measurement of certain types of light vs. the intensity of the light. All turfgrass plants have the following responses to solar radiation. Refer to Figure 2.4.

Temperature is another factor of a plant's growth rate and success. There is a minimum, optimum, and maximum temperature for

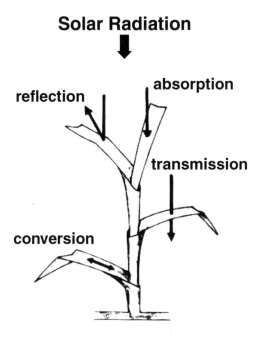

FIGURE 2.4   Turfgrass radiation diagram

every type of turfgrass species. The minimum temperature is the lowest temperature in which a plant can survive during the winter or extreme cold periods. The optimum temperature would be that in which the plant is prone to grow the most vigorously. The maximum temperature is the temperature at which it becomes too hot to grow. Sometimes the maximum temperature will induce dormancy in a grass species, but in other species it can cause death or serious turf injury. Plants have a specific range for optimum growth and also an optimum temperature for seed germination. Seeds of every turfgrass species will usually germinate at one certain temperature range but may thrive at a much higher temperature range.

Moisture is the most important condition for plant cell survival. Turfgrass, as I mentioned before, is 90% water. The functions of water in a plant include maintaining turgidity, transporting nutrients, comprising much of the living protoplasm, helping with chemical

TURFTIP

The functions of water in a plant include maintaining turgidity, transporting nutrients, comprising much of the living protoplasm, helping with chemical processes, and helping to buffer plants against wide temperature fluctuations.

processes, and helping to buffer plants against wide temperature fluctuations.

Wind is often not considered to be an environmental factor but can have direct effects on turfgrass. Air pressure over a specific turfgrass site is directly related to the general weather patterns that are occurring. Topography and geographic location can also influence and change the direct effects on the turf. Air flow can cause plants to wilt or dry out. Wind can also spread weed seeds or other potential pests into a specific area. Pollutants and even pathogens all use

TURFTIP

Air flow can cause plants to wilt or dry out. Wind can also spread weed seeds or other potential pests into a specific area. Pollutants and even pathogens all use wind as a mode of transportation. This is why wind breaks and turfgrass location can directly affect turf quality.

wind as a mode of transportation. This is why wind breaks and turfgrass location can directly affect turf quality.

The importance of light as the basis for all life on this planet has been discussed. I will again mention the importance of light to the process of photosynthesis and why it occurs. Of course, all living plants require a certain amount of light for survival, but once again I want to point out that not all turf plants require the high amounts of light thought to be necessary for optimum growth.

Human factors are simply the effects humans have on turfgrass plants and what relevance this has to plant development and establishment (Figure 2.5). Humans have a detrimental effect on many

FIGURE 2.5 Golf cart damage—cart damage often occurs in areas where people consistently drive over the same areas instead of scattering out.

> **TURFTIP**
>
> From construction damage to walking, it is important to consider the human factor while planning to establish grass in any areas.

elements in our environment, and turfgrass is no exception to this. Plants that are developing are not going to do well when subjected to foot traffic, automobile traffic, or anything that will smash and disrupt the seedling growth. Humans also tend to have a "cattle" mentality when we walk or travel in any sort of motorized vehicles. Golf carts are the prime example of a daily frustration for golf course superintendents. I know I fought the battle many times to try to disperse golfers over a particular area instead of using the same area repeatedly. It was also a huge frustration to have to put up large and obvious barriers to keep people from walking across newly seeded grass. These are just a handful of examples of human damage, and certainly a whole chapter could be spent on this. From construction damage to walking, it is important to consider the human factor while planning to establish grass in any areas.

Chapter 3

Turfgrass Species

TURFGRASS ZONES AND SPECIES

Turfgrass species vary in the areas where they grow in for a variety of reasons. There is usually a specific type of turf that will thrive in a location or at least be suitable to grow under a range of conditions. As turfgrass professionals, it is critical that we learn to identify the types of grasses and the locations suitable for each variety of turf to be planted.

The first factor that should be considered is the zonal requirements or more specifically the difference between warm-season and cool-season grasses. The warm-season grasses for the most part are considered to grow in the southern half of the United States or other warm climates and are susceptible to long periods of freezing temperatures. Some of the grass types are not capable of handling frost or freezing temperatures for even one night, while others can handle cold temperatures but perhaps not below 0° F. temperatures. The basic types of warm-season grasses that will grow in these regions are:

- *Bermuda grass*

- *Zoysia grass*
- *Carpetgrass*
- *St. Augustine grass*
- *Centipede grass*

The cool-season grasses are just the opposite in terms of acclimation to survival. They are usually found in climates in the northern half of the United States and other climates that are cooler and more moderate in temperature conditions. Grasses in these areas usually prefer moderate to heavy rainfall and do not like hot, dry, or arid conditions that would typically be found in the south. The types of grasses used in these climate locations are:

- *Bluegrass*
- *Fescues*
- *Rye grass*
- *Bentgrass*

TURFTIP

There is a large section of the United States that is both hot and cold during the calendar year. This region is referred to as the transition zone and is a sweeping area that extends right through the middle of the United States.

There is, however, a large section of the United States that is both hot and cold during the calendar year. This region is referred to as the transition zone and is a sweeping area that extends right through the middle of the United States. The transition zone often has winters that are too cold for one type of grass and too warm for another type of grass. In many cases there are warm-season grasses planted in the same neighborhood with cool-season grasses. A turf professional becomes a valuable asset in determining when and where a grass species will survive. Even more critical is knowing how to maintain different turf types when they are used in the same locations.

I want to point out that the grasses I discuss in this book certainly grow in other countries. I am mainly referring to the zonal conditions that occur in the United States. Refer to Figure 3.1. I also focus on the transition zone, since that is where many problems occur for turfgrass professionals.

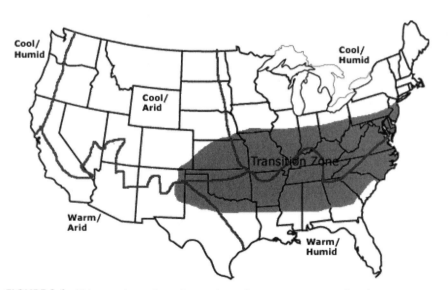

FIGURE 3.1 This map shows the various regions where warm-season and cool-season grasses grow. The colored area is the transition zone which can grow cool- or warm-season grasses.

> **TURFTIP**
>
> When selecting a product, you should also take a look at soil type, temperature, exposure, light, traffic, and, most importantly, the specific function of the grass you are planting.

Growing up in the Midwest myself, I have seen many instances of 15 inches of rain in July and other instances with no rain in July or August at all. You can imagine the difficulty this creates when trying to put together a schedule for maintaining turf in the best manner. When choosing a grass for any climate, there are many factors to consider. Rainfall, as I mentioned, is very critical to growing grass, especially if you recall that plants are mostly composed of water. You should also take a look at soil type, temperature, exposure, light, traffic, and, most importantly, the specific function of the grass you are planting.

Cool-Season Grasses

Most of the grasses planted in the United States fall under the category of cool-season grasses. Some of these grasses are adapted to warmer climates and the transition zone, but for the most part cool-season grasses prefer cool seasons. There is a need for high fertility and a luxurious amount of water for most types.

Kentucky Bluegrass

The most common cool-season grass grown widely in the northern states and also in a large part of the transition zone is Kentucky blue-

> **TURFTIP**
>
> Bluegrass can be grown for use in a lawn or even on golf tees. The grass has a very good color and when fertilized makes a very attractive area of turf.

grass, or Poa pratensis. Kentucky bluegrass is the thin-bladed plant that is used in lawns, golf courses, and many other recreational facilities. It can spread vegetatively by tillers and rhizomes. It is a fairly aggressive spreader and can repair itself when damaged. It can thrive under rigorous mowing conditions and is commonly maintained at a height of anywhere from $1^1/_2$ inches to over 3 inches. Bluegrass can be grown for use in a lawn or even on golf tees. The grass has a very good color and when fertilized makes a very attractive area of turf.

Bluegrass was introduced into the United States in the 1700s and is native to Europe. It was first used in Jamestown. It has been planted in lawns for many years and adapts to a range of growing conditions. I have seen healthy bluegrass from Minnesota clear into the southern states, which shows how adaptable this plant can be. The earlier varieties were considered to be somewhat troubled by disease problems and required high rates of seed to establish. Newer varieties are more suitable for warm weather and less susceptible to disease. (Here is a footnote for someone who might be inexperienced in purchasing bluegrass seed. I always hate to hear of inexperienced turf managers and homeowners who claim to have purchased really cheap bluegrass when they have actually purchased Kentucky 31 fescue instead. This is a fairly common mistake made by many but something to be aware of in case you ever encounter this problem yourself.) Refer to Figure 3.2.

Kentucky bluegrass is best suited for full sun to lightly shaded areas, although tremendous breakthroughs in shade varieties have been developed in the last few years. This grass prefers moderate to heavy fertilization and can be somewhat of a fertilizer glutton. Without moisture, Kentucky bluegrass has the unique ability to go dormant for long periods of time without any long-term damage to

FIGURE 3.2 Although this bag clearly reads tall fescue, do not be fooled by the word Kentucky and assume it is bluegrass.

TURFTIP

Kentucky bluegrass has the unique ability to go dormant for long periods of time without any long-term damage to the turf. Once the rains return, the turf area will usually resume its green luster and appearance. Since it is such a prolific spreader, the rhizomnes on the plant aid in this recovery process and help to assure survival.

---

the turf. Once the rains return, the turf area will usually resume its green luster and appearance. Since it is such a prolific spreader, the rhizomes on the plant aid in this recovery process and help to assure survival.

Grasses often do best when they are put into a blend or mixture. A blend is a combination of two or more varieties of turfgrass

TURFTIP

Grasses often do best when they are put into a blend or mixture. A blend is a combination of two or more varieties of turfgrass within the same species.

within the same species. In other words, you might have the variety of bluegrass called Pennstar blended in with another variety called Flyking. Both of these grasses are in the bluegrass family but have characteristics that make them more suitable when used together. The reason blends are used is due to the unique nature of each individual cultivar. A cultivar is a cultivated variety of a grass that is bred to have a specific trait. Some cultivars are genetically bred for color, while other cultivars have disease resistance.

A mixture of grasses involves two completely different species of grasses that are used to obtain a desired affect. I have briefly mentioned how bluegrass and ryegrass often accompany each other. The idea is very similar to that of a blend. When a couple of varieties are used in a location, each variety can adapt to where it is most comfortable. For instance, one species of grass may thrive in a shady location while another grass may prefer full sun.

Since either a blend or a mixture will contain different seed types and varieties, the consumer or turf professional should always read the seed label tag on every bag in order to identify these types. Every seed package is required to have a tag on it, identifying the type and quality of seed (Figure 3.3). Make sure that the inert material in the bag is less than 2% and that the weed content is less

TURFTIP

Always remember that if a seed bag is opened and not completely used, you should always leave the tag attached until the bag is disposed of. This will avoid costly mistakes and keep the misuse of turf species to a minimum.

Turfgrass Species   41

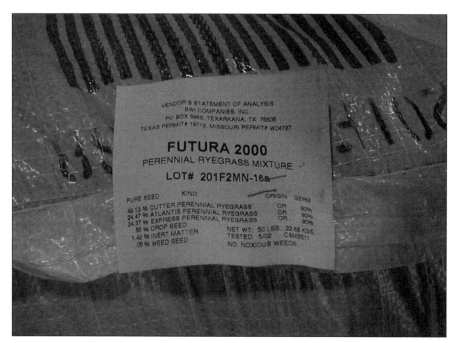

FIGURE 3.3 Notice this seed tag on a bag of perennial ryegrass. It shows important information such as the producer, seed content, and amount of weed seed.

than 0.1%. Good seed companies will produce seed with less than 1% inert and 0.05% weed content in some cases. The tag should also note that there are no noxious weeds. This simply means that you are not going to buy a product loaded with evasive seeds in the mixture. Each variety or type of seed should be marked on the tag along with the percentage of that seed that is contained in the bag. Keep in mind that seeds have different sizes and weights, so the naked eye is a poor judge of content. Also, since seeds are different sizes, the percentage of weeds and inert materials should be much smaller when the seed sizes are smaller. You should always remember that if a seed bag is opened and not completely used, you should always leave the tag attached until the bag is disposed of. This will avoid costly mistakes and keep the misuse of turf species to a minimum.

TURFTIP

One of the few grasses to prefer a shady location is a grass called rough-stalked bluegrass, or Poa trivialis.

Rough-stalked Bluegrass

One of the few grasses to prefer a shady location is a grass called rough-stalked bluegrass, or Poa trivialis. Rough-stalked bluegrass is unusual because it prefers cool, wet, and shady conditions. The only drawback that I have seen to this type of grass is that it cannot withstand drought conditions very well. Hot summers will kill this grass or thin it out very badly. Its usefulness should not be overlooked in blends or mixtures. Rough-stalked bluegrass is also commonly used in the south as a grass to overseed Bermuda grass on golf course fairways and athletic fields. The reason this is done is so that when the Bermuda grass starts to go dormant for the winter, the rough-stalked bluegrass can take over to give the area a playable surface. Rough-stalked bluegrass is not too aggressive so the Bermuda grass can take over in the spring. Remember, as I mentioned, rough-stalked bluegrass cannot handle the heat so it basically dies out as the Bermuda grass reclaims the area for play during the normal season. It can be said that the rough-stalked variety provides "good transition" as the seasons change. A popular variety of rough-stalked bluegrass is called 'Sabre.'

Annual Bluegrass

Another grass that is worth a mention is annual bluegrass, or Poa annua. For the most part this grass is considered a weed. Keeping in mind that a weed is really a "plant out of place," Poa annua is a very pro-

TURFTIP

Another grass that is worth a mention is annual bluegrass, or Poa annua. For the most part this grass is considered a weed. Keeping in mind that a weed is really a "plant out of place," Poa annua is a very prolific plant that spreads very easily. The most common place for this grass to be noticed is in a golf course green when it is competing with creeping bentgrass.

lific plant that spreads very easily. The most common place for this grass to be noticed is in a golf course green when it is competing with creeping bentgrass. Poa annua is capable of handling a range of soil types including compacted and wet soils. It is also capable of withstanding mowing heights down to bentgrass heights of 1/8 inch or lower in height. The worst feature is that it has the capability to seed out at these lower heights. This is the primary reason for its evasiveness. There are also few chemical controls for it outside of growth retardants, which limit the plant's ability to grow and reproduce seeds. Up until just a few years ago, growth retardants were one of the only ways to control this type of grass. The reason it is not desirable for the most part is its light green color compared to that of regular grass species. An important fact to remember is that it is an annual, which means that the grass comes back from new seeds each year. The grass germinates in late fall, grows through the winter, and is often gone at the first signs of hot and dry weather in the late spring.

Perennial Ryegrass

Another very common cool-season grass is perennial ryegrass, or Lolium perenne. Perennial ryegrass can be found in lawns, athletic fields, and golf course fairways. As I mentioned, this grass is usually grown in accompaniment with Kentucky bluegrass in most instances. The reason is that ryegrass is somewhat of a clump-type grass; it does not spread and fill in nearly as quickly as bluegrass does. Ryegrass often has a darker color than bluegrass and can withstand mowing heights as low as a 1/2 inch if irrigated well and maintained correctly.

It is believed that this grass originated in the British Isles and was introduced to the United States many years later than Kentucky bluegrass. Perennial ryegrass thrives in moist soil conditions and is mostly a full-sun grass. It does require good fertility but has one of the darkest green colors of any grass species when it is healthy and vigorously growing. The most useful characteristic of the ryegrasses in general is the ability to germinate very quickly. This is one of the primary reasons that perennial ryegrass is used in seed mixtures along with other grasses. Since many other grasses are slow to germinate or establish, the perennial ryegrass can provide a cover crop, so to speak, until the others catch up. The reason it is so common in athletic fields and golf courses is due to its ability to tolerate traffic and wear at lower growing heights. Irrigation is always helpful in keeping perennial ryegrass looking good, because it does love moist growing conditions.

TURFTIP

The most useful characteristic of the ryegrasses in general is the ability to germinate very quickly. This is one of the primary reasons that perennial ryegrass is used in seed mixtures along with other grasses.

TURFTIP

Lolium multiflorum, which is commonly known as annual ryegrass, is only used a cover crop. This grass has a limited usefulness but can make a valuable cover crop because of its rapid germination. Since it will germinate in seven to eight days in many cases, it can allow ample time for other grasses to establish. If you need a temporary cover or something to prevent erosion, this would be the perfect grass variety.

Annual Ryegrass

Lolium multiflorum, which is commonly known as annual ryegrass, is only used a cover crop. This grass has a limited usefulness but can make a valuable cover crop because of its rapid germination. Since it will germinate in seven to eight days in many cases, it can allow ample time for other grasses to establish. If you need a temporary cover or something to prevent erosion, this would be the perfect grass variety. It is not a useful turf for aesthetic reasons or for a long-term lawn.

TURFTIP

Tall fescue, or Festuca arundinacea, is the most durable, versatile, and adaptable grass species there is. It will grow in any soil condition including the worst clays.

Tall Fescue

Tall fescue, or Festuca arundinacea, is the most durable, versatile, and adaptable grass species there is. It will grow in any soil condition including the worst clays. It is not planted for a desirable appearance, but to fill in where other grasses cannot be used. Highway departments for years have relied on tall fescue to establish medians and prevent erosion on highway projects. It has excellent heat and drought tolerance and is resistant to most insect and fungus problems.

It is believed that tall fescue is native to Europe, although it has become popular all over the world because it is so durable and can grow in such a wide range of conditions. It has been found in Africa, New Zealand, Asia, and of course in the United States. It is considered to be an important forage grass and has been used widely, even in the agriculture industry. Although this grass can be found in many pastures, tall fescue does have some adverse affects on livestock and grazing animals. The reason is that the grass possesses a naturally occurring fungus that lives in the crown area of the plant. This fungus can cause grazing animals to become sick and cattle to abort fetuses. It is toxic to some extent to all animals that feed upon it, so

---

TURFTIP

Although this grass can be found in many pastures, tall fescue does have some adverse affects on livestock and grazing animals. The reason is that the grass possesses a naturally occurring fungus that lives in the crown area of the plant. This fungus can cause grazing animals to become sick and cattle to abort fetuses.

this is important to consider when using fescue in certain locations. The fungus for all other purposes is very beneficial to the plant because it protects it from insects and pests. It is a symbiotic relationship that gives fescue some of its most desirable traits. Since the toxicity of fescue has been realized for years, newer varieties have been bred and altered so that they do not posses the endophyte. This allows tall fescues to be used in fields where grazing animals are present. On the flip side, since the endophyte does allow insect resistance and some other positive features, there has also been an attempt to breed the endophyte into ryegrass and bluegrass types. One more note: in a case study a select group of grass plants were exposed to severe moisture stress; 75% of endophyte-free plants died and all endophyte-infected plants survived. This might raise an eyebrow as researchers continue to tamper with this endophyte trait in grass plants.

This grass is a bunch-type grass, and the normal varieties are not usually used where an aesthetically pleasing effect is desired. It has a clumpy or uneven appearance and does require a fairly high mowing height because of its large crowns. Because of these large crowns it also has a rough feeling when it is walked on or driven over, much like that of old rough road. Tall fescue is not an aggressive spreader and only spreads by tillers. It has no rhizomes, so if an area of turf is damaged, it is difficult for fescue to repair itself. This is also the

TURFTIP

Tall fescue is not an aggressive spreader and only spreads by tillers. It has no rhizomes, so if an area of turf is damaged, it is difficult for fescue to repair itself.

reason that the grass is seeded at a very high rate, because it cannot spread to fill itself in. If a pickup truck were to drive through an area that was covered with tall fescue and make ruts, those tracks would probably fill in with weeds or another grass type. This is again because of the nature of fescue and its inability to spread on its own. For that reason, it is very common for tall fescue to be mixed with bluegrass in many cases.

Kentucky 31, which I mentioned earlier, is a tall fescue and not a bluegrass. The Kentucky appellation has for years troubled homeowners in the way it is promoted in retail trade. This is the oldest and best-known variety of the tall fescues. It is a forage grass and is reasonably inexpensive, which is why it has become so common over the years. Now in more recent years there are many newer varieties of tall fescue that are a little more desirable in appearance and in their growth habits.

More recently dwarf versions of this grass have been developed and are being used widely in the lawn and landscaping industry. These slower-growing varieties are much denser in growth. They are now very common in golf course roughs and out-of-play areas. The only problem with them is that, since they are slower growing dwarf plants, their recovery from injury is much slower as well. One way that researchers are dealing with this problem is by trying to breed rhizome characteristics into fescues so that they will eventually be able to spread out on their own and the seed rates required for establishment will be much lower as well.

Fine Fescues

There are several different fine fescues known as the Festuca spp. This grass type is known for having fine hairlike characteristics and a delicate appearance. They adapt well to low fertility and can even tolerate a low pH. The fine fescues will grow in shade and can tolerate hard, dry soils while still maintaining a decent appearance. These grasses are not really considered a stand-alone grass type and are

TURFTIP

The fine fescues will grow in shade and can tolerate hard, dry soils while still maintaining a decent appearance. These grasses are not really considered a stand-alone grass type and are usually always mixed in with other grass varieties.

usually always mixed in with other grass varieties. They do germinate fairly quickly and can be considered a starter grass.

Creeping Bentgrass

Probably the most popular grass in the industry since its introduction in the early 1900s is creeping bentgrass. Agrostis palustris and Agrostris stolonifera are the two most commonly known varieties in

TURFTIP

Probably the most popular grass in the industry since its introduction in the early 1900s is creeping bentgrass. Agrostis palustris and Agrostris stolonifera are the two most commonly known varieties in the bentgrass family.

the bentgrass family. The bentgrasses were known for their use in putting greens but now have become so widely used that many golf courses have bentgrass fairways, tees, and greens (Figure 3.4). You can also find bentgrass on tennis and bowling courts.

The first grasses to be known as bentgrasses were used in Austria and were called seeded German bentgrasses. They made a reasonable putting area for golfers and were usually seeded in mixtures that contained chewing fescues. The first vegetative bentgrasses were derived from these German varieties.

FIGURE 3.4 A bentgrass golf green in Missouri.

In the 1920s a new seeded variety of bentgrass was discovered that would change the industry forever. A grass called 'Seaside' was discovered in Oregon, and it would become the focus for improvements in bentgrass that would not happen until years later. It was actually in the 1950s that new strides were made in bentgrass development. At that time 'Penncross' was developed, which can still be found in most putting greens today. This was the first seed to be widely available for the golf course industry and was also the first "stable" grass to be considered usable for the purpose.

In just the last few years many varieties have become available. Many of these are adapted to different climates and soil conditions. They are also less susceptible to the many diseases that affect bentgrass. Disease susceptibility is the biggest problem associated with the maintenance of bentgrass. All bentgrass thrives in moist conditions and in full sun. High fertility is a must, and low mowing must be done to keep the grass in the healthiest condition. Unlike some other grasses, if bentgrass is allowed to grow too tall or let go to seed, it will never look the same again. The primary use of bentgrass is on putting greens, and it is found more commonly in the northern states where temperatures are cooler because of its susceptibility to disease and heat stress.

TURFTIP

All bentgrass thrives in moist conditions and in full sun. High fertility is a must, and low mowing must be done to keep the grass in the healthiest condition. Unlike some other grasses, if bentgrass is allowed to grow too tall or let go to seed, it will never look the same again.

Warm-Season Grasses

There are only a few warm-season grasses that grow in the United States. These grasses are found only in southern climates and a portion of the lower half of the transition zone. The warm-season grasses that do grow in the transition zone are often subjected to severe damage during the winter months due to the extreme cold conditions in the Midwest.

Zoysia Grass

Zoysia grass is, for the most part, the only warm-season grass that is grown in the middle or northern regions of the transition zone. Although it is cold-tolerant and will come back from year to year, it does remain dormant for long periods of time when planted in the transition zone. The three types of zoysia that are grown in the United States are Zoysia japonica, Zoysia matrella, and Zoysia tenuifolia. Zoysia, as you might guess from the Latin binomial, originated in Japan.

The original cultivar for American zoysia was 'Meyer.' It is still one of the most popular and readily available varieties on the market today.

TURFTIP

Zoysia grass is, for the most part, the only warm-season grass that is grown in the middle or northern regions of the transition zone. Although it is cold-tolerant and will come back from year to year, it does remain dormant for long periods of time when planted in the transition zone.

It is very drought-tolerant and has a great color in season. It has also been proven to be the one of the most cold-tolerant of the zoysia species. I know of several places in Iowa that have great looking zoysia lawns and fields even though at one time Iowa was considered too far north for zoysia grass. Zoysia is also a very disease- and insect-free type of turf that requires little maintenance. Also, since it is so aggressive and lush in the way that it grows, weeds often have a hard time competing in a zoysia lawn. It requires little water and only minimal fertilization to look good. It does tend to remain dormant for at least six months in the northern regions but stays fairly active in the southern states.

Sometimes zoysia is sold in trade publications and popular magazines as a "wonder grass" that sounds too good to be true. Although this grass has some great characteristics, it is not as perfect as some sales gimmicks make it out to be. If there is any major drawback to zoysia, it is the dormancy period. Not only is the dormancy effect unsightly to look at, but it can also create a fire hazard (Figure 3.5). This grass type is also problematic when it comes to thatch problems. Thatch is the partially decomposed layer of organic matter that is situated above the ground. It can cause drainage problems and even nutrient-availability problems for any and all turfgrasses, but warm-season grasses tend to be worst. I often

TURFTIP

I often jokingly say there are only two types of zoysia grass: zoysia grass that has a thatch problem and zoysia grass that will have a thatch problem. For the most part this is one of the biggest struggles is dealing with zoysia.

54 Turfgrass Installation: Management and Maintenance

FIGURE 3.5 This zoysia grass was only established on the left side of the yard. The right side of the lawn contains bluegrass and is still growing. The zoysia on the left has already gone dormant.

jokingly say there are only two types of zoysia grass: zoysia grass that has a thatch problem and zoysia grass that will have a thatch problem. For the most part this is one of the biggest struggles is dealing with zoysia. The other problem is that once zoysia is established, it can be somewhat evasive and almost impossible to get rid of if you ever decide to change grass types later on.

There is a belief that burning off zoysia can help to reduce thatch problems in the lawn and cause the grass to come out of dormancy faster. This is somewhat true, but there are some considerations before you should ever try to burn off zoysia. First of all, thatch is a reoccurring problem and it occurs at the soil line. Burning off the turf is not a guarantee that your lawn will be thatch-free afterward. I already mentioned the fire hazard of this grass when it is dry. The last fact I want to point out is that, yes, the zoysia might green up faster but only because the black soil surface from burning will gather light rays faster than the normal brownish color. No miracle really, just the effect of black as a heat-gathering color.

Bermuda Grass

Bermuda grass is a warm-season grass that is very popular in the southern transition zone and almost all parts of the southern states. In the northern parts of the transition zone and other regions of the country, Bermuda is considered to be a weed. Cynodon dactylon, as it is called botanically, is a very useful grass for stadiums and golf courses in many parts of the country (Figure 3.6). It has a very adaptable mowing height and can be cut anywhere from $1/8$ inch up to 3 inches.

It is believed that Bermuda grass actually originated somewhere in Africa although its name would suggest otherwise. This grass type has excellent heat and drought tolerance, even better than that

TURFTIP

A major disease problem that exists with Bermuda lawns in stadiums and athletic fields is a disease called "spring dead spot." It is still somewhat of a mystery exactly why this disease occurs. What happens is that large dead areas develop in the turf through the summer months, can continue into fall, and possibly even into the next season. The spots are ringlike areas where all or most of the turf dies in the center.

FIGURE 3.6 Busch Stadium has Bermuda grass turf that was installed in 1996. The use of Bermuda grass is very common on athletic fields and stadiums.

of zoysia. It adapts well to close mowings and can tolerate a wide range of soil types. The grass is very aggressive and recovers from damage quickly. It is not a cold-tolerant grass, however, and is very susceptible to winter injury. This is why it is not considered a good grass to use in northern regions.

The reason Bermuda grass is considered a weed in the north is due to its winter injury problem. Instead of thick, thriving lawns like those seen in the south, small areas of Bermuda pop up here and there in the lawn. They are not really that noticeable in well-manicured areas until first frost when these areas become brown long before the rest of the lawn goes dormant.

A major disease problem that exists with Bermuda lawns in stadiums and athletic fields is a disease called "spring dead spot." It is still somewhat of a mystery exactly why this disease occurs. What happens is that large dead areas develop in the turf through the summer months, can continue into fall, and possibly even into the next season. The spots are ringlike areas where all or most of the turf dies in the center.

The other problems with Bermuda are the same as with most of the warm-season grasses. The grass is very evasive and can get out of control. Once established, it can be hard to eradicate. This is also a grass that can have problems with thatch. And when Bermuda is planted in the transition zone, one can expect dormancy for six months of the year.

The newest and most successful forms of Bermuda that have been planted in recent years are referred to as the "tiff" series. There is Tifgreen, Tifway, and Tiflawn. You can guess by the names that these grasses are marketed for golf courses, fairways, and lawn usage.

St. Augustine Grass

St. Augustine grass is an extreme southern grass. Stenotaphrum secundatum is the botanical name for St. Augustine grass. Although it is the most common grass in the state of Florida, it is not really found in many other parts of the United States. It is used on both golf courses and residential lawns. The reason it is not found in any other regions is because it is not cold-tolerant at all. This has limited its progression even in northern Florida. There are newer varieties of St. Augustine that can now be found as far north as North Carolina.

TURFTIP

One of most distinctive features of St. Augustine grass is the plush carpet feel it has when you walk on it and the aggressive way that the grass spreads using stolons. St. Augustine is the most aggressive spreader, in my opinion, that exists. I have seen this grass cover a 3- to 4-foot sidewalk in a little over a growing season.

One of most distinctive features of St. Augustine grass is the plush carpet feel it has when you walk on it and the aggressive way that the grass spreads using stolons. St. Augustine is the most aggressive spreader, in my opinion, that exists. I have seen this grass cover a 3- to 4-foot sidewalk in a little over a growing season. Since this grass does spread so rapidly, edging sidewalks at the time of each mowing is a must. St. Augustine also has the capability to thrive in the poorest growing conditions. It will grow in salty conditions, which are common in Florida, and even in full shade. St. Augustine is a glutton for water and does require a little fertilizer, but when it is healthy, "watch out" because this stuff will really grow fast. Mowing is often needed more than once a week when temperatures are right and moisture is abundant.

There is a disease worth mentioning that seems to be more common every year. It is referred to as SAD or St. Augustine decline. It is typically a disease that shows up four to five years after establishment of a yard and is now being fought by simply using varieties that are resistant. 'Raleigh' is a variety of St. Augustine that is showing great promise in resistance to SAD and also expanding the use of the grass into North Carolina and other more northern states.

Centipede Grass

Eremochloa ophiuroides, or centipede grass, is a grass I at least have to mention because it has selective usefulness. It does make a reasonable turf for areas that are unfertilized and not watered. It is grown for the most part in the southernmost United States, and it's not really that common in lawns. This could be considered the honorable mention of the low-maintenance warm-season turfs. It you were looking for something to plant in an area that was not mowed or maintained that heavily, then this grass might at least be a consideration. I am not impressed by its growing habits nor its appearance, even when it is properly maintained.

Buffalo Grass

The last of the grasses I am going to mention is buffalo grass. It is a very adaptable native grass that can be used as a turfgrass. There are really very few native grasses grown in the United States that would qualify as both a native grass and a turfgrass. It also has the unusual characteristic of being monoecious, which means that it has both male and female plants.

I had to mention this grass because it is a good low-maintenance turfgrass. It requires little water or fertilizer, but the appearance is very rugged, as you would expect from a native grass. It only tolerates a high cut and is difficult to harvest seed from, making it less available in the industry. This is another plant that is not very useful for a manicured turfgrass, although I have seen some beautiful areas in Colorado. It still can be seen in large areas of prairie and open plains in some western states (Figure 3.7). In the two instances where I have dealt with it in the central midwest, it was already being used and the client was trying to figure out how to eradicate it. Anything native, of course, is hardy and can be somewhat evasive to control.

FIGURE 3.7 Buffalo grass is still common to prairies in the West.

Chapter 4

Careers in Turfgrass

I hope that many of you who are reading this book are already turf professionals or at least active within the turfgrass industry. The multibillion-dollar field of turfgrass maintenance encompasses a wide range of occupations and areas of expertise. People of all ages and economic backgrounds have a prime opportunity to take advantage of the numerous career opportunities that exist in turfgrass management. There are many other major areas of study closely related to turfgrass, which include agronomy, horticulture, greenhouse management, nursery crop production, floriculture, and landscape design.

The major area of study described in this book is known as turfgrass management. Professionals in this career can range from lawn care providers to athletic and sports field managers, and of course golf course superintendents. Levels of pay in the industry, as with any occupation, are based on experience and practical knowledge of the topic. I have found from my personal experience that many people do not even know what horticulture is and certainly do not understand the potential for making a living in lawn care or turfgrass. Horticulture is the scientific term for the study of plant science pertaining to landscaping and turf maintenance.

The first horticulturalists go back to grass developers and breeders in England long before the United States even existed. The Chinese and Japanese were developing and breeding plants even longer ago. Turfgrass was not introduced into the United States until the early 1700s. As I mentioned before, the types and varieties of grasses that were brought here were somewhat crude, but all served a purpose. I do find it interesting that some of the grasses that were introduced here are considered to be weeds in the turfgrass industry today. As the study and development of grass progressed both here and abroad, the need for expertise in the field increased. The first people who studied or worked with grass were referred to either as gardeners or as greens keepers. This label actually lasted until the late 1900s. Even today there are some people who do not think of turfgrass management as a profession. This is one of the main reasons for people like you to prove that there is a professional status in this line of work.

Turfgrass has now become one of the fastest growing industries. Revenues are in the billions, but the number of educated professionals in the industry is still very low. There are many career opportunities available, and the numbers increase every year. Thousands of jobs go unfilled in agriculture and hundreds of thousands in horticulture

TURFTIP

Turfgrass has now become one of the fastest growing industries. Revenues are in the billions, but the number of educated professionals in the industry is still very low. There are many career opportunities available, and the numbers increase every year.

each year. There continues to be a gap between educated high-paying jobs and uneducated low-paying jobs. I remember when I received my bachelor's degree in plant science there were only ten other graduates in the program. There were 15,000 business majors, many of whom struggled to find employment. I am mentioning this to show the need for educated professionals and how easy it is to find employment with a formal degree.

The major area of study is now called plant science, which is the study of horticulture. All areas of study, such as floriculture, turfgrass, landscape design, interior plant design, and nursery production, fall under the rubric of plant science. There are other career options besides the ones I have just listed, but these are certainly the more common choices that people will take.

GOLF COURSE CAREERS

The most common career choice that people in turfgrass usually pursue is golf course management. I spent nearly 10 years in the golf course industry and can understand the struggles and challenges that a superintendent can experience in the course of a year. There are career opportunities ranging from a laborer all the way up to the superintendent position. The superintendent is the person who is usually responsible for overseeing and planning all course maintenance

---

TURFTIP

The most common career choice that people in turfgrass usually pursue is golf course management.

---

activities. The superintendent should make sure that everything is budgeted for and that all course work is performed in accordance with board or owner directives. In most cases a golf course is operated by a board of directors. The superintendent must make the decisions of the board become a reality while still maintaining the integrity of the course. What usually happens with a board of directors is that many of the people who sit on the board have no practical knowledge of turfgrass and what it takes to actually run a golf course. It is one thing for a board member to suggest "speeding up the greens," but the reality is that the superintendent must be educated enough to explain if that can happen and why. Then the superintendent must relay to the crew how to make it happen. The biggest challenge for a superintendent is often the public relations part of his or her job in dealing with both a board of directors and the public.

The assistant superintendent is the next person in the chain of command. On many golf courses this person's job may be just as important as or more important than the superintendent's job. Although the pay is usually significantly lower, the assistant superintendent in some cases becomes the "scapegoat" for the superintendent and the employees. I hesitantly use that term, but the reality is that the assistant is a mediator between employees and the superintendent. If something is not carried out correctly by an employee or foreman, it is usually the assistant's responsibility to explain why. This is also the case if everything is done correctly but the end result is not what the board had intended. This means that often the assistant is getting the "buck" passed at him from both ends. I have seen many cases on larger courses where the assistant's job was more stressful than the actual superintendent's job. Keep in mind, though, that it is very rare to become a superintendent without having been the assistant superintendent first.

The next in command at a golf course is the crew foreman, who usually is in charge of the laborers. This person usually follows the instruction of the assistant and makes sure the final tasks are completed as planned and on time. He or she might be responsible for some scheduling and making sure that the equipment is in good shape. The equipment should be one of the biggest concerns for everyone at the course, though, since the work cannot be finished without it.

A full-time mechanic might be required to make a course run efficiently. This is not only a good time to mention how important a mechanic's role is in the golf course industry but also that it is becoming a good career option for many associated with the turfgrass industry. This is an example of the ripple effect that the turf industry creates as it continues to grow.

These are the most common chains of command at a golf course of average size; however this is not always the case. Many courses are large enough to have two assistants just to handle the numerous work orders that are performed each day. On the other hand, a course might be so small that there are only three people responsible for keeping up an entire 9-hole course. The responsibility burden may be quite high on all of the workers, and the need for productive people and equipment is quite high.

I now want to focus on why any of us have a career in any profession, and this is of course the money aspect. Of course, there are numerous other rewards for being involved in the green industry, but financial reward is why most of us work at all (Figure 4.1).

If any of you become major players in the golf course industry, you will probably become involved in an organization known as the GCSAA or Golf Course Superintendents Association of America. This is an organization I have been involved with over the years and can tell you that you should be a part of if you are in the golf course business on any level. The GCSAA is responsible for numerous changes in the turf industry and is constantly trying to improve the level of professionalism that is present on golf courses. The GCSAA provides many educational seminars as well as quality literature and information for people who are interested in learning more about the turfgrass industry. One of the pieces of literature the GCSAA provides to its members is a statistical guide called the Compensation and Benefits Report, which shows the level of pay for all levels of

National Average Wages for Maintenance Crew Positions

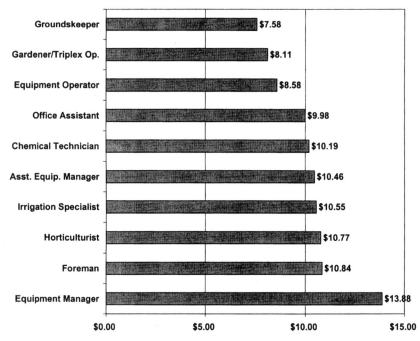

FIGURE 4.1 Average wages by position for golf course employees.

employment. Many of the statistics and diagrams in this chapter have been provided by the GCSAA.

One statistic that I would like to share with you is the average salary of a golf course superintendent in the United States, based on data compiled in 2000. It shows that the average superintendent made $57,057, which is up considerably from the $53,205 average in 1998. What is impressive to me is that right now half of all superintendents make over $50,000 dollars a year annually; 25% make more than $68,000 a year annually; and 10% make more than $88,000 a year annually. As you can see, these are impressive figures. Most people who are superintendents start this occupation sometime around the age of 30: 77% of superintendents are between 30-49 years old (Figure 4.2).

TURFTIP

One statistic that I would like to share with you is the average salary of a golf course superintendent in the United States, based on data compiled in 2000. It shows that the average superintendent made $57,057.

Age

	Number of Responses	Mean Age	Less than 30 Years	30 to 39 Years	40 to 49 Years	50 to 59 Years	60 years or More
Total	3,361	41	9%	37%	40%	13%	2%

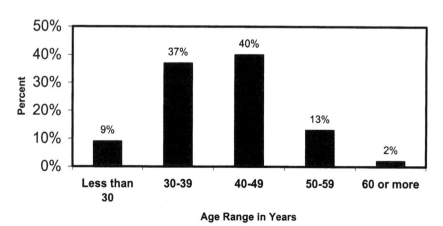

FIGURE 4.2 Average age of a superintendent.

I mentioned the importance of education within the field and how relevant it can often be to a person's success in the turfgrass industry. One of the more interesting statistics is a comparison of the level of pay for educated professionals vs. noneducated professionals with practical experience (Figure 4.3). Base salary by demographics is shown in Figure 4.4.

LAWN CARE PROFESSIONS

Another common occupation that a turfgrass professional might be involved in is a career as a lawn care operator or LCO. Lawn care companies vary from one-person operations up to large-sized companies that have hundreds of employees. It is certainly a respectable

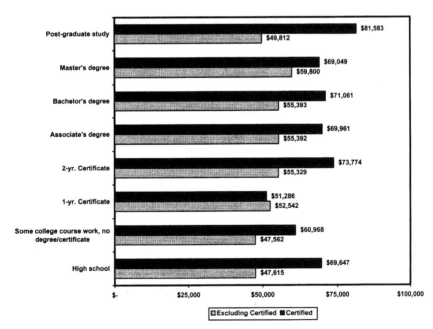

FIGURE 4.3   Average base salary of golf course superintendents by level of education.

	Number of Responses	Less than $35,000	$35,000 $49,999	$50,000 $64,999	$65,000 $79,999	$80,000 $94,999	$95,000 $109,999	$110,000 $124,999	$125,000 or more	Mean	Median
Country											
Canada	100	11%	21%	38%	19%	7%	1%	0%	1%	$55,757	$54,250
USA	3,424	10%	32%	28%	14%	8%	4%	2%	1%	$57,410	$52,000
Facility Type											
Private	1,452	4%	18%	27%	21%	15%	8%	3%	3%	$68,367	$65,000
Daily Fee	1,345	17%	41%	28%	8%	3%	1%	1%	0%	$48,467	$45,000
Municipal	425	11%	45%	33%	8%	2%	0%	0%	0%	$49,191	$47,000
Other	245	7%	33%	27%	19%	9%	2%	2%	0%	$57,505	$54,000
Military	33	18%	42%	27%	12%	0%	0%	0%	0%	$46,752	$43,500
University	18	17%	56%	22%	6%	0%	0%	0%	0%	$45,550	$42,000
Number of Holes											
9	280	46%	36%	13%	3%	1%	0%	0%	0%	$37,294	$35,000
18	2,580	8%	33%	29%	15%	8%	4%	1%	1%	$57,529	$52,000
27	278	4%	28%	33%	16%	12%	3%	2%	1%	$59,986	$55,000
36	241	5%	23%	29%	16%	12%	5%	2%	7%	$67,517	$60,000
45+	130	4%	18%	22%	23%	13%	5%	6%	6%	$71,443	$67,390
Annual Maintenance Budget											
Less than $100,000	372	45%	43%	8%	1%	0%	1%	1%	0%	$36,867	$35,400
$100,000 to $249,999	1,208	11%	43%	32%	10%	3%	1%	0%	0%	$50,443	$48,000
$250,000 to $499,999	1,076	3%	28%	34%	18%	10%	4%	1%	1%	$60,422	$56,000
$500,000 to $749,999	399	2%	18%	30%	23%	16%	7%	2%	3%	$67,618	$64,045
$750,000 to $999,999	157	2%	10%	22%	25%	21%	11%	6%	2%	$75,342	$75,000
$1 million or more	235	1%	6%	18%	22%	20%	14%	8%	10%	$84,412	$80,000

FIGURE 4.4 Base salary of golf course superintendents by demographics.

career and one that people seldom thought about just a few years ago. Today's homeowner seldom has an abundance of free time to spend "cutting the grass." This has spawned one of the largest growing segments of the turfgrass industry. People are becoming more conscious of the appearance of their yards and certainly are willing to spend good money to have them look good. I have found that it is hard for many lawn care businesses not to be involved in either pesticide applications and/or landscaping. Consumers are driving companies to be all-inclusive in taking care of outside work on a property. From a business owner's perspective, this allows a business to diversify and to pick up additional revenues that might have not been considered otherwise. If you find yourself in this situation, you can either expand the business to include those services or rely on reputable subcontractors to do the work for you.

LCOs and especially small lawn businesses are flourishing in all parts of the United States and even in countries abroad. The reason that many start in the lawn care business is because they think that the work can be done in the evenings or weekends while still maintaining a full-time job. In some cases many part-time "weekend warriors," as I like to call them, grow into multimillion-dollar operations. In some instances people start lawn care because they find themselves in a situation where they need employment and look to this occupation as an area where they can have a future with little to no training. Although there are such successes, it is hard to compete and grow in the lawn care profession without any training, experience, or formal education.

Employment opportunities exist outside of running your own lawn care business. You can be a foreman for a larger, more reputable company that already exists. Most LCOs struggle to find high-quality employees just like other businesses do. I have said many times that there are really only two types of jobs for any of us no matter what career we choose. We choose to be either an employee or an employer. There is a lot to be said for either side of this scenario. Certainly with leadership comes the penalty of responsibility and stress. However, these stresses are often compensated with significant economic rewards. A lawn care business requires many long hours of

TURFTIP

A lawn care business requires many long hours of hard work and a dedication to taking pride in a field that most will refer to as "cutting the grass."

hard work and a dedication to taking pride in a field that most will refer to as "cutting the grass." I mentioned earlier that until just recently many occupations associated with lawns and turfgrass maintenance were considered to be held by people of low economic status or of little education. This is simply not true. People who survive in the golf course and lawn care businesses are usually hard-working well-educated people. By educating ourselves and the public we raise the bar of income and create a general improvement in the industry as a whole. After all, turfgrass and plant management is a science. Through good public relations skills and education of employees the industry will continue to grow and prosper. The level of the wages will continue to increase, and turfgrass will become more well known as a profession.

I would like to share with you some statistics that show the level of progress the lawn care business is experiencing (Figure 4.5). It is unfortunate that the recent slowdown the economy has experienced reflects some negative trends and percentages; however, the turfgrass industry is one segment of the economy that has fared better than others during the last couple of years.

A variety of different services were offered by most of the companies that were included in the data pool. Most companies are diversifying their businesses to compete with other companies. Below is a breakdown of services rendered (Figure 4.6).

It is always interesting to see how companies fare from year to year. Considering the problems the economy has suffered in 2000 and 2001, here are the results (Figure 4.7).

I constantly focus on the importance of education of the workforce of turfgrass professionals. Figure 4.8 shows what experience the people in the lawn and landscaping industry have.

Although many companies provide respectable wages to their employees, salaries are dictated by geographic location and the economic level of the area the business is located in. If you own a lawn care business or are thinking of starting one, consider the level of competition in your geographic area. If you live in a large city and there is a LCO on every corner, then you can expect that margins might be low. By the same token, if you are in a more rural area where fewer people live, the price to do business might be lower and competition could be less. The big city might pay big dollars, but the

	Industry Average	Entry-level (<$250,000)	Mid-level ($250,00-$1M)	Established (>$1M)
Years in business	18	12.5	18	24
Revenues rising	72.4%	62.0%	72.6%	84.6%
Revenues holding	21.0%	25.8%	23.2%	10.7%
Ave % of revenue rise	18%	24%	14.5%	18%
Price rising	66.3%	62.0%	73.2%	71.6%
Price holding	31.7%	36.2%	26.7%	25.0%
Ave % of price increase	8%	10%	7.5%	5.5%
Residential customer %	57%	72.5%	60.0%	35.5%
Commercial customer %	38%	26.5%	37.5%	55.5%
Government client %	4.0%	1.0%	2.0%	5.0%
Miscellaneous client %	1.0%	0.0%	0.0%	4.0%
Residential profit margin	28%	39%	24%	18.5%
Commercial profit margin	23%	38.5%	20%	13.5%
Specialty services margin	30%	50.5%	39%	14.5%
Landscape maintenance margin	26%	47%	18.5%	14.5%
Chemical applications margin	26%	44.5%	25.5%	17%
Landscape design/install margin	24%	38.5%	22.5%	14.5%
Irrigation services margin	23%	40%	24%	16%
Tree care margin	23%	44.5%	34%	15%
#Full- time employees	34	3	7	131.5
#Part-time employees	13	2	6	41.5
#Family members employed	3	2	2	6.5
#Workers of foreign origin	36	5.5	7	85.5
Average hourly wage; best supervisor	$16.01/hour	$14.20	$15.03	$18.56
Average hourly wage; entry-level crew	$8.24/hour	$8.45	$8.18	$8.

FIGURE 4.5 Industry averages for growth in 2001 for lawn/landscaping.

Industry Service	Average
Landscape installation	86.3%
Mowing	72.6%
Landscape design	71.7%
Turf fertilization	69.3%
Ornamental care	66.0%
Turf weed control	62.3%
Turf aeration	59.0%
Turf insect control	54.7%
Turf disease control	49.5%
Irrigation maintenance	49.1%
Landscape lighting	48.1%
Irrigation installation	46.7%
Tree care/ Arbor services	45.3%
Hardscape/ paving/ patio installation	43.9%
Snow removal	42.9%
Pond/ lake maintenance	24.1%

FIGURE 4.6 Services offered by companies in the lawn care industry.

overhead could be much higher than what the rural setting would provide. With any business there is no guarantee for success. It is up to the individual to balance profits vs. overhead to create the ideal balance for growth.

OTHER AREAS FOR EMPLOYMENT

An area that has been overlooked by turf professionals is college and municipal maintenance. There are many colleges and schools that

74 Turfgrass Installation: Management and Maintenance

In 2001 we...	Industry average
Had more sales	76.6%
Found it more difficult to compete in the market	62.5%
Had more customers	64.4%
Used more suppliers & distributors	48.7%
Had more discerning customers	59.4%
Found it harder to be profitable	64.8%
Had financing/ cash flow problems	58.6%
Found it harder to recruit good employees	64.0%

FIGURE 4.7 Comparison to 2000.

	Industry Average	Entry Level <$250K	Mid-Level $250K - $1M	Estd. >$1M
Average working hours/week	56	54.5	58	55
Average years in industry	19	16	18.5	20.5
Average Age of worker	44	43	44	45.5
High school or less	12.0%	12.0%	6.2%	17.3%
Some college	30.4%	30.3%	39.3%	17.3%
Vocational/tech school	4.3%	6.0%	3.0%	4.3%
Associate's degree	10.9%	15.1%	9.0%	8.6%
Bachelor's degree	35.9%	36.3%	33.3%	43.4%
Post graduate degree	6.5%	0.0%	3.0%	4.3%
Very satisfied with career (well rewarded)	82.7%	71.6%	84.9%	90.4%
Somewhat satisfied w/ career (or rewarded)	14.9%	25.0%	12.3%	7.9%
Not satisfied with career (or rewarded)	2.4%	3.3%	2.7%	1.3%
Ave. personal take home pay	$73,351	$45,884	$67,149	$106,666

FIGURE 4.8 Hard-working landscape professionals.

TURFTIP

An area that has been overlooked by turf professionals is college and municipal maintenance. There are many colleges and schools that maintain large campuses, which require expertise and a good understanding of turfgrass.

maintain large campuses, which require expertise and a good understanding of turfgrass. Many of these campuses contain athletic fields that are considered to be full-time responsibilities in some cases. At larger universities there is usually a maintenance staff that is responsible for college grounds and another staff that is responsible for the athletic fields. The business of maintaining sports fields is a very high-paying job. Many schools and colleges would not even consider a grounds superintendent who did not have a degree or extensive knowledge of the industry.

It is obvious that the higher-paying jobs for athletic field maintenance are with professional sports teams or big collegiate schools. A degree is now almost a must, and the people who do this work can expect $60,000 to as much as $150,000+. As stadiums continue to make the change back to regular grass instead of artificial turf, the job openings continue to increase. At one time it was believed that artificial turf could eliminate major expenses, but it was later concluded that the increase in injuries was not worth the change. Several fields that were once famous for the condition of their turfgrass went to artificial turf back in the early eighties. This was thought to be the ideal condition for field managers. By eliminating grass, equipment, and crews, the turf complexes and stadiums believed that they could save big bucks. The reality was that artificial turf was

> **TURFTIP**
>
> It is obvious that the higher-paying jobs for athletic field maintenance are with professional sports teams or big collegiate schools. A degree is now almost a must, and the people who do this work can expect $60,000 to as much as $150,000+.

nothing more than glorified carpet on top of concrete. This was why the injury level was so high and also why many fields now have gone back to regular grass. There are certainly other factors involved the transition back to grass. For instance consider just the temperature reduction that grass can create in a closed area like a stadium. There was also a belief that artificial turf could extend the playing season. What it comes down to, though, is that there is just no replacing the job and benefits of an experienced turfgrass manager.

Athletic fields can consist of a wide range of sports. As the popularity of soccer continues to increase in the United States, more jobs open up. There are baseball, football, rugby, lawn bowling, tennis, and other types of fields. All of these types of facilities provide numerous career opportunities. It is hard to find a geographic area that would not have at least some type of golf course or athletic field.

There are also teaching and research jobs available in the turf industry. With so many new advances being made in turfgrass science each year, there are new openings available. Many chemical companies have research facilities. There continues to be an effort to make grass types greener, more hardy, and less susceptible to disease. It is also a continuing goal to reduce the need for chemicals and to breed disease- and pest-resistant grasses. Although most researchers have graduate degrees, there are still possibilities for assistants or maintenance people. Almost any college or chemical research center will have a test plot or research farm that requires a crew. Although students and researchers do a lot of the work on a turfgrass research farm, there is still a need for someone to do the maintenance work.

The last area that I want to mention as a career in turfgrass is seed and sod production. Although this can often be closely related to research because of the nature of improving seed or sod quality, there are different avenues for work. A person could own and operate an entire seed production facility or farm. There is a need for experience and expertise. The equipment used for seed production is also very specialized and often quite expensive.

Sod farms are also on the increase as construction contractors and commercial landscapers look for the immediate effect that sod creates. Sod also requires very specialized equipment and very high standards for turf quality and maintenance to provide a desirable turfgrass crop. Sod farms are held in many cases to some of the highest-quality requirements. A career in sod production can be a good career choice.

TURFTIP

Sod farms are on the increase as construction contractors and commercial landscapers look for the immediate effect that sod creates. Sod requires very specialized equipment and very high standards for turf quality and maintenance to provide a desirable turfgrass crop.

There are certainly many jobs that are closely related to turfgrass but do not directly require much experience or an education. Who can put a number to the numerous industrial jobs that exist due to the thousands of commercial mowers that are sold each year? As the turfgrass industry has progressed over the last few years, new types of mowers and specialized equipment continue to be developed. The idea is to create power equipment that can maximize efficiency and reduce work hours. One of the main reasons for this is because business owners cannot find enough quality employees to work on crews and do maintenance work. If a business owner has an opportunity to buy a piece of equipment that can potentially replace three members of a crew and reduce overhead in the long run, he or she will probably make that choice. A mower or tractor does not get lazy, call in sick, or need a vacation. As the need for employees has increased, the demand for unique and efficient equipment has increased as well.

Chemical companies are another large component of the industry where manufacturing jobs exist due to turfgrass maintenance. Not only is it necessary to have test farms and research plots of grass, but someone has to actually manufacture and produce the chemicals that are used. Thousands of pounds of fertilizer and liquid chemicals are produced each year to keep up with the needs of turfgrass professionals.

There is a snowballing effect in how the greens industry continues to create jobs (Figure 4.9). There is something to be said for the person who makes the rake that fits in the toolbox that goes on the truck and so on and so forth. What I am saying is that there is no way to assess the true effect turfgrass has had on the American economy and the jobs that have been created from lawn care. I just hope that I can stimulate you to get interested or excited about turfgrass and the many career choices that it offers.

Service Segment	Fastest growth (Industry average)	Rate of growth (Industry average)
Landscape maintenance	45.0%	30%
Landscape design/ installation	41.9%	23%
Chemical application	11%	47%
Specialty services	8.9%	23%
Irrigation design & installation	6.3%	18%
Tree care / arbor services	4.7%	28%

FIGURE 4.9 Fast-growing services.

Chapter 5
Cultural Practices

MOWING

Since mowing is the primary maintenance practice associated with turfgrass, it should be the one you are most familiar with. All grass grows and therefore needs a routine mowing to keep it looking and performing its best. All turfgrass is stimulated by the practice of mowing, therefore it can improve turf quality and condition. There are optimal techniques for mowing and optimal times that it should be done.

TURFTIP

One of the primary variations in mowing a golf course is to create the different playing surfaces.

Golf Course Mowing

I will start first by focusing on golf course mowing. Golf course mowing is performed the most frequently and to some of the highest standards of any turfgrass. The mowing is done daily in many cases on a majority of the course whenever possible. This is true primarily for larger facilities that have the crew and equipment to provide that degree of maintenance. Variations in mowing height for golf courses can be measured in as little as $1/32$ inch. One of the primary variations in mowing a golf course is to create the different playing surfaces. Different mowing heights create the actual play areas and the differences between the "roughs" and the "greens" Without the mowing frequency or differences in height, a golf course would look like a large open lawn. This is also the reason that smaller courses do not often provide the highest-quality level of play. If you do not have all the different types of equipment required for creating the variable heights in grass, you end up with fewer types of playing surfaces.

Mowing on a golf course does require specialized equipment of all sizes. It usually requires the use of larger equipment that is not commonly used for lawn care operations. There are three types of

TURFTIP

There are three types of mowers that are used on a golf course: the flail mower, the reel mower, and the rotary mower. The most common types of mowers used are the rotary and reel mowers, but in recent years I have seen a handful of courses using flail mowers.

mowers that are used on a golf course: the flail mower, the reel mower, and the rotary mower. The most common types of mowers used are the rotary and reel mowers, but in recent years I have seen a handful of courses using flail mowers. A rotary mower is the most common type in general grounds maintenance and is the one that most people are familiar with. A rotary mower has a rotating blade or blades that actually tear the grass off as the machine spins in a circle. I know this may sound elementary and you might be thinking, "I know how a mower works," but do you realize that the grass is being literally torn off and not actually cut? This leaves a jagged edge on the grass tips. The only mower used on golf courses that truly cuts grass is a reel mower (Figure 5.1). The reel mower provides an actual scissors action that pinches the grass between a blade and what is called a bed knife. This type of mowing produces the best appearance and cut quality. This is why most mowers that are used on golf courses are reel mowers. It would be impossible for a rotary mower to cut anything shorter than ½ inch in most cases without scalping. Scalping occurs when the mower engages the soil along with the turf, causing injury and leaving a somewhat unattractive appearance. Scalping is very harmful to the turf because it can cause damage to the plant crowns. I have mentioned before how the crown is a very critical point of growth in a grass plant, and damage to this area can take some time to recover from if it does not kill the grass.

I do want to mention that a sickle mower also provides a true cutting action. This mower has sharp slicing knifes that reciprocate back and forth over one another. This action actually chops the grass off, and it will fall over at the point on the plant where it is cut. This mower is more commonly used in hay production or other agricultural uses. The large pieces of grass that it leaves are desirable for hay production but not for golfers. I have seen sickle mowers used in golf courses for outlying areas or pond bank maintenance, but those are about the only instances in which they would be useful.

The last type of mower I mentioned for golf course use is the flail mower, which has also been more commonly known for its use

84 Turfgrass Installation: Management and Maintenance

FIGURE 5.1 This wide area reel mower leaves wonderful striping that is very desirable.

in agriculture. It uses an odd-shaped triangular blade that spins on a reel. I do not want to say that it is a combination of the reel and the rotary mower, but it does combine some of those characteristics. For all practical purposes I myself have never used a flail mower for a golf course because I do not consider the mowing quality to be up to golf course standards.

Mowing Areas

When mowing on a golf course, there are several areas that require different styles of mowing. The areas are as follows:

- *Greens are the areas that golfers use for putting. The grass heights are less than a ¼ inch on most courses and are the highest points of maintenance on the course. Mowing is always performed daily during the season of play.*

- *Collars and fringes are the areas surrounding the greens. Typical mowing heights range from ¼ to ¾ inch in height on most courses. Often these areas contain irrigation heads and create a protective border between the greens and taller turf. Mowing is performed here daily if possible or two to three times per week.*

- *Tees are the areas where golfers start each hole. This grass is maintained at an average height of ¼ to 1 inch. Mowing is usually performed here daily or at least twice a week.*

- *Fairways are the areas that lie between the tees in greens. This is "the playable route" that should be followed to get from tee to green. The height is variable but can range from ¼ inch up to 2 inches in height depending on the mowing equipment available and the type of grass present. Mowing can be performed as often as daily or as little as twice a week depending on these factors.*

- *Roughs are the areas that outline all other playing areas. They encompass a majority of most course mowing. These areas are covered with the tallest grass on a course that is still considered to be "in play." Mowing heights are commonly 2½ up to 5 inches in height and mowing occurs only one or two times a week, depending on desired height.*

These are just the most common playing areas that are found on a majority of golf courses. Mowing heights vary a great deal based on desired playing conditions, equipment present on the course, and turfgrass species present. Some courses may break their maintenance into primary and secondary areas. For example, roughs may be mowed at two heights and referred to as primary rough and secondary rough.

Mowing frequencies vary a great deal on a golf course based on the specific requirements of play for that area. All mowing heights tend to be raised slightly during hot and/or dry periods to reduce stress. The reason this is done is because all turf can develop a puffy

> **TURFTIP**
>
> It is critical when mowing at extremely low heights to mow early in the morning and not during the hot parts of the day. Heat causes more stress to the grass and can increase scalping and turf injury. Evening mowing is not recommended because it can lead to increased disease occurrence.

appearance and be much more susceptible to scalping or damage during these times. In extreme drought some areas of the course may not be mowed for long periods of time to reduce the stress to the grass further. We hopefully have learned by now that grass is composed of water, which is why reducing heights and mowing frequency helps to reduce the plants' water loss.

It is also critical when mowing at extremely low heights to mow early in the morning and not during the hot parts of the day. Heat causes more stress to the grass and can increase scalping and turf injury. Evening mowing is not recommended because it can lead to increased disease occurrence. If the turf has just been mowed, the freshly cut shoots are an ideal area for disease to invade.

Another practice that is common on golf courses is that of bagging a majority of the turf that is mowed on the course (Figure 5.2). This takes away moisture from the turf that is being cut but is often offset by the irrigation that is almost always provided to golf course turf. Irrigation plays a critical role in golf course maintenance, and without it, mowing could never be performed to such low heights. Recycling of clippings or nonbagged turf occurs in the larger areas, commonly just the fairway and rough. The reason the clippings are collected in the other parts of the course is because they can actually interfere with course play by causing the ball to roll or bounce off-track. There is the more obvious reason of the aesthetic benefits from the removal of the clipping debris. There have been some studies that compare disease resistance on recycled vs. bagged turf. Recycling clippings does reduce certain disease instances; however, other diseases were not affected and in some cases actually increased in occurrence.

FIGURE 5.2 Greens mower picking up clippings.

FERTILIZERS

I talked a great deal in Chapter 1 on soil fertility and the importance of a good fertilizer program. Fertilizer is the key to having healthy dark green turf that is free from stress and disease. Fertilizer comes in many different forms, and each type is best suited for different plants. I am not going to mention every type of fertilizer, but I do want to focus on some of the more common types you will encounter and what I think you should know about them.

Fertilizer Types

I want to start off with the most common type of fertilizer that for use on turf. It is either urea-based or ammonium nitrate-based and has no slow release properties at all. These types of fertilizers have been around for many years. They are still used today and they are very effective. Any combination of the elements Nitrogen-Potassium-K can be found in these fertilizers, but a micronutrient package is fairly uncommon unless it is a more reputable brand. Urea-based fertilizers are still very popular in the agriculture industry, where they are used in fields and pastures. They are also used on golf courses

> **TURFTIP**
>
> Urea-based fertilizers are very popular in the agriculture industry, where they are used in fields and pastures. They are also used on golf courses and in the lawn care business for turf applications.

and in the lawn care business for turf applications. The main reason that urea-type fertilizers are used is because of cost. Quick-release urea and ammonium-nitrate fertilizers are still some of the cheapest types available. The prices of these fertilizers are dictated slightly by fuel prices because heat is required to manufacture them, but in most cases they are much cheaper than some of the alternatives.

The close cousin so to speak of regular urea-based fertilizers are SCUs, or sulfur-coated ureas. These are urea-based fertilizers, but they also have a coating that gives them a slow-release quality that makes them very desirable. There is not a huge price increase over the traditional urea fertilizers. SCU is usually mixed in with a regular urea fertilizer, yielding a fertilizer that is partly readily available and partly slow-release. The slow-release portion can be expected to last three to six months depending on many other factors such as the brand of the fertilizer, the zone it is applied in, and environmental factors such as rain or temperature.

There are other types of slow-release urea and ammonium nitrate fertilizers that use different technologies for the release mechanism. For example, there are methylene ureas and other coated urea fertilizers. They share the desirable slow-release properties. I would suggest that any fertilizer for use on turf should be a slow-release type. This is especially true for applications that are put down in the fall.

There are other types of slow-release fertilizers that are coated with a poly coating or some sort of plastic coating. One example of this is a product called Osmocote™ (Figure 5.3), which is a ammonium nitrate-based fertilizer with a plastic-type coating that makes it slow-release. As water, temperature, microbes, and other factors work on the coating, the fertilizers are released a little at a time. These types of coating advertise six to nine months of slow-release properties. The only problem I have encountered with them is the effects of freezing temperatures, which can cause the coatings to break down prematurely and immediately dump the nitrogen. This problem can occur with other coated fertilizers, but the results from the freezing are not usually as drastic.

TURFTIP

I suggest always using rotary spreaders as opposed to drop spreaders for applying fertilizer. Rotary spreaders ensure a more even distribution and leave less chance for overlap and turf burning.

FIGURE 5.3 Even though this particular bag is not for lawns, this is an example of a coated fertilizer that offers 120 days of feeding. This would be a type of coated ammonium nitrate.

All granular fertilizers come in small particle form and are dry for easy spreading with some type of drop or rotary spreader. I suggest always using rotary spreaders as opposed to drop spreaders for applying fertilizer. Rotary spreaders ensure a more even distribution and leave less chance for overlap and turf burning (Figure 5.4). Another way to help improve distribution and consistency is by using fertilizers that are well blended and have good particle sizes that spread easily. There are several types of fertilizer blends that are available. Some companies use what are called homogenous blended fertilizers. Homogenous blends ensure product consistency and more evenly sized fertilizer particles so that all of the turfgrass areas you are spreading receive even coverage and the same amount of each element.

There are many technologies and features of fertilizers. The purpose of fertilizer is to enhance turf quality, be available when needed, and provide a well-balanced release mechanism. Some fertilizers are immediately available and provide the plant with an instant jump start. A very popular choice is liquid fertilization. This is a common way for lawn care operations to be able to supplement fertilizer needs along with a weed treatment program on the lawn. The use

TURFTIP

The use of liquid fertilizer is only possible for operators who are using sprayers that apply 3 to10 gallons of water per 1000 square feet. Liquid fertilizing is done to give lawns a quick burst of green color or to rejuvenate the grass along with a weed treatment.

FIGURE 5.4 When spreading fertilizer, a rotary spreader is best for distributing the fertilizer evenly. It also the best device to control waste of any of the material.

of liquid fertilizer is only possible for operators who are using sprayers that apply 3 to 10 gallons of water per 1000 square feet. Liquid fertilizing is done to give lawns a quick burst of green color or to rejuvenate the grass along with a weed treatment. It usually only provides a short-term effect, and I am not a fan of using liquid fertilizer for the entire growing season.

Liquid fertilizer provides a temporary release into the plant or soil. The problem is that once the plant has used up this form of nitrogen there is little else available for turfgrass consumption. Liquids move quickly through the plant and do give it an instant burst of color. However, they often are being applied to turfgrass plants at times and at rates that are not conducive for fertilizer treatments. There are companies that claim to offer slow-release liquid fertilizers. I still have not seen any liquid that compares to the results provided by high-quality coated fertilizers.

The last fertilizer I want to tell you about is a throwback to the past. Organic gardening and lawn care have created a great demand for manure or organic-based fertilizer. These fertilizers are often made from ingredients such as recycled human and pet waste, fish and plant byproducts, or composted and recycled materials. The idea is to create a totally organic product that is still easy to spread. Some fertilizers offer a valued natural side effect of disease resistance and color enhancement. Turfgrass responds very well to the organics, but typically availability and price has limited their usage. As the demand continues for organics, prices have fallen and newer improved products are being created. I personally had unbelievable results when spreading a chicken manure product to my golf tees years ago. There was an odor issue, but I did have a tremendous period of dark green color, disease resistance, and improved turf quality. The problem came when I walked into the clubhouse after I finished spreading the material.

Organics are very popular because of methods such as integrated pest management and other practices that encourage less fre-

TURFTIP

Organics are very popular because of methods such as integrated pest management and other practices that encourage less frequent use of chemicals.

quent use of chemicals. IPM, as it is known, would suggest that you treat the golf course on a have-to basis and use preventive practices and chemical alternatives. In other words, chemicals are only used once thresholds are reached. The threshold is measured based on a damage factor, financial factor, or aesthetic factor. The deciding commonality among these three issues is when the damage from any pest or disease reaches an intolerable point. Since organics rely on natural materials found in the environment, they are more desirable for people with a low appreciation for chemicals.

Application Methods

The use of fertilizer is a necessary practice associated with turfgrass growth and production. In most cases the existing soil that is being utilized for grass establishment is deficient in some or all of the elements required to keep turf looking its best. Turfgrass prefers to have an abundance of nitrogen, phosphorus, and potassium available for growth. The grass may also need several minor elements. I want to point out once again that proper fertilization and soil fertility are the key to turfgrass success. I could grow grass on a slice of bread by supplementing the nutrition that plants need along with water and light.

TURFTIP

Turfgrass prefers to have an abundance of nitrogen, phosphorus, and potassium available for growth. The grass may also need several minor elements.

> **TURFTIP**
>
> Grass has a specific time frame when fertilizer should be applied, and it is different for warm-season and cool-season grasses. The biggest mistake I see home-owners or private individuals make in their yards is to fertilize at the wrong time of the year.

Grass has a specific time frame when fertilizer should be applied, and it is different for warm-season and cool-season grasses. The biggest mistake I see home-owners or private individuals make in their yards is to fertilize at the wrong time of the year. Part of the problem has come about from the marketing practices of retailers and manufacturers who sell fertilizers and lawn care products. By the time you as an individual are seeing these products on the shelf, half the application window has passed in many cases. The key to good fertilizing methods is to promote root growth and not top growth.

Cool-Season Fertilization

Let us first start with the cool-season grasses and the times in which they should be fertilized. It is very common to see fertilizer in the store around late April, May, and even into June. This is great if you are fertilizing warm-season turf but not so good for cool-season grass. I want to start in the fall and work towards the summer, because this is how your fertilizing should work. The best time to apply fertilizer is in early fall, which in the Midwest is sometime around the first of September and on into October. At this time you would ideally apply one pound of slow-release nitrogen per 1000 square feet. I like to see one application of one pound in September and a second in late October or early November of ½ to 1 pound. You can use up to three pounds of nitrogen per 1000 square feet in one application, but anything over that amount is almost always a disaster for turf. Even two to three pounds per 1000 square feet can result in turf injury or burning if the application is done incorrectly or applied on a hot day. The reason why fall is the best time of the year to fertilize is because you are promoting root growth through the entire winter months instead of top growth. It is a given that in the spring all grass will green up, grow like crazy, and often be hard to keep up with when mowing. The reason why the homeowner feels the need to apply fertilizer to the lawn at this time is because that is perceived to be the best time to do it and that is when the stores sell it. This practice and mentality needs to be changed.

As the winter progresses and the grass becomes dormant, there is a lot going on in the turfgrass plant that people do not see. Even though the grass on the surface of the soil is brown and dormant for the winter, the grass is still very actively growing under the soil. This can be seen by digging into the sod with a knife. You will still

TURFTIP

As the winter progresses and the grass becomes dormant, there is a lot going on in the turfgrass plant that people do not see. Even though the grass on the surface of the soil is brown and dormant for the winter, the grass is still very actively growing under the soil.

see white roots that actively grow and develop under the soil all through the winter months. The grass will utilize the nitrogen that has been applied to the soil for the most part by late winter or early spring. If you are only going to fertilize once a year, fall is the time to do it. However, if you are going to fertilize more than once a year, then you will want to start thinking about it very early in the spring of the following year.

I always suggest that people fertilize in the spring when grass is still partially dormant or just starting to green up. This will not only "kick start" the grass for the spring growing season, but it will also still allow some positive root growth early in the year. Grass that has a good root system is able to survive disease and drought much better than grass without it. Roots are the key to good turfgrass growth; without good roots grass will never appear healthy. If you apply one pound of nitrogen per 1000 square feet around March 1 in the Midwest, you have met most of the fertilizer needs for the season. I do like to follow up the spring fertilizing with a light fertilizing sometime in May. This application may be accompanied by a weed treatment or even a grub control, but the idea is to ensure that the grass has that last burst of energy to get through the hot summer months.

TURFTIP

I always suggest that people fertilize in the spring when grass is still partially dormant or just starting to green up. This will not only "kick start" the grass for the spring growing season, but it will also still allow some positive root growth early in the year.

All the fertilizer you apply should have some slow-release characteristics to ensure that some is available for the turfgrass after rain so it can quickly rejuvenate.

The problem with fertilizing in the late spring is that the grass is already growing and probably at a very fast rate. Once you dump fertilizer down late in the spring or summer, the grass is only encouraged to grow more. What you are doing is encouraging top growth at the expense of the root system and therefore weakening the plant's long-term survival capabilities. This is especially true for grass that may not have received any fertilizer in the fall. What happens is that the plants shoot out a burst of growth and then often bake from hot summer temperatures as the growing season progresses. If the grass plants have no root systems, they are very slow to recover from heat or drought stress. Disease is also much more likely to damage turfgrass with no root system.

One misperception of fertilizing in general is that it just causes you to mow more often. Properly fertilized turf with an established root system will grow just as slowly or more slowly as turf that has not been fertilized. The reason is because the plant is developing and growing in ways that benefit the turf quality instead of just creating shoot growth.

TURFTIP

All the fertilizer you apply should have some slow-release characteristics to ensure that some is available for the turfgrass after rain so it can quickly rejuvenate.

I mentioned the use of liquid fertilizers on turfgrass plants and how they are a quick fix for recovery. Plants can and will utilize liquid fertilizers in the summer if they are applied. They can be effective if the plants have a well-developed root system from fall fertilization practices. A little bit of fertilizer is fine to use in midsummer if you are a believer in that practice, but a lot of fertilizer does nothing more than to cause high stress and turf injury. I do not suggest fertilizing cool-season grasses in the Midwest for any reason from the first of June until late August, unless it is with low amounts of nitrogen in a liquid fertilizer. I do stress the low nitrogen content for this type of fertilizing.

Ideally turfgrass would prefer to have small "spoon feedings" of nitrogen throughout a majority of the year and not a large amount at any one time. Since this is impractical, I suggest the use of slow-release fertilizers in a good program to have basically the same effect. Following is a one-year schedule for cool season turf in the transition zone:

- *One pound of nitrogen per 1000 square feet in 50% or less slow release form can be applied the last two weeks in February or the first two weeks in March, possibly 28-3-10 (10-50% SCU). It can be used as a carrier for a pre-emergent treatment.*

TURFTIP

Ideally turfgrass would prefer to have small "spoon feedings" of nitrogen throughout a majority of the year and not a large amount at any one time. Since this is impractical, I suggest the use of slow-release fertilizers in a good program to have basically the same effect.

- In May or June ½ pound or less of nitrogen per 1000 square feet, possibly 12-2-10; slow release is not a necessity. This could be a carrier for a post-emergent treatment for crabgrass or other annual weeds or for grub control if one is needed.

- In June to August supplement liquid fertilizer if you use it. Low nitrogen and high water volume would be best. Use only one treatment in combination with weed or late grub treatment or avoid all together. I recommend no dry fertilizer from June 1 until late August.

- In August to September use one to three pounds of nitrogen per 1000 square feet in a slow-release fertilizer with 50% SCU or higher. I suggest one pound in August and the other in a second treatment in October, possibly 34-10-18 (50% SCU). It can be used as a carrier for fall weed treatments but this is not required.

- In October to November use one pound of nitrogen per 1000 square feet, possibly a 20% or less SCU. Use as a second fall treatment if 1½ pounds or less were applied in early September.

This a just one example of a recommended program for cool-season turf in most of the Midwest. Each region can be different, and certainly the monthly temperatures can vary from state to state. This may not be the best program for your area. This program also discourages the used of liquid fertilization.

Warm-Season Fertilization

Warm-season grasses have a different fertilization timetable. This is especially true for warm-season turf grown in the transition zone due to the long periods of dormancy that most such grasses have.

Fall fertilizing is not as big a necessity, but it does allow some positive root production. The most critical fertilizer applications for

> **TURFTIP**
>
> The most critical fertilizer applications for warm-season grasses are right before and during the actively growing periods in the summer months.

warm-season grasses are right before and during the actively growing periods in the summer months. I suggest a treatment in late April or early May of one to two pounds of nitrogen per 1000 square feet. This will give the grass a good start after a long period of dormancy. There can then be follow-up treatments of ½ to 1 pound of nitrogen in June and again in August.

Warm season-grasses like to have a high amount of nitrogen available and plenty of moisture while they are actively growing. In the winter months while they are dormant little growth activity occurs. You are not going to notice a patch of zoysia spreading out or growing in December but rather in July, when many of the cool-season grasses have gone dormant from the heat. Warm-season grasses prefer warm temperatures and an abundance of food and water while they are growing.

IRRIGATION

This is a good time to focus again on the water requirements for turfgrass and the best way to provide grass with water. Water is so critical to plant functions and health that irrigation is always used when top-quality turfgrass is demanded. You can always find irrigation systems on a golf course where grass greens are used. Irrigation is also becoming more popular in homeowner lawns as the desire for that perfect yard grows in suburbia. Irrigation is also commonplace in most athletic fields and sports complexes (Figure 5.5).

FIGURE 5.5 Rotary irrigation system in use.

TURFTIP

Most irrigation systems use a natural water source such as a lake, stream, or river due to the cost and availability of water in general, especially municipal water.

Irrigation does not always utilize pipes and sprinkler heads, although that is the most common method. Irrigation simply refers to the use of a supplemental water source to take care of the needs of grass or crops. Water is usually moved by some external source such as a pipe, person, or tank. Irrigation can come from many different sources. Most irrigation systems use a natural water source such as a lake, stream, or river due to the cost and availability of water in general, especially municipal water. Use of city water is almost always cost-prohibitive, and in most cases the city may already be using the same source you are considering for your irrigation needs for area-wide drinking water needs.

There is only a limited supply of fresh water available on the earth's surface, and if one field is fortunate enough to be receiving rain, the chances are there is another badly in need of it. Rain is a cyclical event that moves from region to region depending on ocean currents and the jet stream. With the exception of desert and arid regions, a specific location can only rely on the estimated average rainfall per year. It is possible that one year could be above average and the following year could have drought conditions; the average is only statistical.

The reason irrigation exists is to take care of the time periods when moisture is not available through rainfall. The period of time is irrelevant in most cases because many varieties of turfgrass are dependent on almost daily rainfall. This is at least true for a wide variety of cool-season grasses in warm summer months. What irrigation does is allow for growth of healthy turf throughout the growing season and also for turf production in locations where grass could not be grown otherwise. Certainly the golf courses in the southwestern part of the country would not even be possibilities without the help of an outside water source to keep the grass green in the hot, dry summer months. Even warm-season grasses need a certain level of moisture, but many of them are less susceptible to damage compared to the cool-season grasses, which are acclimated to cooler and moister conditions.

A typical irrigation system is put into place to simply supplement the existing rainfall that occurs in that area. I did mention the Southwest as an area that receives so little rainfall that most of the turfgrass is completely dependent on irrigation for its survival. In the Midwest, where I live, the inconsistency of precipitation creates the need for irrigation. It is common to have very wet years where irrigation is seldom needed followed by years where it hardly rains at all.

Golf courses have the most intricate and advanced irrigation systems. Current systems are run by high-tech satellite links with cell-phone electronic controls that barely require a human's presence for much more than pushing a button. Some of the first irrigation systems that were used on golf courses utilize ponds and lakes located in a central location on the course. The lakes and ponds are pumped to fill a complex layout of pipes that run throughout the course. The pumps used today are electronic and most run off computers and switches. The older systems, however, still rely on a human to turn a valve to release that water to the greens or other areas that are irrigated. Water is a must on golf courses due to the high mowing frequency and the extremely short mowing heights. It does not matter what type or variety of grass you have on a green when it comes the need for water. All grass is kept at maximum stress lev-

---

TURFTIP

Golf courses have the most intricate and advanced irrigation systems. Current systems are run by high-tech satellite links with cell-phone electronic controls that barely require a human's presence for much more than pushing a button.

els during most of the year to provide golfers with so-called "ideal playing conditions."

Most golf course layouts are complex in design and often are created by engineers who specialize in irrigation. The concepts used to create these designs are very simple. The idea is to simply create a sprinkler layout that will efficiently allow water coverage over an outlined area. There should be little overlap in the coverage and a minimized amount of waste water. It is also important to make sure that all areas are completely covered. I will be the first to admit to walking over greens with high-priced irrigation layouts where the water was not covering all of the designated areas.

There are many factors in irrigation layouts: size and type of sprinklers, water pressure, size of pumps, nozzle types—and this is just the hardware. Then take into consideration environmental factors such as wind or sun that could affect water distribution, and you can see how the complexity of the system increases.

The biggest and most troublesome problem with irrigation is actually keeping everything in working order. Maintenance is over half of the battle, and repairs to a system can be nightmares. Freezing temperatures create one of the largest challenges for anyone who deals with an irrigation system. There is also the electricity factor, which creates an endless variety of scenarios.

Sprinkler-head problems are the easiest to fix because you are dealing with an isolated problem in one specific spot. The heads can usually be removed and cleaned out or simply replaced, and the problem is cured. Water leaks or pressure problems can be a much larger issue, and most of the time indicate a major leak in the water system. A leak means digging into the soil down to the leak and fixing broken pipes. A landscaper or golf course employee may need to have some plumbing experience. Leaks can also be quite costly to repair (Figure 5.6, Figure 5.7).

The reason leaks or major repairs are such a problem is due to the short period of time that is available to fix the irrigation system.

106 Turfgrass Installation: Management and Maintenance

FIGURE 5.6   Pipes can have several tee fittings or angled connections making repairs more difficult or complicated.

The grass will need water in the immediate future, and it is up to you to get those repairs done fast enough so that the turf is not damaged. Be prepared to spend big money for the timeliness and convenience of getting a repair done. It is much cheaper to pay subcontractors or an excavator than it is to replace a lawn or especially a golf green. It is important that watering is performed almost daily during hot and dry periods and that the grass receives a specific amount to keep it at its best.

Golf greens should be watered almost daily when temperatures are above 50 to 60 degrees F. and there has not been any precipitation. Daily watering could be needed even with cooler temperatures, depending on how dry conditions are or on the level of stress the grass is showing. In my area a green should receive $^1/_{10}$ to ¼ inch of moisture during the summer. This is not a definite statistic, though. Moisture in a green should be based on desired playing conditions and the level of stress the grass is under. Taller turf requires less moisture, and if you want a firmer playing surface you can use even less

TURFTIP

Golf greens should be watered almost daily when temperatures are above 50 to 60 degrees F. and there has not been any precipitation. Daily watering could be needed even with cooler temperatures, depending on how dry conditions are or on the level of stress the grass is showing.

FIGURE 5.7 Major water leaks may need extensive digging to reveal the broken pipe for repair. Heavy equipment is often necessary and repairs can be quite costly.

water. It is important that all the environmental factors are taken into consideration when using irrigation.

An example of environmental factors could be a situation where you want to apply $2/10$ inch of water to a green on a windy day. If you placed a rain gauge on the green on a given day, the result might show that the green is receiving $1/10$ inch per 10 minutes of use. On a windy day the same gauge might show that the green only received half that much moisture because it either blew off-site or evaporated before it reached the ground.

When considering lawn irrigation the following aspects should be addressed. If below-ground irrigation is installed in a lawn, the pipes are often more shallow than on a golf course and the control mechanisms are much less complex. Usually the operation and maintenance is a little easier, but there is still a lot of maintenance. Lawn systems operate on considerably lower water pressure, and the size of the pipes and sprinklers is smaller than those used in other applications. Timers are very simple if they are used at all, and almost all systems operate off city water systems.

What is used for a majority of lawns are lawn and garden sprinklers on the end of a garden hose. This is the cheapest and most inexpensive way to keep a yard looking green. Most lawns require one inch of moisture a week in the transition zone to retain a green color in the summer. This can be applied in $1/10$-inch increments per

---

TURFTIP

Most lawns require one inch of moisture a week in the transition zone to retain a green color in the summer.

day or in ½-inch increments every two to three days, which is preferable. The desired result is water penetration without runoff. This means that the water that is applied should penetrate the soil but not run off down to the street after being left on for too long or unattended. Don't get me wrong; someone who at least tries to water his lawn in midsummer will have better results compared to not watering at all. There is a point, however, where a yard could become so saturated that it may show mower tracks or minor erosion. There is a balancing act to determine when and where irrigation should be used and how much. Many of these answers will come, unfortunately, through trial and error.

I do want to mention a type of irrigation that was once common but that is only seen for the most part on sod farms today. This is a surface irrigation system that has above-ground pipes on large rolling wheels. These systems are quite spectacular and are more common in the Northwest, where much of the nation's grass seed crop is harvested. These systems allow constant movement of the entire system so the seed or sod can be harvested. Then the system is rolled back into its desired location so that watering can continue when needed.

The most efficient type of irrigation is drip irrigation or soaker hoses. Drip irrigation uses hoses with small holes in them or a permeable

---

TURFTIP

The most efficient type of irrigation is drip irrigation or soaker hoses. Drip irrigation uses hoses with small holes in them or a permeable hose that allows water to seep out.

hose that allows water to seep out. The reason this type of irrigation is best is because the water can be directed to very specific locations and the amount of water lost to runoff or evaporation is very minimal. All the water a drip irrigation system puts out is usually quickly absorbed by the soil and since the hoses lie on the ground, very little water can evaporate or run off-site. This is not a very common type of irrigation to be used for turf even though it is very efficient, because the hoses can be troublesome to move around and this system is useless on a large-scale operation such as a golf course. I just want you to be aware that this is an option for irrigation and that it is potentially quite useful for smaller irrigation needs.

INTEGRATED PEST MANAGEMENT

I briefly mentioned IPM or integrated pest management earlier and how this practice can help to eliminate excess chemical usage and improve overall turfgrass quality. The effectiveness of IPM is really only as good as the person who is implementing it. What IPM does is maximize productivity through cultural practices to eliminate the need for excessive chemical usage.

When using IPM all the outside factors affecting pest control need to be thought out and calculated as to the outcomes they will have on the turf quality. I gave an example of how wind affected irrigation amounts and distribution. Most people do not consider IPM practices at all when chemicals are fast, easy, and require little planning. What makes IPM work is the planning process. Water, for instance, is a very good growing medium for diseases and fungus. When temperatures and other factors such as humidity are figured into the scenario, disease occurrence can increase. I will talk in more detail in the pest control chapter of how and why IPM works and how you can implement these practices.

The key to IPM is to treat disease based on the threshold level that each pest problem may present. The most important threshold point is economic in nature. The question you should be asking is, at what time is the economic damage worth the treatment? What is the cost to repair or replace? Is the unsightly appearance worth the expense of controlling the pest?

By answering these questions, safe and successful control measures can be implemented. IPM does not suggest banning chemicals or not using chemical controls; however, it does prevent excessive chemical usage. Using IPM in combination with good cultural practices such as water management and mowing frequency allows us all to become better turf managers.

Chapter 6
Turf Establishment

It is a challenge in itself to have to do all the things that keep turf looking its best. An even a bigger challenge for most of the people who grow turfgrass is getting it established. When I refer to establishment I am talking about actually getting grass to start growing in an area that may not be that suitable for growing much of anything. There are two main ways of establishing turfgrass: seeding and sodding. There are also a couple of other ways to establish turf that are used for certain types of warm-season grasses. Establishment of turfgrass is the most critical element in ensuring the long-term health of the grass because at this point a basis is created for the rest of the plants' life span.

SEEDING AND REESTABLISHMENT OF TURFGRASS

Usually the need for turfgrass establishment occurs after a couple of different scenarios. The first might be that you have an area that has never been seeded in grass before and now you want to turn this site into a grassy area. This could be perhaps an area on a golf course or perhaps a new lawn around a home. The other scenario might be that there is already existing turf in a particular area and you want to start over with new grass due to the condition of the existing grass.

When assessing the situation regarding new or existing turfgrass you must determine what will need to be done to the soil, how it should be done, and what will ensure the long-term success of the grass you are trying to grow in this location. This could be compared to a site analysis that a landscape professional might perform when assessing the landscaping around a new home. In both instances all success factors for the new grass should be considered: soil type, drainage, light, wind, and all other environmental factors that exist in the location. By doing a site analysis the most effective methods can be used when starting the establishment process.

Is Renovation the Answer?

When looking at a location for reestablishment you need to assess whether it is an area to improve or just simply start over. That is the biggest determination for either working with the existing turf or completely renovating the area and starting over with all new grass. There

---

TURFTIP

When assessing the situation regarding new or existing turfgrass you must determine what will need to be done to the soil, how it should be done, and what will ensure the long-term success of the grass you are trying to grow in this location.

> ## TURFTIP
>
> When looking at a location for reestablishment you need to assess whether it is an area to improve or just simply start over. That is the biggest determination for either working with the existing turf or completely renovating the area and starting over with all new grass.

is a threshold for weeds in the lawn that will help to make that determination. Arguably, everyone has a percentage of too many weeds vs. the amount of healthy grass in a lawn. The percentages could vary based on existing factors, possibly limiting renovation, or by what type of turf is desired for a particular area. I feel that any lawn that is 50% weed-infested is probably at a point where starting over is going to be the cheapest and most effective way to establish new grass. I am not saying that this is set in stone and that you cannot work with what is already there, but in many cases it is just more time- and labor-consuming to fix the existing turf vs. starting with all new grass. There are also exceptional benefits to being able to start over because of all of the soil improvements that can be made vs. trying to improve soil with grass already on it.

If the area you are planning to establish grass on is questionably deficient in an element or appears to have a nutrient problem, the first thing to do is a soil test. I mentioned earlier the importance of soil quality and improving soil to provide the ideal growing medium for the turfgrass. A soil test will reveal the answer to successful turf establishment. If the pH tests low or a nutrient is lacking in the soil, being able to incorporate this element before seeding will create much more desirable results. This is why I say that you are at an advantage when tilling up old grass instead of trying to add elements to the soil over existing grass. By tilling in an element or nutrient the improvement to the soil is more immediate (Figure 6.1). When all these nutrients are mixed into the soil, there is an instant availability to the grass as well as a fluffing of the soil that can be gained by tilling. The tilling action of the soil loosens soil particles and gives seedling grass a better chance to root into the soil.

When trying to improve soil content with established grass, the process can be a frustrating one. Soil improvement is limited by the

FIGURE 6.1 Tilling the soil allows all elements to be incorporated which will increase plant availability later on. *(Courtesy of BEFCO)*

time that is required for certain elements to become available. A lot has to do with what is called the cat ion exchange (CEC) capacity in the soil and all the other environmental factors that are present. I did not mention the CEC before because it involves some of the more scientific aspects of soil science. What is important is to recognize the limitations of certain element availability and address these issues in an appropriate manner. Lime, I mentioned, is very slow to affect soil conditions and so are some other crucial elements like potassium. This is why complete turf renovation is a much better answer to improvement in some cases.

If erosion is a possible problem to your project, then leaving the existing grass will certainly be the better choice. It is not difficult to improve an existing lawn into as good a condition as if you had started over. The difference is that sometimes it takes more time to do this. Erosion requires the use of sod or other methods of establishment that I will discuss a little later. What I am trying to emphasize here is that in many cases improvement of existing turf is often a better route for establishment.

If you indeed have a situation where starting over is not the right choice, then you might be wondering how elements can be incor-

TURFTIP

There are mechanical devices that are specifically designed to renovate existing turfgrass. One of my favorites is an aerator. An aerator is a machine that either pokes holes into the turf or pulls out cores of turf and distributes them onto the soil surface.

porated in the soil. There are mechanical devices that are specifically designed to renovate existing turfgrass. One of my favorites is an aerator (Figure 6.2). An aerator is a machine that either pokes holes into the turf or pulls out cores of turf and distributes them onto the soil surface. The hole-puncturing method is not nearly as effective as the core pulling method. Core aeration is very effective in that it churns the soil without tilling. The reason I say this is because you are pulling soil from below and putting it back onto the soil surface. At the same time you are allowing nutrients and moisture to seep into the soil, which then become readily available to the turfgrass roots. By doing this, of course, the movement of lime and other elements is more effective and immediate for plant uptake.

Another machine that is useful for seeding work is called a power rake (Figure 6.3). A power rake is a gas-powered machine with slicing knives that shred through the existing grass, pulling up thatch and debris while slightly loosening the soil surface at the same time. This unit does not allow quite as good soil disturbance as a plugger or core aerator, but it does create a much better seed bed on the surface. Often when aeration is used the grass only comes up in the core holes, whereas a power rake will leave many little grooves where the seed can germinate. Seed to soil contact is important with either machine, which is why an aerator or power rake will greatly improve your germination results.

TURFTIP

A power rake is a gas-powered machine with slicing knives that shred through the existing grass, pulling up thatch and debris while slightly loosening the soil surface at the same time.

FIGURE 6.2 Aerating is often the easiest thing to do to improve a lawn without getting into any major renovation. *(Courtesy of Ryan)*

TURFTIP

Rooting grass requires a good quantity of phosphorus for root stimulation but does not require a lot of nitrogen because roots are what are desired at this stage.

If you have just finished using a power rake, then all of the debris will need to be removed from the yard. In the case of aeration, nothing will have to be done to remove the cores. These cores will break

FIGURE 6.3 A power rake will remove thatch and debris in the lawn and create a great seed bed for overseeding. *(Courtesy of Ryan)*

down on their own eventually. For golf courses, however, the process is much different and core removal is very common. After aeration, lime and fertilizer are usually added to the soil. The lime and fertilizer will penetrate best when applied after a mechanical renovation of any sort. It is always a good idea to use a starter fertilizer or one that is not horribly high in nitrogen. Rooting grass requires a good quantity of phosphorus for root stimulation but does not require a lot of nitrogen because roots are what are desired at this stage. If the young grass plants get too much nitrogen, then they can become leggy or spindly in appearance.

The next step is to spread the seed out over the area that has been renovated. This process is referred to as overseeding. It is possible to overseed a yard with a spreader of some type, either a drop spreader or a rotary style. You will probably have just applied a fertilizer of some kind and possibly lime with a spreader of this type. If no mechanical renovation has been performed to the turf, overseeding with a spreader will have very limited success. The seed will be left on the surface of the ground, where it is subject to many possible scenarios. The seed may be tracked off by human or pet foot traffic. In some instances birds will actually consume a portion of the seed before it germinates. A large portion of the seed simply will not germinate at all because it never had that opportunity for the seed to soil contact that is required for it to grow.

I mentioned the threshold of 50% weed infestation as being a startover point for established grass. The reason this is the case is because the new grass often does not have the ability to compete with already established weeds in the area. Perennial weeds are often very good at stealing water and nutrients from existing grass. Some perennial weeds are such prolific spreaders that sometimes tilling them while they are alive is not the best way to deal with them. It is often better to use a nonselective herbicide such as glyphosate, which is known by the trade name of Roundup™, to kill all the vegetation in an area before any seeding is done. The reason this should be done is because some more evasive species of weeds actually thrive from cultivation practices such as tilling (Figure 6.4). This is definitely not what you want to happen. By simply spraying the area first, everything is killed out for a clean start. Won't the Roundup™ stay in the soil and affect the germinating grass? The answer to this is "no," because it has no systemic or residual properties. The reason it (or any glyphosate chemical) works so well is because it only kills what it comes into contact with and basically it is safe to proceed in an area once it has dried. In other words, if you sprayed an area and tilled the same area a day later it would not affect the new grass at all. I do suggest allowing a complete burn-down effect or the browning of all of the vegetation, which will take a few days. This allows for easier tilling and a better chance for a kill on all the weedy species. This is not always a possibility due to time constraints, so you might have to till the day after the area is sprayed.

FIGURE 6.4 Ground Ivy is one particular weed that spreads from cultivation and tilling. This would be an example of a weed that, if present, should be sprayed and killed before seeding or tilling.

TURFTIP

Organic matter is very helpful in breaking up clay particles and improving characteristics of the soil. Organic matter also helps to improve moisture and nutrient-holding capabilities.

Now that you have eliminated the competing species of weeds and old grasses, the new grass will have a much better chance for survival. You have also assessed any possible fertilizer or lime needs. There may be one more consideration that will need to be addressed when working with exceptionally poor soils. The incorporation of some type of organic matter might be the difference between success or failure especially when working with very poor soils or clay soils. Organic matter is very helpful in breaking up clay particles and improving characteristics of the soil. Organic matter also helps to improve moisture and nutrient-holding capabilities. Using organic matter will help to improve the turf quality short-term and long-term. Old leaves and peat moss make an exceptional organic product that can be tilled right in with the fertilizers and lime if they are needed.

> **TURFTIP**
>
> Using organic matter will help to improve the turf quality short-term and long-term. Old leaves and peat moss make an exceptional organic product that can be tilled right in with the fertilizers and lime if they are needed.

There are only certain times of the year when any type of seeding should be considered. This will vary somewhat based on the particular seasons in your area as well as the type of grass you are trying to establish. Spring and fall are certainly your best choices for grass establishment, be it by renovation or improvement of existing turf. Fall is the better of the two because turfgrass will tend to develop better roots in the fall than in spring. Weed control is much more difficult when you seed in spring. In Missouri I like to start on spring projects around the first of March and be seeding no later than the first of June. Then I resume seeding around the end of August and seed up until the beginning of November. This is certainly not an exact schedule of what happens every year. In many cases there is snow on the ground in March or snow flying in October. I have experienced unusual droughts in both spring and fall where conditions were just not suitable for doing any seeding at all. The reason that spring and fall work best is because rain is more likely and soil temperatures are the most suitable during these time frames. As I mentioned, everyone's spring and fall are slightly different so the dates I mention are just rules of thumb that would apply to a large majority of the Midwest primarily for cool-season grasses. Warm-season grasses are usually established by different methods and are done at different times. They are discussed later in this chapter.

124 *Turfgrass Installation: Management and Maintenance*

Seeding work sometimes has to be done in off-season times to meet contract requirements. Seed can lie on the ground for a reasonable length of time and still come up later on. The percentage of the grass that actually will come up falls off very drastically as time progresses. You might have seen your local highway department or state transportation commission working on a project in the middle of winter. Usually when spring rolls around there is that surprising green grass still poking up out of the layers of straw (Figure 6.5). The same is true for seeding in midsummer, although I have experienced significantly low germination rates from extended periods of no moisture.

FIGURE 6.5 This seeding project was done along a major highway during the winter. You can see how the straw has prevented water from washing the soil away. Although there is no grass showing yet, this will be quite lush by springtime.

New Installations

Getting back to completely new seed installations, what is the next step? You now know what soil assessments should be made and how the soil can be improved. The next factor you should look at is the type of seed you are using and how it will be applied.

Many people unfortunately look first at the price of the seed. This is unfortunate because seed comes in a wide variety of types and qualities. I mentioned earlier that blends and mixtures will work best because there are several types and/or varieties of seed present. Each seed type is more likely to find a suitable location to grow in. Look for at least a middle of the line seed quality if price is the only thing you are concerned with or if you are not familiar with grass seeds. I often use a combination of 70% bluegrass and 30% ryegrass. This gives a nice fine-bladed turf that will germinate somewhat quickly under good conditions.

Once you have figured out what type of turfgrass you are using, you must next look at how this seed will be spread. I mentioned earlier the use of a rotary spreader when renovating a yard after the use of a power rake or aerator. A rotary spreader is still a good choice to use when tilling and starting a new lawn. The seed should make good seed-to-soil contact because at this point there should not be any turf to interfere with the germination of the new seeds.

The most common method for seeding is with some sort of a seeding device that is engine-driven or that fits on a tractor. The size of the project at hand will have a lot to do with what type of machine is being used. There are many types and styles of seeders available. There are roller-type seeders, which I refer to as non-powered seeders (Figure 6.6). A non-powered seeder usually has a drop box for the seed that is followed by some sort of large roller that will help to press the seed into the soil. These are almost always tractor-mounted seeders because of the weight needed to be effective.

A powered seeder can have either an engine or a Power Take Off (PTO) to operate a slicing device of some sort. Much like a power

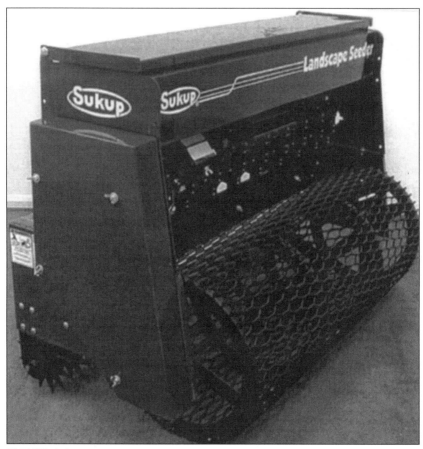

FIGURE 6.6 This is a non-powered seeder made by the company Sukup™. This seeder works on ground drive principles and does not have a PTO. The seed is dispersed out of the seed box and rolled in by the large corrugated roller on the rear. (Courtesy of Sukup)

rake, there are usually knives involved that will slice the soil while depositing seed from a seed box at the same time. A powered seeder works much like a crop drill, the only difference being that the grass is the crop that is being planted. There are many small walk-behind types of seeders as well as larger tractor-mounted versions (Figure 6.7, Figure 6.8). Good brands of large powered seeders will usually have a large roller on the back to roll the seed into the soil

TURFTIP

Some companies claim to have one-step seeding machines that till, drill, and roll all in one pass. From my experience their effectiveness has not been as remarkable as claimed, and they are very expensive to own or rent.

to maximize seed germination. Some companies claim to have one-step seeding machines that till, drill, and roll all in one pass. From my experience their effectiveness has not been as remarkable as claimed, and they are very expensive to own or rent.

FIGURE 6.7 This powered seeder made by Befco™ has slicing knives that penetrate the soil and disperse seed into the grooves. *(Courtesy of Befco)*

128 Turfgrass Installation: Management and Maintenance

FIGURE 6.8 Walk behind seeders are great for working with small areas like residential yards. However, they are not that effective on larger scale projects. *(Courtesy of Ryan)*

By using a seeder to drill the seed in some way the soil is also rolled out as the seed is being deposited. The only problem with using a rotary spreader to spread the seed is that it does nothing to level the freshly tilled soil. If you are walking on tilled ground, then there are foot tracks to worry about; and if a tractor has been used for the project in any way, then there are usually tractor tire ruts to contend with. You can use a turf roller or pull-behind roller to level out the soil if you are using a rotary spreader. These devices are somewhat effective for leveling on small projects where walk-behind equipment is used but not so good for removing large tractor ruts or tracks.

The seeding should be drilled in a way to minimize tracking to provide the smoothest possible result. Once the lawn has grown in and is ready for mowing, it is often hard to correct unlevel conditions at that stage. I also want to emphasize that the more you can do to reduce traffic on newly seeded lawns, the better off the grass will be. This goes for before, after, and during a seeding project. This is why I discourage the use of lawn rollers and other equipment that will cause overcompaction of the soil to the point that the tilling provides little benefit. Keep in mind also that moist or wet soil may compact more than dry soil.

I like to run my seeding projects by covering the ground in a bi-directional pattern to ensure good coverage. This is really the best way to spread fertilizers, dry chemicals, or seed. By cutting the rate in half and going over the area in two different directions, you help to reduce skipped areas or places that might not be reached while

TURFTIP

I like to run my seeding projects by covering the ground in a bi-directional pattern to ensure good coverage. This is really the best way to spread fertilizers, dry chemicals, or seed. By cutting the rate in half and going over the area in two different directions, you help to reduce skipped areas or places that might not be reached while going over the area the first time.

going over the area the first time. In the case of tractor seeding this might help to further cover up any tire tracks. Let's take, for instance, a football field as an example You are pulling a seeder that only has slicing knives on 3-inch centers. This will mean that a majority of the seed will end up growing in those grooves but at a distance of 3 inches apart. Now think if you cut that rate in half and put down half of the seed running in one direction and the other half running in the other direction (much like a tic tac toe board). This will provide the best result when it grows in and make a much denser surface to start out with. In the case of fescue, which if you remember is a nonspreading turf, you may need to go over the top of the area with a rotary spreader even after the seed is drilled. This will help to eliminate bare or skipped areas from the seeder spacing and allow a more dense coverage.

Hopefully at this point you are looking at a dirt area that has been nicely leveled and is ready to grow. There is one last determination that needs to be made to finish a new seeding area. Should this area be covered with straw and when should it be covered? First of all, straw has several functions for a seeding project. The first and most important is erosion control. If the seeding project you are working on is on a steep incline or in an area where runoff is a concern, then you will probably want to cover the seeding with straw when it is completed. This will help to slow runoff and minimize damage from rainfall while the grass is establishing itself. The straw will also help to retain any moisture the seeding receives while keeping the ground from drying out. The straw shields the soil surface from intense sunlight and because of its porous nature tends to retain a tremendous amount of moisture. This moisture will then be available for the new grass as it grows. Straw will also help to increase germination times dramatically. The reason is due to the canopy effect it creates. The grass seeds germinate and, due to the natural phototropic nature of plants, the grass shoots up for light. This will make the grass seem to appear magically overnight because it can happen so quickly. Once the grass seeds germinate, the straw continues to provide moisture and protection to the young plants as they continue to mature. Once it is time for the first mowing and the straw is shredded up with a mower, there is also the slight benefit of the organic matter added to the soil from the mulched-up straw. The combination of all these factors is why straw is usually used.

There are reasons and situations where straw is not needed and a couple of disadvantages of using straw. The biggest disadvantage is the amount of weeds that you may be introducing back to your newly seeded area. Straw is usually cut from mature wheat after the seed heads are harvested or sometimes from other agriculture crops like barley. The straw should always be purchased from someone who has a reputation for providing high-quality straw from well-maintained fields free from noxious weeds. The weeds are not as big a concern as many people think, because all turf needs to have a weed treatment program to continue to look good even after it is renovated. If there is a situation where the seeding is being done in a level location or in an area where weeds are a concern, such as a golf course or athletic field, then you may not want to use any straw. The grass will come up if good seed to soil contact is established, although it might take a little longer than if the area were covered with straw.

The amount of straw used is another factor that varies with each project; 50% coverage is a good middle point. This does not mean to only cover half the seeding but rather to apply the straw in a way

TURFTIP

Straw should always be purchased from someone who has a reputation for providing high-quality straw from well-maintained fields free from noxious weeds.

132 *Turfgrass Installation: Management and Maintenance*

so that light can still reach the ground surface. If there is a moisture or erosion concern, then more straw may be required. Wind is also a factor to consider when applying straw.

Many people still "shake out bales" by hand. This is just the old-fashioned way of picking up a chunk of straw and shaking it out into smaller pieces when covering the seed. There are now many types of straw application equipment. These "straw blowers," (Figure 6.9) as they are called, have a shredding device with a blower that shoots the straw out onto the seeding. Some of the units have long hoses to disperse the straw, while others mount on trucks and shoot bales out as if they are coming from a cannon.

There should be some consideration given to weeds while seeding, which is why the use of straw is often contemplated. Weeds are usually more of an issue after the grass is up and established. When seeding is performed in late spring, there is always a concern with summer annual weeds, especially crabgrass. This is one of the many

FIGURE 6.9 With the use of the straw blower, the labor intensity of covering a seeding project with straw has been reduced considerably.

reasons why grass should be sown in the fall vs. the spring. Weed treatments while seeding are almost impossible because chemicals will directly affect seed germination. A good rule of thumb when seeding should be to wait four to six weeks or after two mowings before any chemical treatment is put down. This is especially true for broadleaf herbicides, which tend to be the most harmful to young grass plants. The chemical known as Tupersan can be used while seeding in late spring and is often found in many starter fertilizers. Tupersan is the generic name for one of the few known grass herbicides that can be used while seeding. It is primarily used as a crabgrass preventive and is fairly effective when used in this way. Most other common chemicals that are used on turf should be avoided until the new grass has had time to establish.

Once a seeding project has been completed, the first concern to think about is water, whether or not any fertilizers and/or chemicals were used. All seeds need water to actually germinate and to continue a healthy level of growth. Remembering once again that water is the main component of grass, you can imagine that grass does not do very well when lacking water. A new seeding ideally would receive $^1/_{10}$ to $^2/_{10}$ inch of moisture per day for at least the first couple of weeks. Since most grasses germinate within 14 to 21 days or less, this at least gets that process started and hopefully ensures a successful start. I like

---

TURFTIP

Once a seeding project has been completed, the first concern to think about is water, whether or not any fertilizers and/or chemicals were used. All seeds need water to actually germinate and to continue a healthy level of growth.

to continue to see the ground receive ¼ inch of moisture every other day at that point. Watering can be done with a garden hose and sprinkler or by many other methods. The idea is that the grass should receive a steady rate of moisture over a long period of time vs. a deluge of water for just a few minutes. This will help to minimize runoff and erosion while allowing the grass to grow. By watering lightly for longer periods of time the water will have time to soak down to the roots. I like to see a project that is covered with straw watered as quickly after completion as possible to get the straw saturated so it will retain moisture and not blow away (Figure 6.10).

There is a specialized machine that will supposedly do all the things I have talked about in one simple step. Hydro-seeders are now becoming more popular than ever. The problem with hydro-seeding is that it often creates an attractive lawn for the short term but fails in the long term. The concept of a hydro-seeder is that a mulching component is mixed into a large water tank along with fertilizer and all the "elements necessary for a beautiful lawn." I have seen absolutely horrible results from using these machines and am not really a believer that they are the best and most effective way

TURFTIP

The concept of a hydro-seeder is that a mulching component is mixed into a large water tank along with fertilizer and all the "elements necessary for a beautiful lawn." I have seen absolutely horrible results from using these machines and am not really a believer that they are the best and most effective way to create a lawn.

FIGURE 6.10 Although straw was applied to this seeding at a church, much of it blew away from lack of rain or protection from wind.

to create a lawn. There is always an inconsistency in the seeding because everything that is mixed in the tank is of different sizes and textures. There is also no chance for incorporating elements into the soil unless this is done prior to using the machine. I have seen cases

> **TURF TIP**
>
> I will say that hydro-seeding is a valuable tool for seeding highway projects or medians because these machines can cover large areas of ground without ever having to drive over the top of the surface at all.

where the weather was just right and a hydro-seeder provided an excellent stand of turf, but it is often short-lived because of the lack of soil improvement that is usually undertaken. In most cases these machines are used right after construction and just cover up hard, nasty soils with a colored mulch with a little grass in it. I will say that hydro-seeding is a valuable tool for seeding highway projects or medians because these machines can cover large areas of ground without ever having to drive over the top of the surface at all. When steep banks are a concern and you have miles upon miles of seeding to perform, a hydro-seeder is a much more economical choice.

Any lawn you have established should be able to be mowed within a few weeks. Do not believe the old wives' tale that grass should seed out before the first mowing. That is simply nowhere close to the reality of what should occur. Mowing should start when the majority of the seedling grass has reached a height at which normal mowing might occur or just slightly higher. I usually mow around 4 to 5 inches of height and cut the grass no lower than $2^1/_2$ to 3 inches. I recommend 3-inch mowing for a period of time with most cool-season grasses because it will not be such a shock to the new seedling grass. Mowing is very important at this stage in order to cause the grass to tiller and spread. Once a grass plant is mowed, it begins to explore new avenues of spreading out on its own. Be cautious not to mow the first time when the ground is too wet. You may have spent a lot of time on your project and you do not want to mess it up with ruts from a heavy mower.

I mentioned the difficulties of controlling weeds while establishing grass, but there needs to be a weed control effort put into effect as quickly after establishment as possible. A good weed and fertility program is well worth it when compared to repeated costs of renovation or replacement. A well-maintained lawn will last for years if all treatments are done in a correct timeframe and the other maintenance is kept up. If the weather gets dry, then it is time to get out the sprinklers. Retest the pH of the lawn a minimum of every three years and make sure that those weeds are kept under control. I have seen so many instances where a newly established lawn that looks absolutely beautiful is let go within the first year or two because the homeowner does not control the weeds. Establishment is half the game; the other half is keeping it looking that way. Evasive weeds only need a short window to come into a lawn and ruin its appearance. For instance, one crabgrass plant might produce 10,000 viable seeds in one season. Now think of your own yard and do the multiplication over three growing seasons to see what could happen if left untreated.

TURFTIP

Evasive weeds only need a short window to come into a lawn and ruin its appearance. For instance, one crabgrass plant might produce 10,000 viable seeds in one season.

SODDING

The fastest and most effective way to establish grass in a short period of time is through the process of sodding (Figure 6.11). Sodding is done by laying carpetlike pieces of turfgrass that have been harvested from a sod farm over an area. These pieces of sod are cut with specialized pieces of equipment called sod cutters. The sod may range from $1/2$ to 2 inches in thickness and often comes in 3-foot lengths for easy handling. Sod can be bought in large rolls, but these require addi-

FIGURE 6.11 New sod on this golf course tee looks great after only three months.

tional specialized equipment to lay the sod. Rolls are only used for athletic fields or large-scale projects. Sod is used for immediate effect for the most part. Many contractors and landscape professionals find that sod is very good for providing the instant finished look that many people will pay good money to have. Sod also eliminates the concerns for immediate weed control. It does not eliminate the need for water.

Probably the biggest need of sod is for gluttonous amounts of water for a few weeks after being installed. Sod experiences a shock reaction for a while because it has the task over the first few weeks it is laid out to try to find a source for water and to grow roots. Since its roots were severed in the field, it takes time to establish a good basis with which it can start taking up water again. I like to give sod ¼ inch of water or more a day for two to four weeks or until established. I do find that spring tends to be a better time to sod than fall because of the generous spring rains that we have in our area.

Although sod has soil with it when it is purchased, you often may not know a lot about the location or history of the soil where this sod came from. Consider the source of the sod and its potential for weed or fertility problems. You are only getting the top $1/2$ to 1 inch of ground typically, and many sod farms have been harvested on for years. In some cases the topsoil the sod actually contains is low

TURFTIP

Although sod has soil with it when it is purchased, you often may not know a lot about the location or history of the soil where this sod came from. Consider the source of the sod and its potential for weed or fertility problems.

in nutrients or at an undesirable pH because it has been stripped out of the same fields for so long with nothing added back to it. This is why there should still be some consideration for site preparation even before sod is laid.

Ideally the same factors are looked at as would be for a seeding job. The soil is still one of the most critical areas. This means that tilling may still be desired and all the proper elements incorporated into the soil prior to laying the sod out. If tilling is done, you must make sure that the area is raked or rolled out very smooth prior to the actual installation of the sod. It would be like trying to roll a carpet out over a floor covered with golf balls and baseballs otherwise. By doing the proper site preparation, problems of uneven turf can be eliminated.

Sod should be mowed at about the same time as you would normally mow grass or shortly after that. As with a seeding project, you might want to mow a little higher than normal at first to eliminate stress to the turf. Sod usually takes a while before the grass tips grow tall enough to be mowed. It takes a while for the sod to root well enough to establish top growth. Once the sod roots itself, then top growth seems to happen very quickly. Make sure the subsoil is

TURFTIP

If you have laid the sod out without doing any soil improvements, make sure that a light amount of a starter fertilizer is applied over the top when you are finished. Once again, fertilizers high in nitrogen are not necessary because ideally you are trying to establish roots and not top growth at this point.

not too wet to get mowing equipment onto the sod and cause damage to the plants. It is easy to scalp the new sod if the ground is uneven, and recovery is very slow and unsightly. If no soil improvements are made prior to sod installation, fertility should be addressed at some point after it is laid.

Sod can be laid right over the top of existing grass and/or undisturbed ground. This will provide the most level surface but maybe not the most ideal conditions for rooting. It you have laid the sod out without doing any soil improvements, make sure that a light amount of a starter fertilizer is applied over the top when you are finished. Once again, fertilizers high in nitrogen are not necessary because ideally you are trying to establish roots and not top growth at this point. Top growth, as I said, will come soon enough.

ESTABLISHMENT OF WARM-SEASON GRASSES

With warm-season grasses the methods for establishment are quite different. Sodding is done for any large-scale grass establishment just the same as it would be for any cool-season grasses. One way to establish warm-season grasses on a smaller scale is through a method

TURFTIP

One way to establish warm-season grasses on a smaller scale is through a method referred to as plugging. Plugging is probably more commonly known when used for zoysia grass.

referred to as plugging. Plugging is probably more commonly known when used for zoysia grass. Zoysia, like other warm-season grasses, is only available in sod or plugs and cannot be seeded. The plugging I am referring to in this case is different than the references made to core aerating. Plugging is done through the use of a hand tool that is actually called a plugger. This tool is a hollow tube that has a handle of some sort on the end. The idea is to drive 2 to 3 inches into the soil surface to pull a core that can be transplanted into a new location. Cool-season grasses can also be plugged so they can be spread or replaced with warm-season grass. A plug of cool-season grass is pulled out and a new one of warm-season grass is put back in.

Plugging can be useful for more than establishment. Golf course professionals have used plugging to repair damaged areas on greens and tees for years. In this case the damaged area is removed and the plug of healthy new grass is put back in. Plugging warm-season grasses can be done when the grass is actively growing, although dormant plugs will root just fine once temperatures reach a point desirable for growth. It is a crude and simple process to establish grass by plugging, but it is the cheapest and most effective way to establish many warm-season grasses. Once these grasses are plugged in along with cool-season turf, they take over after a period of time. Plugs are certainly much cheaper than sodding an area. It requires little skill or knowledge of turf to do, and zoysia is marketed in popular gardening magazines for that very reason.

Another way to establish warm-season grasses is through a process of sprigging. Sprigging is commonly used for larger-scale projects and more specifically for golf course fairways. Many golf courses in the Midwest in recent years have made a transition to the warm-season zoysia or Bermuda grass. This has been done due to the demand from golfers for higher playing standards. The warm-season grasses can be maintained at much lower heights and have other more desirable characteristics for golfing as well. What the sprigging process consists of is using small pieces of tillers and rhizomes and simply broadcasting them out over an area and heavily watering them. It is like spreading shreds of actively growing turf out over the top of other turf and watching them take over. The reason this method is used is once again cost. From my experience it is an effective method for establishment even though it may seem a little unorthodox at times.

These are the most common methods for establishing most kinds of grass. There are ways to establish warm-season grasses by seed as well as ways to plug and sprig cool- season grasses, but for the most part the methods that are done stay the same. There are arguably advantages and disadvantages to all seeding and establishment methods. It is up to the professional to assess each circumstance to determine which ones are the best.

Chapter 7

Residential and Homeowner Lawn Care Business

Lawn care is a very profitable industry, and it is growing in size every day. The old days of getting the local neighborhood kid to "cut the grass" or mom and dad having the time to mow on the weekend are over. Times are changing and the lawn care industry is changing right along with it. When everyone in a home is working, there is little time to go out and take care of the lawn. Children seem to be busier than ever with school activities, and there is less time to take on extra weekend chores. This has created a great demand for lawn care businesses of all sizes and almost every demographic.

TURFTIP

Most people now can hire a service for about the same amount of money as it takes to purchase and maintain the equipment. That is a very critical point of sale if you are in a lawn care business and trying to promote yourself and your services.

It used to be that having the lawn maintained by an outside company was considered a luxury and thought to be affordable only by people in higher-class areas. Most people now can hire a service for about the same amount of money as it takes to purchase and maintain the equipment. That is a very critical point of sale if you are in a lawn care business and trying to promote yourself and your services. Because of the continued increase in cost of mowing equipment it becomes less affordable to actually own a mower each year. In some cases it is even cheaper for homeowners with larger lawns to pay to have the lawn maintained.

I talked earlier in this book how lawn care and landscaping used not to be considered professional businesses. The gardener in a fine neighborhood was often looked at as a necessity much like the maid or the butler. Gardening, landscaping, and lawn care are now very respectable occupations and require highly educated people to do them. Getting into residential work of any kind requires skills and a certain repertoire in dealing with people. You must also have a diversified company that can offer a quality, dependable service at a fair price. There are a lot of competitors, so each business will face daily challenges of competition and the usual headaches of a business operation. It is important to do what the market requires in your area

TURFTIP

There are numerous year-round services that most lawn care companies provide to their customers. This allows a lawn care operator (LCO) to be able to make a living on a 12-month calendar year as well as take care of all the outdoor needs of the customer.

and provide a diversified business that can handle all the services requested by the customers. This is why so many companies offer full-service outdoor work of all kinds (Figure 7.1).

There are numerous year-round services that most lawn care companies provide to their customers. This allows a lawn care operator (LCO) to be able to make a living on a 12-month calendar year as well as take care of all the outdoor needs of the customer. Your customers do not need to be commercial for you to achieve year-round profit from them. Too many times I see lawn care companies that overvalue commercial accounts. The problem with a lot of commercial accounts is that you can end up the next "low-baller" to get the job. What I am saying is that the biggest struggle for a lawn care company is to be undercut by someone else who is working for little to no profit. Many who complain about these practices end up doing the same thing just to get what they consider to be a large client.

FIGURE 7.1 Pruning shrubs

No client, large or small, is worth having if the company is not retaining any sort of profit from providing its services. This is one of the biggest benefits I can see from soliciting more residential accounts vs. commercial accounts. It is often a much easier sale to convince one homeowner that you offer a superior service compared to that of some large-scale company that might be looking to save a buck.

FINDING AND SERVICING THE CUSTOMER

Homeowners may be one of the best, yet most overlooked, resources for a lawn care company. My personal experience has shown me that if we find ourselves in a neighborhood where we are able to take care of three to four residences, we are able to maximize profits by eliminating road time. I personally feel that road time and travel are the most overlooked expenses that a lawn care business has on a day-to-day basis. Fuel, equipment, and labor can all be figured out very accurately in a budget, but when the travel expense is added in, it actually works as a multiplier on the other factors. If an employee

---

TURFTIP

Homeowners may be one of the best, yet most overlooked, resources for a lawn care company. My personal experience has shown me that if we find ourselves in a neighborhood where we are able to take care of three to four residences, we are able to maximize profits by eliminating road time.

spends three hours a day riding in an expensive truck that burns gas and takes up his time, his cost to the company is harder to figure per hour. The reason homeowner accounts work so well is that if you are a reputable company with good sales and marketing capabilities your services will sell themselves. By picking up one account in a nice neighborhood and providing that customer the best work possible, others in the area will take notice.

"Keeping up with the Jones's" is the generic phrase for people who refuse to be outdone by their neighbors. You will find that this mentality will raise the bar for the appearance of the property in an entire neighborhood. If you are fortunate enough to be the person or company that has started this trend, then you may very easily find yourself with many new properties to maintain. This is when good work will reward you and "word of mouth" advertising comes into play. People driving by will take heed of your performance whether you do good or bad work.

"Word of mouth" advertising is one strategy that I find to be the fastest track to either failure or success in any type of business. Unfortunately, it does not cross some business owners' minds. Job performance is not thought of by some companies at all. A good job will sometimes be noticed but a bad one always will be. The positive effect of neighbors watching high-quality work can be a great benefit to your company, but if you are doing below-standard work, people will pay even closer attention. Remember that those who pay attention to good work are much more likely to notice poor work. If a customer is extremely satisfied with the performance done on a project, they will tell up to eight other people about your work. If that same customer is unsatisfied, they will tell around 30 people—that is nearly four times is as many people. It takes a lot of "atta boys" to make up for one "oops" when it comes to the service industry in general.

If you are fortunate enough to land that customer of a lifetime who happens to be in a prominent neighborhood, do not be afraid to interact with that customer or with the neighbors. It may be a

> **TURFTIP**
>
> Education of the customer and the public is the best way to solve problems even before they start.

good idea to do some door-to-door interaction with these customers. This is the first step to what I consider to be the most challenging skills any business owner can possess—customer-relations skills. Interaction with the public is the most difficult aspect of being in any business but especially lawn care. Many people have emotional attachments to plants and even more so to plants that might have been memorials for a loved one. When dealing with the public, always tread lightly on emotional subjects or when problems arise on a job site. By dealing with problems quickly and effectively, even bigger problems can be avoided later on. If you do not handle the public very well, then lawn care might not be the right career for you. I know I have had a lot of instances when the homeowner was completely wrong about something, but I never started out a conversation by telling them that. Education of the customer and the public is the best way to solve problems even before they start.

By explaining to the customer what are you doing beforehand, why you are doing it, and when you are going to do it, you start off on the right foot before providing a service. Make sure to stick to what you say in this conversation. Homeowners do not mind waiting a reasonable amount of time for something to be done. However, do not tell a person you will be there next week if you already know that you are a month behind. It is hard to get rid of a reputation of being days or weeks late in starting and completing a task. If you find that your plans are forced to change from weather or any other outside factor, take the time to call the customer and let them know why this is occurring.

If changes develop on a project with a homeowner, you should let them know beforehand if possible. If something unexpected is encountered during a project or an accident occurs, let the homeowner know as quickly as possible. Even when doing something as simple as mowing, there is always the possibility of property damage. If this happens, make sure to tell the client and offer to take care of it as quickly as possible. Homeowners can accept accidents, but liars are a different scenario.

Take advantage of any interactions with your customers. Every interaction is a potential for additional revenues with that client. Door-to-door flyers might help to get your foot in the door, but you need to be at least a good enough salesperson to get hired for the job. I like to look at every interaction with the customer as a time to talk about future work or projects that may lie ahead. Even the customers with small, less lucrative projects have the potential to provide a lifetime of income on future work.

TURFTIP

When your company is on a property weekly, for instance, to mow the lawn, you have an opportunity to give yourself as much new work as you can sell. Each customer is a new opportunity and provides different resources than the one before. By talking to your customers at every opportunity you can educate them and explain to them the many services you offer and what you think they need to do next.

When your company is on a property weekly, for instance, to mow the lawn, you have an opportunity to give yourself as much new work as you can sell. Each customer is a new opportunity and provides different resources than the one before. By talking to your customers at every opportunity you can educate them and explain to them the many services you offer and what you think they need to do next. A simple interaction for instance might start out as, "Sir, I was noticing while mowing that.....(fill in the blank)." This conversation might lead in any number of directions at that point. A successful lawn care professional will shine because of two reasons. The first reason is that you are utilizing good human resource skills. The second reason is that you know what you are talking about and the customers will respond to your suggestions.

By knowing exactly what you are talking about and passing this knowledge on to the customer you are solving future problems and creating new work for yourself. If you have few answers and little skill, chances are you will have a hard time making a future for yourself. This is why in most city phone books five out of ten lawn care companies change every year. For some reason many people still think that anyone with a lawnmower can call him or herself a landscaper or lawn care professional.

Even if you consider yourself to be educated and experienced, you will encounter questions you cannot answer. If you are new to the business, this happens more often than not. It is fine to say, "I don't know," but you should follow with, "I will find out for you, though." This is an easy way to deal with awkward situations that may arise in an interaction with a client. It is better to find out the correct answer to a question than to give incorrect information that you'll need to correct later on.

You will find that many conversations concern topics that a client has read about in the local paper or a gardening magazine. Just because something appears in print does not mean that it is an absolute or that there are not alternative methods to performing such a task. You also need to take responsibility yourself in reading publications and articles about the lawn care industry. I have mentioned how this business continues to change on a day-to-day basis. If you find yourself outdated by your practices or not willing to change with the times, more conflicts can result with individuals. Customers often will "do their homework" before hiring someone to perform a service for them. This is not usually the case, but when it does occur that person will take every opportunity to prove you wrong.

There are many other ways to generate new business or create new opportunities. I think that the easiest or most obvious ones are often overlooked. Business with homeowners can come about in a lot of ways. Word of mouth is always the best way, but there are always options available no matter how small your area or operation is. Versatility is an attribute that might get you in the door and provide ways to find customers that you might not have thought of before. That aeration job that the "other guy" didn't have the equipment for could land that account for all outdoor services permanently. If you are doing the snow removal or other nonseasonal work but nothing else for a client, you have an opportunity in front of you. I had a particular instance myself where I was performing all the maintenance duties except for the mowing, because "the other guy had been doing it for years." These customers will let you become involved in their lawn care maintenance if you do good work for them and establish a good relationship. Eventually the "other guy" will quit doing it or get tired of your presence on the property. If you have done a good job on the other services you have provided, the mowing will come along eventually.

MAINTENANCE SCHEDULE

Once you have secured a client for any sort of lawn care work, you need to think ahead to scheduling services that will get you onto that property on a repeated basis. Like I said, the more often you are there and the more interaction you have with the home-owner,

> **TURFTIP**
>
> Versatility is an attribute that might get you in the door and provide ways to find customers that you might not have thought of before.

the more chances you have for doing future work there. The most common service to be provided is mowing the lawn. This will create the most frequent trips to the residence, which can be once a week is most cases and sometimes more than that in the spring. Do not let the customer control the schedule if at all possible. You will always have a select few customers that will want to call you when they think the lawn needs to be mowed. This will create numerous long-term problems when dealing with that particular client. First of all, when they do call they will expect you to "be there yesterday," which is often impossible. They will also expect the lawn to look as good as if you were mowing it when it actually was needed instead of when they thought it needed it. There may also be scheduling issues when they do call if your plate is already full for the week. You cannot bump one good customer on your schedule to make room for a part-time customer who only needs you when it is convenient. Finally, you will face the problem that they expect you to charge the same low price when sometimes it takes twice as long to do the property. This is why it is important for you to remain in the driver's seat when dealing with clients. I am not saying not to accept the customer as a client, but consider the possibility that you may lose money. Be prepared to let the client know that if the property gets out of control you will not be able to provide your services anymore. It is possible that the client could be comfortable with paying an hourly rate or a variable cost for mowing, but this is highly unlikely and never has been the case for me.

Another great way to get yourself back to a property, even if you do not mow for the customer, is by putting the client on a chemical application schedule. Chemical treatments are done in most cases on a monthly basis during the active growing season.

I use a minimum of four treatments in a summer for the less particular clients and five to six treatments for the very particular clients.

Common application schedule for cool-season grasses in transition zone:

- *Late February-mid-March: pre-emergent treatment on fertilizer for annual grasses*
- *Mid-March-mid-May: post-emergent broadleaf application*
- *Early May-late May: post-emergent treatment on fertilizer for annual grasses*
- *Mid-May-early August: grub control (sometimes with fertilizer if performed early)*
- *Late August-late November: fall fertilizing; can be in two applications. (possibly lime)*

Chemical treatment schedules can also be set up to allow fall fertilizing to be combined with aeration or other types or services. This is just one example of how these treatments can lead to additional work with the client.

Take opportunities to combine services for several reasons. You can tell a customer that one particular task is scheduled but you noticed that there was also a need to do something else. Then explain what that is and why you are doing it. By doing several tasks at the same time you can increase profits and eliminate road time. A good scheduler can get more profit per hour by lining up several services to be performed at one time.

BILLING AND INVOICING

Billing is the next task to consider. It will make or break a company because without the actual collection of the revenues you are unable to continue to be in business. Billing in a small operation should be handled in a very personable way. If you hand-write (Figure 7.2) or even print invoices you may want to take the time to deliver the first few bills in person. This creates more opportunity to interact with the client and to see if the services you are providing are satisfactory. The client will usually spend some time with you talking about possible future work they would like you to do. An in-person meeting also means that you will be paid promptly. By addressing the person on a one-on-one basis the customer is forced to pay you on the spot unless you arrange otherwise. This may be your first indication of a problem if your client makes an excuse for not paying you at that time. Many customers will appreciate the personal contact, whether you have one client or several hundred clients. Sometimes you can correct a problem that you did not even know existed by interacting with customers.

As a company grows or as time becomes more limited, the billing becomes more complex. Hand-written invoices will become

TURFTIP

Some companies bill on a third-third-third method, which is a good way to keep cash flow coming on a larger project. That system typically would have the customer pay one-third up front, one-third halfway through the project, and the last third upon completion.

very overwhelming with a large number of bills to send out. Should you fall behind, you are only hurting yourself because you will not collect on services that are not billed. It also becomes very easy to accidentally forget to send a bill. A computer program will help you manage the business in a better way (Figure 7.3). This will allow tracking of services and will keep records in order from season to season.

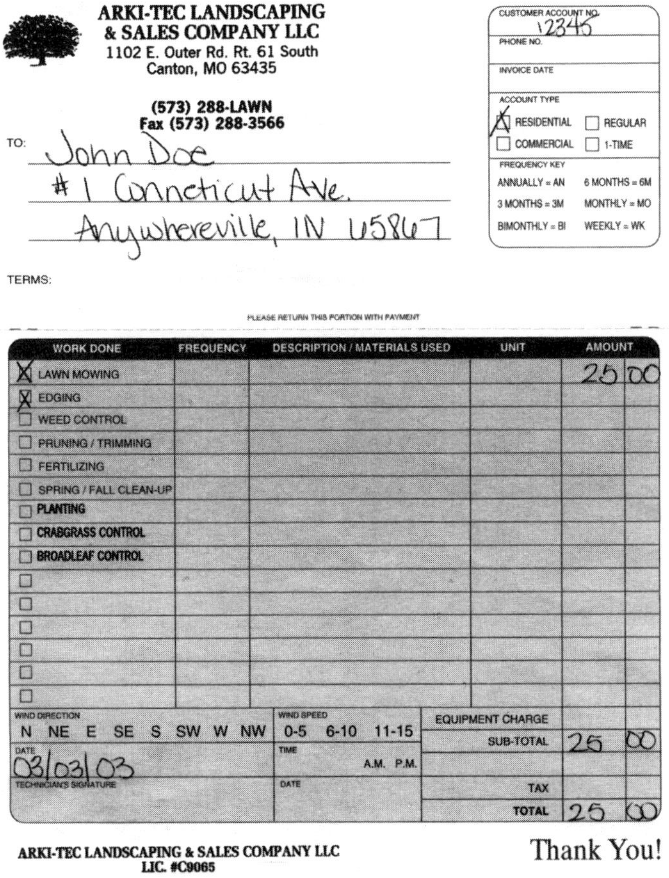

ARKI-TEC LANDSCAPING & SALES COMPANY LLC
1102 E. Outer Rd. Rt. 61 South
Canton, MO 63435

(573) 288-LAWN
Fax (573) 288-3566

CUSTOMER ACCOUNT NO. 12345
PHONE NO.
INVOICE DATE
ACCOUNT TYPE
☒ RESIDENTIAL ☐ REGULAR
☐ COMMERCIAL ☐ 1-TIME
FREQUENCY KEY
ANNUALLY = AN 6 MONTHS = 6M
3 MONTHS = 3M MONTHLY = MO
BIMONTHLY = BI WEEKLY = WK

TO: John Doe
#1 Connecticut Ave.
Anywhereville, IN 45867

TERMS:

PLEASE RETURN THIS PORTION WITH PAYMENT

WORK DONE	FREQUENCY	DESCRIPTION / MATERIALS USED	UNIT	AMOUNT	
☒ LAWN MOWING				25	00
☒ EDGING					
☐ WEED CONTROL					
☐ PRUNING / TRIMMING					
☐ FERTILIZING					
☐ SPRING / FALL CLEAN-UP					
☐ PLANTING					
☐ CRABGRASS CONTROL					
☐ BROADLEAF CONTROL					
☐					
☐					
☐					
☐					
☐					
☐					

WIND DIRECTION: N NE E SE S SW W NW
WIND SPEED: 0-5 6-10 11-15
EQUIPMENT CHARGE
SUB-TOTAL 25 00
DATE 03/03/03
TIME A.M. P.M.
TECHNICIAN'S SIGNATURE
DATE
TAX
TOTAL 25 00

ARKI-TEC LANDSCAPING & SALES COMPANY LLC
LIC. #C9065

Thank You!

FIGURE 7.2 Simple handwritten invoice. When you are a small company simple hand-written invoices are a good way to bill customers; make sure to include all the data you need to refer to.

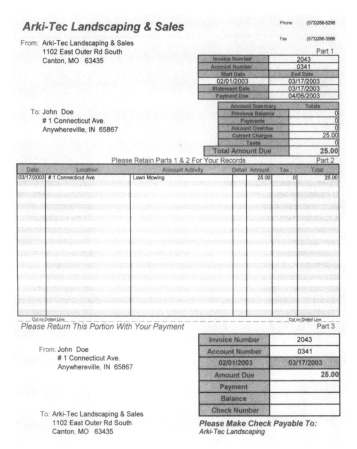

FIGURE 7.3 Computer invoice. A computer system can help you to track customers as your business grows, allowing more detailed billings and keeping track of numerous accounts.

Make sure when you are billing a client to not only be consistent from client to client but also with the same client from year to year. Clients will talk about your services and also keep statements to refer to from time to time. Price increases are allowed in any business, but make sure they are justified or at least explainable. Try to mail bills on at least a monthly basis. This will keep cash flow coming in regularly. On larger projects that can tie up large sums of money, ask for half down and half on completion. This practice is more common with landscaping work, but seeding jobs can get into big dollars in a hurry. You will only lose part of your money of the customer fails to pay you the other half.

Some companies bill on a third-third-third method, which is a good way to keep cash flow coming on a larger project. That system typically would have the customer pay one-third up front, one-third halfway through the project, and the last third upon completion. Whatever system you use, make sure not to allow too much money to be outstanding on the books. Cash flow is the most common failure for business owners. Many companies fail because business was too good and billing issues were overlooked.

CHARGES AND FINANCE

The amount you can charge for any service will vary a great deal based on your location and the population of the city you live in, as well as on the competition. The average in 1995 was still only $15 to mow a residential lawn. This has increased only slightly in the last few years but takes into account many economically poor regions. In larger cities the averages are around $30 to $50, which sometimes does not include trimming. I hate to tell people what I think they can charge because it varies so much from state to state. It even varies from town to town based on the level of competition and the condition of the economy. You will need to do some checking with people who are paying for services to see what the average in your location is. Do not hesitate to ask in a tasteful manner, if you are meeting with a new client, what he or she has been paying. If you find yourself losing bids with residential customers, ask the clients why and what bids they accepted. This will be more of a concern when dealing with commercial clients.

Do not feel that you need to do work by the hour. Hourly work should be a consideration in your mind but not on paper when you are billing. Know exactly what you need to charge before approaching someone about work. Know if you have any room in your price and what you need to have to make any money at all.

I do not want to get into a discussion of overhead because of the numerous factors and variables involved. Overhead is an important aspect of business, and it should always be considered when doing business. If you meet with a customer and he or she beats you down to a price where you are not making any money, it is usually not worth your time to have the client in the first place. It is fine to have different rates for different tasks. This is just part of what overhead actually is—the cost of doing business. Consult an accountant and keep close tabs on spending when starting a business. An accountant will be able to advise you on your choices. Set up good relationships with lenders or banks to ensure quality relations for years to come. Almost anyone can borrow money, but many cannot pay it back. That is why good credit is the key to good business. Without credit or the access to money a business cannot grow.

I like to think of money as nothing more than a tool. It is no different than a hammer, mower, or automobile. The reason is that it takes money to buy these tools to make additional money. There is the old saying that "It is easy for someone to make money when they already have it." This is really true because that person may have a better access to cash. Someone who has money does not have to worry about interest on the things they buy. Interest is an underlying cost that exists for almost any business.

Make sure to keep detailed records of all your customers and the services you provide for them. Also track addresses and phone numbers so contact with each client can be done quickly and easily. Tracking all these aspects will help to establish routines and maintain consistency. By being consistent you ensure that all customers feel important to your operation and are considered to be a priority. By keeping up-to-date addresses and phone numbers, interactions can be improved through mailings or calls to clients. Door-to-door calls are not always possible, but picking up the phone to check with clients once in a while makes them feel like they are still a concern.

One way to create awareness is through advertising (Figure 7.4). Advertising for business is almost a necessity at some point, but there are many different ways to approach it. The first thing to establish is a specific budget. This budget should be a portion of money you can afford to use. It is hard to measure a response from advertising, because you never can tell exactly how a customer has found out about you. Even when you ask a customer, he or she may not be able to give you an accurate answer. I have had clients tell me that they heard about me on the radio that week when I might not have run a radio ad for months. Consider are which advertising media will reach your target audience. What is your target audience and what age are the people in this audience? There are many choices available, including newspaper, radio, TV, Yellow Pages, and signage. Each has a particular usefulness. Ask questions and meet with the various media sources to find out what audience their services reach.

Budgeting is an important aspect of your business. Set strict budgets each year and try to follow them as closely as possible. This goes for advertising, purchases, overhead, expenditures, and maintenance. Look back at all the fixed and variable costs you had from previous years and try to anticipate what will happen in the current season. A margin of error of 10% could be expected for some of your figures just because of economic and business trends.

APPEARANCE

Make sure that when you conduct business, you always maintain a certain integrity and appearance. This goes for both your personal and professional appearance when interacting with the public. Try to make your equipment look as good as it can. This does not mean that it has to be new but only that you should take reasonable care

FIGURE 7.4 Some advertising media examples.

of it and not bring junk onto a homeowner's property. Also try to have a vehicle that does not appear trashed or unsightly. Even if you drive a 20-year-old truck you can use signage on the doors or open spaces on trailers (Figure 7.5) to put up your company name and logo. Signage on your equipment is a cheap way for you to get your name out to people just by parking on the street while you are working. This might be a way to meet a new customer after hours as you go about your business.

FIGURE 7.5 ArkiTec landscaping trailer.

SPECIAL RESIDENTIAL CULTURAL PRACTICES

It takes a well-thought-out program to transform a customer into a regular client. The key is to provide a well-kept lawn at an affordable price. Lawns are the basis and main focus for building a successful lawn care business. The practices for growing and maintaining turf are discussed throughout the book, but there are some special needs for maintaining a residential lawn.

Most lawn care customers will want to have the lawn mowed on a weekly schedule throughout the growing season. Typical height for a residence for cool-season grass is 2½ to 3 inches. There is always a mix of customers who will want to have the lawn bagged (Figure 7.6) and others who will prefer not to have the lawn clippings collected. In more recent years it has become a trend to mulch the lawn. Each customer will have a preference or a general idea for what he or she feels will provide the best results. It is up to you as an individual to sell your particular service and explain what your company offers. Some larger companies may have the advantage of offering the customer the choice of any of the mowing practices because of the range of equipment they own. I personally like the appearance of a mowed lawn with the clippings collected. Supplemental irrigation and fertilization can be performed to offset any disadvantages to vacuuming the lawn. I just think that it looks nicer.

For warm-season grasses, more frequent mowing may be required and at lower heights. Many Bermuda, zoysia, and St. Augustine lawns are cut at heights ranging from 2 inches or less. The warm-season grasses are more capable of handling the lower heights that are expected by customers in some of the southern states.

Chemical treatments should always be performed on a lawn to maintain its long-term integrity. An evaluation of the soil quality and pH should be performed at least every three years if possible. Try to sell the idea of aeration once a year. Aeration is not used by all lawn care businesses, but I find that it is one the most beneficial sales you can make to the customer. It will keep the soil from becoming compacted and reduce the thatch in the lawn, and the homeowner will

FIGURE 7.6 Bagging mower.

be much more satisfied long-term. Another task that I see as a necessity at least every other year, if not yearly, is to overseed the yard. Power raking or aeration along with an overseeding program is the best, but any type of overseeding once a year is better than none at all. This will make sure that, if there is turf injury, reestablishment can occur. Without some overseeding to a lawn area, bare spots may be slow to recover and the lawn may decline in appearance. By introducing new varieties of grass into the lawn new species may adapt and find areas more suitable for growth. This will ensure that overall turf quality will stay at its highest point. It is much less expensive to do little bits of maintenance along the way than to let a yard go until there is little that can be done except to start over with new turf.

The combination of all the maintenance practices makes a lawn look its best. Remember that the lawn is composed of living material that we call turfgrass. Nurturing and manicuring this grass creates the most attractive results. Take out any one component of maintenance, and the overall quality of the lawn will decrease. Maintenance is the key to success.

It is important for you to be involved in continuing education to keep up with the highest standards and latest technologies of lawn care maintenance. Much like the medical field the practices associated with the treatment and care of plants are constantly changing. As technology improves we continue to learn more about the life cycle and science of turfgrass. With each new discovery the practices and maintenance habits that have been done for many years may have to adapt and change. It is through these changes and by utilizing this knowledge that we can all become better turfgrass managers.

Chapter 8

Commercial Lawn Care

One of the fastest growing segments of the green industry is commercial lawn care services. Whether the focus is on residential lawn accounts or business accounts, anyone who has a mowing service that provides lawn care falls into this category. There are a handful of companies that specifically focus on residential accounts and numerous commercial mowing businesses that do both residential and commercial mowing.

The specific focus of most lawn care businesses tends to gravitate towards more commercial work. Owners of lawn care companies feel that large profits and a huge gross sales volume can only come by having numerous commercial accounts. I mentioned earlier that this is not always the case due to the bidding process that occurs for most commercial contracts. As with any business there will always be competition, but in lawn care the number of competitors in one geographic area seems to be much higher. I believe that this trend results from a mentality that if you have a mower you can start a business It has also been proven that in periods of economic recession or high unemployment people resort to lawn care because they feel it is something that requires little skill to do. I am not going to go into the many reasons why this is not so. Any business takes a high amount of skill and know-how to succeed.

Commercial mowing started to accelerate in growth sometime during the 1980s but did not boom until the early to mid-1990s. Many feel this was due to the economy in general and that people could afford to have their lawns mowed who may not have been able to afford it before. Businesses became more inclined to hire out services than to keep everything in-house as was done in the past. The idea behind contracting work out was to cut costs and maintenance expenses. Mowing had become one of those areas that it seemed more practical to hire out vs. keeping a piece of equipment on hand and trying to pay an operator to use it (Figure 8.1).

I mentioned in the last chapter how much cheaper it is for some homeowners to pay to have mowing done compared to the costs associated with keeping a lawn mower. This is also the case for a majority of commercial businesses, which have large areas to mow. The problem most companies have with turning someone loose on a mower is that the person may not be suited for doing that type of work. There is also the problem of someone damaging the equipment because they have blatant disregard for other people's prop-

---

TURFTIP

Commercial mowing started to accelerate in growth sometime during the 1980s but did not boom until the early to mid-1990s. Many feel this was due to the economy in general and that many people could afford to have their lawns mowed who may not have been able to afford it before.

erty. By adding the repair costs in with the hourly wages of the employee, you can see why some companies feel that hiring the job out is a much better alternative. Another reason mowing is hired out is because an outside company can be held more accountable for the work that has been performed. There may also be a need for additional outside work, which requires even more equipment to be on hand at a business, which is why many commercial lawn care operations have diversified their services to provide a complete maintenance package. With snow removal, leaves, and weed treatments being a concern for some companies, it is not practical for someone within the company to do these procedures.

FIGURE 8.1 Small storage facilities at a large business; it is not practical to keep much equipment around.

> **TURFTIP**
>
> Do not overlook schools or municipalities for mowing contracts. Many cities are looking to management companies and subcontractors to meet the needs of shrinking budgets. There are potential customers in schools, parks, and city organizations. There is also great potential with churches and other religious facilities.

Lawn care companies that diversify often have better success with year-round cash flow instead of trying to survive on mowing income alone. Let's face it, unless you live in the extreme South, there are going to be periods of the year when it is just impossible to get any mowing done. It also may be a good idea for companies to consider having equipment that is versatile enough to be used during the whole calendar year. For instance, buying a mower, skid steer, or tractor that has the versatility for use in all seasons just makes sense. This will reduce the need for equipment that might sit around a majority of the year and also will allow some use in the off-season. If you are going to have payments for the equipment, it sure would be nice to be able to get some profit from those items in the winter also. Mowing is a great occupation that provides high returns in the season, but with the unpredictability of the seasonal weather factored in, your return is not always guaranteed. I find that more diversified companies have a better survival rate than those who only perform mowing services. This is the key to lasting through hardships that are caused by any unforeseen circumstances including the economy.

> **TURFTIP**
>
> The bidding process is the biggest key to success for a commercial lawn care company.

BIDS AND CONTRACTS

The bidding process is the biggest key to success for a commercial lawn care company. The only sure way to find your price range is unfortunately through trial and error, but there are some tips for success when bidding. First of all, find out as much about the potential client as possible. This can be done just by asking someone you know at the company or someone who knows a little about the history of the company. You can also or try to ask questions of your contact, such as, "Why they are bidding out the mowing in the first place?" This might provide answers to a majority of your questions and also let you know if this is a client you need or want to be doing business with. The company could respond with negative comments about every single person who has ever worked there, which might indicate difficulties in keeping this client happy. They might also indicate that they always bid the work out and are simply looking for the lowest bid. This can also be a bad situation, because you might simply be the next "low ball Joe" who happened to respond to a bid request.

The questions you ask could also reveal something very positive, such as a situation where the current mowing company had gone out of business or moved out of the area. This might be a situation where a long-term business relationship had been going on and this was the first time in years that the mowing had even had to come up for bid. Some companies will lock in people providing services as long as their prices do not change; some even allow a reasonable increase over a period of years as long as it is justified and the work you provide is satisfactory. Once a price increase is requested from most clients, the bidding process may be opened back up. I have seen this to be the case with many cemeteries. This is often due to their limited resources or strict budget restrictions.

Cemeteries are a special segment of the commercial mowing industry. Many try to get into cemetery accounts because they often pay pretty well if they are of any size. These accounts are usually controlled by a board of directors and often run off what is called a perpetual fund. This fund is in place to pay for the mowing and maintenance of the cemetery. The idea is that there is an account set up that is supposed to cover the maintenance costs of the facility. It is preferred that the cemetery run off the interest without dipping into the principal of these accounts. This may not be possible during slower periods of the economy when interest rates are not very high.

There are many rural cemeteries that may provide an unknown source of income for your lawn mowing company. Many times there are few people who know the cemeteries exist and do not think of this market for potential customers. It is important when bidding a

---

TURFTIP

Cemeteries are a special segment of the commercial mowing industry. Many try to get into cemetery accounts because they often pay pretty well if they are of any size.

cemetery to consider the tremendous amount of trimming that will be involved with the upkeep of the property. My personal experience with cemeteries has been that you need to allow as much time for trimming as for the actual mowing. There is a trend in many newer cemeteries to allow only flat tombstones to create easier maintenance and mowing; however, many of these cemeteries still have an older section that may have larger above-ground stones. You need to take special caution when mowing around tombstones because this can be a very sensitive area to the people who have family buried there. If the property becomes unsightly or a tombstone is damaged, you might find yourself out of work very quickly.

No matter what the focus of your mowing operation, there are many untapped resources for you to consider as new customers. I find that in some rare instances there is even the potential for golf course maintenance by private companies. This is more common in rural areas where there might not be enough money to provide the equipment and people to maintain an entire course. Resourceful companies need take a look and see if they are capable of offering the course the entire maintenance package. Be careful to consider the specialized equipment and chemicals that are required for making such an offer.

Do not overlook schools or municipalities for mowing contracts. Many cities are looking to management companies and subcontractors to meet the needs of shrinking budgets. There are potential customers in schools, parks, and city organizations. There is also great potential with churches and other religious facilities.

When bidding a mowing contract for one time or for the entire season, be sure to factor in all the potential expenses that your company might endure for the season.

List of potential expenses:

- *Employees*
- *Equipment*

- *Insurance*
- *Products and supplies*
- *Licenses*
- *Fuel*
- *Tools*
- *Taxes*
- *Maintenance and upkeep*
- *Repairs*
- *Advertising*
- *Unforeseen expenses*
- *Insurance*
  - *property*
  - *workman's compensation*
  - *vehicles*
  - *liability*
  - *medical*
- *Facility*
  - *phones*
  - *utilities*
  - *upkeep*
  - *additional insurance*
  - *office supplies and expenses*
- *Other*

There are other factors you might be considering and maybe some in this list that you have not thought about. The important thing to remember and my main point here is that all these factors are your costs for doing business, which is known as overhead. Think about them carefully no matter how big or small your operation might be. Without factoring in overhead, your bids will end up being too cheap and may eventually end up breaking your company.

When placing a bid, make sure to gather as much information ahead of time as you possibly can. Then set up an appointment with the owner or contact person within the facility you are bidding. Have that person actually walk around with you to inspect the property. Maps or directions can lead to confusion or bidding problems because they may not be clear and you cannot differentiate the property lines. As you are walking around with this person, you are also interacting with the customer much like what I described in the residential lawn care chapter. This bidding process is again an opportunity for you to sell yourself, your company, and the services that you provide. Even if you do not receive the bid for the work you are looking at, you may have a prime opportunity to show your expertise and talk about other features that you noticed that might need some attention. They might keep you mind for other work down the road if someone else is hired and they do not have the resources to do all the future work.

When you submit a bid, make sure that all the information contained in the bid is very specific. It helps to create a letterhead or document that looks professional, even if you own a small business. Bids should always show the physical address and contain detailed specifications as to who, what, when, where, and how. Be as descriptive of the work as possible and outline all the services that will be performed. This will also give you a reference point later if someone questions your price or the scope of work that has been done on the property. I like to put a disclaimer in the document that shows terms and gives a 30-day deadline to respond to your bid. This will ensure a more prompt response and not allow the company to leave you hanging until the last minute.

Have a game plan together very early in the spring each year or even get started on bidding in midwinter. Keep in mind that sometimes having too much work is worse than not having enough. When you send out bids, try to get quick responses so you know exactly how many customers you will have for the season. It is ok to say "no" once in a while to avoid being overloaded with customers. If your company gets to a point where you have too much business, you may be lucky enough to "weed out" the clients that you are not making much money on. A few high-quality clients are better than twice as many bad ones. Once you reach a point where you can be a little more selective with your clients and bids, your company's headaches will decrease.

I am not suggesting that you base clients on an economic level but more on attitude and reliability. You should seek out customers that are going to be understanding, appreciative, and dependable. They are depending on you as much as you are depending on them, so payment reliability should always be a consideration. Large-scale jobs and contracts are great, but when companies drag out your payments 60 to 90 days you had better have alternative cash flow to keep your company going. It is difficult to survive slow paying clients no matter how good the project may be. When you are bidding work make sure to ask questions about the payment schedule and terms. If it is a major concern to you, then let the client know ahead of time that you expect payment in less than 30 days of billing and that is the only way you will consider taking on the job.

Make sure to read the guidelines of any contracts very carefully before signing up your company to do any work. It is better for you to be the one providing the mowing contract instead of the person who is hiring you. This will give you the best opportunity to set payment terms and lay out the details of the scope of work. If you are a savvy business owner you may be able to convince a company to provide you with a three-year contract to ensure longer term success with that customer. This will keep you from repeating the hassle of the bidding process each year. I still find that when you provide top-quality services for clients, the bidding process seems to disappear. Most of the time the people soliciting the bids do not like the hassle of bidding any more than you do. If you provide a good service at a fair price, chances are you can have that company as a client for as long as you want to. This does go back to the premise of selecting top-quality clients who will appreciate the work you are doing for them.

There are two types of contracts when it comes to the mowing bid process. Seasonal contracts were one type that became very popular with lawn care companies for a while (Figure 8.2). These types of contracts would allow companies to set up a 12-month pay schedule that would cover all the services needed for the year. For example, a company might need snow removal and mowing services provided. The business could pay "x" amount of dollars per month for your services and know exactly what they pay each month. What you might want to do is set up a plan that will figure the total dollars needed for all your work and divide that by the number of months in the year.

By using seasonal contracts you are allowed the benefit of seasonal cash flow and a guaranteed amount of money each month from those clients. On the other hand, it is very easy to come out on the losing end of this type of agreement if you get a rainy or wet season. In a time of drought or a lack of snow in the winter, for instance, your company could stand to make a high return for little to no work. Remember, you have to base your bids on averages, so you could mow 24 times in a season or end up mowing only 10 times. The same scenario could happen with snow removal where you might

---

Cost of service x Number of weeks in an average mowing season ÷ 12 × monthly cost

(Example: $30 per time mowing × 22 weeks ÷ 12 months = $55 per month year-round)

8.2 How to figure a seasonal contract.

plow 2 times or 30 times. This is why you might be better off with a per-time contract.

I prefer to use per-time contracts, which take a lot of the guesswork out of the bidding but can leave a lot of questions as to how much you might make in a year. With a per-time bidding process you only get paid for the work you do, which means in a dry year times could be very tough. The seasonal bid does give some cushion when it comes to dry weather periods. Many companies use a balance of seasonal and per-time contracts so they have some level of guaranteed income, but I still personally prefer the per-time style of contracts.

I am sure some of you are thinking that you just want to get started with this type of business. You may be concerned about scaring away clients or feel that some of my suggestions seem too assertive. Remember that business is business and that your goal is to make money. If you cannot make money, there really is no point is spending your time on a commercial mowing business at all. By being prepared for growth and understanding long-term problems, the short-term issues will solve themselves. For many people just purchasing equipment is a struggle, and for others finding ways to be efficient and effective are the biggest challenges. I believe that sound business practices are the keys to success.

ON THE JOB

Mowing for the most part, just as with residential lawn care, will need to be done at least once a week. I do find that commercial clients are more likely to let you mow more than once a week in the spring if needed when the grass is growing very fast. For most commercial clients there is a higher level of maintenance expected so the appearance of their business is always top-notch. There are some exceptions to that rule—people who want things done cheap and fast. Those types of customers should be avoided at all costs unless you are really struggling to get established.

> ## TURFTIP
>
> When mowing for commercial clients there is a preferred height of at least 2½ inches. This is really the low end for mowing heights for cool-season grasses in this type of account, and many commercial clients prefer somewhere from 3 to even 4 inches in height.

When mowing for commercial clients there is a preferred height of at least 2½ inches. This is really the low end for mowing heights for cool-season grasses in this type of account, and many commercial clients prefer somewhere from 3 to even 4 inches in height. This will give the lawn a better appearance and reduce the amount of weeds in the lawn. Grass does provide a canopy effect to the ground much like that of an agricultural bean field. For instance, when beans are planted as a row crop for production of grain, they are spaced so that, once the beans reach a certain height, they provide shade for the soil surface, which keeps other weeds from growing. Turfgrass can be very similar in the way it can crowd out competing weeds if it is allowed to grow at a higher mowing height. Grass will provide a shading affect to the soil, which can and will eliminate as much as 50% or more of the weeds in the lawn without using chemicals. This can be a critical sales point for suggesting a higher mowing height for commercial and residential customers. Mowing at the higher height is not always an easy sale. It has been somewhat traditional to cut grass at 2½ inches, and suggesting something different can always be hard to sell no matter how convincing you are. You must at least convince the client that the 2½-inch minimum height is necessary. Anything less in cool-season turf for lawn care is damaging to the turf quality and weed control program. Warm-season grass can tolerate much lower mowing heights and in the southern states less than 2 inches is fairly common for lawn care. Zoysia and Bermuda lawns are the preferred turf types, and these grasses thrive under lower mowing heights and conditions.

Another advantage to the higher mowing height in cool-season turf is the reduction in stress to the turf. Taller grass will have much higher moisture content and also benefits from the additional cooling of the soil that the grass canopy provides for the root zone. These two factors in combination with the added weed control do tend to recommend the higher mowing heights for commercial clients. In comparison to residential lawn care the higher mowing heights are not usually allowed and the customers tend to be much pickier about how high the lawn is cut. I wish that I never had to hear the expression "If you mow it shorter you won't have to mow it as much." The reality is that no matter how low the lawn is cut, there will be a height that the customer will deem as acceptable whether it agrees with your plans or not. It is up to you to educate the client and explain why shorter heights are unacceptable.

When mowing commercial lawns, pay close attention to the details and future concerns that are occurring on the property. While you are there on at least a once a week basis, you have plenty of time to scope out new work and look for tasks that can improve the property. Take the time to pick up sticks and debris so the property or business owner notices that you pay attention to the little details. Litter pickup should be a service that is provided and sometimes might have to be done for free just to maintain the integrity of the job. If you are the type of person who mows over obstacles and debris without a care in the world, you are likely to create a trashy appearance.

SAFETY

Safety is a major concern that should be high on your list if you have to deal with the possibility of mowing over sticks, debris, and

> **TURFTIP**
>
> Safety is a major concern that should be high on your list if you have to deal with the possibility of mowing over sticks, debris, and garbage. However, you should be thinking of the safety issues every second of the day when you are involved with lawn care.

garbage. However, you should be thinking of the safety issues every second of the day when you are involved with lawn care. People should always be a major concern when operating a mower, trimmer, or other piece of equipment that could cause flying debris. There is a big risk when it comes to the safety of the people you are mowing around. A rock could be traveling several feet per second when it is projected out from underneath a running mower blade. This can cause severe injury to people in the path of any of kind of flying objects as well as a risk for property damage.

When mowing a lawn, always be on the lookout for potential obstacles or things that could be damaged from a lawnmower's presence. By the same token, also be aware of things that you could come into contact with that could also be damaging to your equipment or yourself. There is always the potential for windows or automobiles to be damaged from something being thrown from a mower or even by the mower itself. This is why, if you have employees workplace safety should always be a priority and the workers should receive instructions and training on the safety and operation of any powered equipment. You should always train the people who work for you and educate them on the in and outs of what they might encounter while mowing. Those employees in turn will be better aware of safety issues and should know how to handle a problem if it occurs.

Worker safety in the lawn and landscaping industry is always a concern and should be a part of your daily thought process for both your workers and yourself. Always be sure to wear appropriate attire when using mowers or powered equipment and make sure to also wear safety equipment when needed. If you are an employee at a facility, ask for safety equipment from your employer, especially if he or she is not responsible enough to provide some for you in the first place. Make sure that ear and eye protection is always worn. I know this seems like it should be common sense, but really safety is not a major concern for many lawn care professionals. I think that safety should be even more of a concern when working at a commercial business where there are more likely to be workers or passers-by.

There is something to be said for the use of safety equipment, and I can attest first-hand that even brief periods of engine noise can cause some hearing loss that cannot be replaced. When you spend time day in and day out in the presence of power equipment, the noise can take its toll on your ears. Because of the nature of power equipment, always be prepared and aware because some accidents will happen. You should also take into consideration some of the less common safety precautions such as the sun and the effects it can have on the body and eyes over a period of years. Melanomas and other skin disorders are at record numbers and continue to rise each year. This happens even with the cautions and knowledge we now have of how the sun can affect a person's body and skin. There is also a significant increase in the risk of cataracts for people who do not wear sunglasses or eye protection. It is the combination of all the risks that can make mowing and landscaping one of the highest-risk occupations that there is. When a company applies for workman's compensation, it is usually pretty obvious just how high-risk the mowing occupation is considered. Much like farming, the primary risks come from the being exposed to the dangers of moving parts on power equipment such as PTO attachments and mowing blades.

There is also a pollen and allergy risk that I find many people overlook when considering the effects a career can have on the human body. I myself suffer from mild allergies to pollen and grass and can imagine that many who do this type of work share similar problems. Do not be afraid to consult your doctor about what treatments are available if you find yourself working in situations where you are exposed to such allergens. Do not be afraid to wear a protective mask, even if it makes you look like the weird person in the crew. Safety should be a more important factor than appearance.

A mask is a very important necessity with the use of pesticides and chemicals. Pesticide fumes and drift can cause a variety of ailments in the human body. Some of the problems associated with chemical use can be sneezing, rashes, and nausea. Chemical exposure can also result in more serious problems and can even cause death if it is not taken seriously. Be aware of the many protective garments that are required for you to safely apply pesticides and chemicals.

LICENSES AND TAXES

This is a good time to focus briefly on the licensing your company should carry for chemical applications and for doing lawn care busi-

---

TURFTIP

There are a number of licenses and laws that lawn care operations need to be aware of.

ness in general. There are a number of licenses and laws that lawn care operations need to be aware of. The spraying or ornamental lawn care license is probably the most common. This license is governed by state law and requires a written test. I will discuss this in more detail in the pesticide section.

First you need to recognize what type of business you are or want to become. There are corporations and limited-liability corporations (LLCs), for example, but many people do not know enough about these types of companies. There are many differences between a corporation and a sole proprietorship. Sole proprietorships are the most common type of business among lawn care professionals. This is due to the large number of one-person operations. Another reason is that there are still many part-time operations. There is nothing wrong with being a sole proprietorship, especially if you are a small company. The taxes on your business are figured in a very simple way, and the deductions come out of your gross profit. You usually will be "doing business as (DBA)," which means that "John Doe" is doing business as "John's Lawns". Your business operation is held directly to your Social Security number, and any late, delinquent, or default payments can greatly affect your credit. I am not giving you tax advice and would suggest that anyone consult an attorney and an accountant to see what type of business is best for your operation. I do want to give you a basic list of the more common types.

The business type that is becoming very popular these days is the LLC. I would call this a hybrid business that has characteristics of a partnership, a corporation, and a sole proprietorship. Taxation is much like a sole proprietorship but with the umbrella protection of a corporation. It is at least something you should consider if you are new to the lawn care industry or to business practices in general. You do not have to be a large company to have this business type, and it does provide some of the benefits of an actual corporation. There is also a split risk among partners.

A regular partnership can be set up in many ways but usually falls into the same format as a sole proprietorship. The only differences are that the bills, profit, and liability are divided among more than one individual. This is often a way for someone with limited financial resources to get started in the lawn care business without having to assume all the liability responsibilities.

The true corporation is not that common for lawn care companies because it is usually a business type for large companies with several owners and high gross profits. The liabilities only exist through the corporation, and no one individual assumes that risk alone. The taxation is often doubled because the profits can be taxed against the company and the owners, who are paid individually through dividends. I again want to point out that I am not an accountant and that the business type best suited for you can be determined by you and your accountant.

No matter how big your operation is, the use of a fictitious name should always be done in a legal fashion. Your company name should be registered with the state to make sure that you are complying with regulations and not violating any trademarks. An example would be using the name "Quality Lawn Care" to represent your company. Usually there is a form that needs to be filled out, which can often be picked up from the Secretary of State. It is important that the state have a real person's name to tie to your business

---

TURFTIP

No matter how big your operation is, the use of a fictitious name should always be done in a legal fashion. Your company name should be registered with the state to make sure that you are complying with regulations and not violating any trademarks.

operation. This could also prevent an accidental violation of the use of an existing name, especially a trademark name or the name of a business that may have used fraudulent practices.

If your company is selling products or has a storefront of any kind, it may be necessary to have a merchant's license to conduct business in your state. This is a very inexpensive license to purchase.

By setting up a business the right way the first time, it will be easier to grow and expand your operation. The laws and rules that are in place are there to protect you as an individual and the customers that you do business with. A lot of information can be found out about lawn care businesses and the laws that govern your state by contacting the agriculture department or extension offices in your state or county.

PESTICIDE TREATMENTS

Many of the commercial businesses that you are seeking out for customers will ask you to provide a majority of their outdoor maintenance services. Once you have established a working relationship with clients, you will want to address their pesticide needs and offer a program of lawn treatments for their facilities. I have already outlined a typical pesticide program that I like to use on homeowner lawns in the prior chapter. This schedule can also be used for any commercial clientele you would have. I have added a couple of treatments in italics that may be requested by particular commercial clients or by picky residential customers.

Application schedule:

- *Late February-mid-March: pre-emergent treatment of fertilizer for annual grasses*
- *Mid-March- mid-May: post-emergent broadleaf application; may want to use one application late March and follow-up application mid-April*

- *Early May-late May: post-emergent treatment of fertilizer for annual grasses*
- *Mid-May-early August: grub control (sometimes with fertilizer if performed early)*
- *September-early October: post-emergent broadleaf application if no seeding work is planned*
- *Late August-late November: fall fertilizing; can be in two applications (possibly lime)*

Warm-season grasses will have a different schedule from the above cool-season schedule. It can be approached in two ways, with a couple of high nitrogen applications or several low nitrogen applications.

Typical schedule for warm-season turf in lawn care:

- *Mid-March- mid-May: post-emergent broadleaf application*
- *Early May-late May: post-emergent treatment on fertilizer for annual grasses;*
- *Higher nitrogen content than with cool season grasses (Nitrogen ideally is applied in what is referred to as spoon feeding, which is the use of ½ pound of nitrogen per 1000 square feet per month over three to four months. If this is not a possibility, you can use two to three pounds of nitrogen in the May application as long as a majority of the nitrogen source is the slow-release form such as an SCU or IBDU.)*
- *Mid-May-early August: one grub control application (sometimes with fertilizer depending on your warm-season program)*
- *Late August: last fertilizing of season, no more than one pound of nitrogen per 1000 square feet (possibly lime)*

Treatments for commercial clientele should be planned to avoid the pedestrian traffic that visits that business or facility. Some cities have regulations that lawn care companies provide notice when pesticide treatments are performed with some sort of marker, flag, and/or a notice. You should always consider the safety of the people that may work at the facility. Pesticide drift is always a concern, and you should limit the exposure of the people that may come into contact with your applications. Chemical drift is one of the biggest reasons why I suggest granular applications rather than sprays. The other reason is because the dry forms have a much higher quality of formulation. Remember that liquid fertilizers are short-lived and have much higher risk of turf injury.

When mowing or treating a property of any kind, always imagine that you are performing the work on your own property. Would you want to come into direct contact with chemicals or would you want the loud noise of a commercial mower operating in peak business hours? I like to time chemical applications and mowing after business hours or even on weekends for commercial clientele. This creates a more positive relationship with that particular business and shows them that you are also willing to go the extra mile to meet their needs. Noise with power equipment can be a big problem, and the state of California, for instance, has recognized that problem with legislation. There are now decibel ratings on equipment to restrict blowers and other hand-held equipment to a certain noise level. Commercial lawn care businesses became so prevalent that the early mornings in many neighborhoods were disrupted with loud noises of power equipment at the crack of dawn. California also made provisions to clean up the air with improved emissions standards on engines sold in that state. It is important to be courteous to all your clients and try to be the least disruptive to their work environment as possible. Many states have considered following in California's footsteps, but if we as lawn care professionals make efforts to run quieter, more environmentally-friendly equipment, such legislation will not have to be passed.

Always remember that the lawn care business is no different than most other businesses and is as competitive as any other. By using good lawn care practices and combining them with effective management decisions, anyone can have success in the lawn care profession. Make sure that you realize that every business has room for improvement and that anyone starting out has the potential to succeed. There is no market that cannot be entered, and there is also no golden rule to success. I would encourage you to look at the lawn care profession as an opportunity to have a fun, exciting career that will allow you to work outside helping to beautify your community.

Chapter 9

Turf in the Landscape

Turf has an important place in the landscape and should be included in landscape designs for both aesthetic and functional reasons. It can accent landscaping plans and buildings. It can also control erosion, reduce runoff, and help reduce pollution. This is especially true where there are large parking lots, buildings, and city streets. Turfgrass can have many positive effects for the environment.

Turfgrass is actually the easiest maintenance feature in the landscape. Many people think they will reduce maintenance by getting rid of grass, but turfgrass can be very inexpensive to grow, establish, and maintain compared to other landscaping. This is especially true for lower maintenance areas where chemical treatments may not be a concern and mowing frequency is not that important. Turfgrass stays green, prevents erosion, and really only needs mowing to prolong its development. The cost to put in grass is considerably lower than landscaping a comparable area of equal size. Mowing is also cheaper than maintaining elaborate flowerbeds, mulched areas, or large landscaping beds. Turfgrass can be one of the most cost-effective features in terms of material and labor in the landscape.

There are some situations where eliminating a turfgrass area can simplify a landscaping problem. This is usually a scenario associated

> **TURFTIP**
>
> The cost to put in grass is considerably lower than landscaping a comparable area of equal size. Mowing is also cheaper than maintaining elaborate flowerbeds, mulched areas, or large landscaping beds. Turfgrass can be one of the most cost-effective features in terms of material and labor in the landscape.

with a large bank, steep incline, or troublesome area that does not allow for easy access or maintenance with a mower. However, the cost to replace that area with an alternative to turf could be quite high. When mowing an area is unsafe or not practical, landscaping would be a sensible alternative.

Turfgrass provides significant erosion control. Erosion reduction is one of the biggest uses for turfgrass, as is evident when driving down the highway or even building a new house. Grass is a very versatile plant that has an extensive root system if it is healthy. This makes it a very inexpensive solution to keeping soil in its place and generally improving appearance. Grass is used for levees, lawns, prairies, pastures, and many other locations. The commonality among all these uses is that turfgrass will keep wind and rain from removing the soil when it is used for cover.

Prairies were covered with grasses even before people developed the areas of the Wild West, and grass played a big role in keeping the land intact. There have always been grasses on the North American continent, and we continue to adapt and modify them for new uses all the time. It is the great versatility that grass provides that makes it so functional. Turfgrass is currently found at nearly every home across the United States, yet people never stop to realize all the positive effects that it has in our everyday lives.

Did you ever stop to realize how much mud would be tracked in our homes without the lawn to walk across? What about a place for your kids to roll around or play? Did you ever think how unsightly our buildings and houses would be without turfgrass surrounding them? Stop and think about how many sports would not exist if it were not for the development of grasses for playing surfaces (Figure 9.1). Our culture would be completely different if we had no green grass to play on.

Turf should be balanced in scale when used in landscape designs. It is becoming much more popular now to design projects with landscaping in the lawn areas. To much or too little turf can distract from a really great landscape design, and it should always be used in scale with a building. If a building is very large, then much larger areas of grass can be used around it without appearing to overwhelm it. This is not possible downtown, but grass does have a use in the cities also. The biggest reason for overuse of turf is probably due to the

FIGURE 9.1 Turf grass comes into contact with most of us on a daily basis yet some people think it has no value.

higher cost associated with landscaping compared to that of grass establishment.

In city locations where turfgrass is very scarce, it nonetheless serves important functions. It provides a natural filter for runoff water. In cities where pollution is high, water carries many pollutants as it runs off from parking areas, walks, and streets. When the water passes through grass, it is slowed and cleaned at the same time. Grass areas provide the absorption capacity that concrete areas do not. This natural cleaning of our water system is very important to our environment in general.

Another factor that is often forgotten, even by turf managers, is the cooling effect a lawn provides to an urban environment. With buildings and large areas of concrete come hot summer temperatures. When turfgrass is added into some of those environments, the temperature is reduced by several degrees. The buildings and parking areas can add additional stresses to the grass from the added heat, but it is a nice symbiotic relationship that makes turf so useful in city locations.

Since cities continue to expand and the areas of turf continue to be replaced by concrete landscapes, we as turf managers should do

TURFTIP

In city locations where turfgrass is very scarce it nonetheless serves important functions. It provides a natural filter for runoff water. In cities where pollution is high, water carries many pollutants as it runs off from parking areas, walks, and streets.

our job to make sure that grass is still used. Many states are providing guidelines and building codes that promote or enforce the use of turf and landscaping. For instance, there may be a requirement that a certain number of trees be implemented into a construction project based on the square footage of parking lot used. It also may be law that grassy areas or landscaping be left for buffers between residential and commercially zoned properties. The underlying factor in any instance is that builders and engineers do not always see landscaping as a necessity, but it should be figured into every building project that is constructed in a city.

Turfgrass used to be a buffer between city streets and city sidewalks. The turf was generally planted for cosmetic reasons, but really that area of grass is quite functional. The grass provides heat reduction to the hot concrete sidewalks and streets, for one thing. It also creates an area that slows runoff of dirt or other movable materials into the city streets. This area of turfgrass is also an area that can be torn up at any time by the city for maintenance of water or power lines. It is easier to repair areas covered in turfgrass than those covered with elaborate landscaping. It is important that these areas be accessible in case of a water main break or some other catastrophe that would require immediate access to the repair area.

Typically this grassy area that exists between the street and sidewalk is also a good area to plant shade trees, which will further the

TURFTIP

Many states are providing guidelines and building codes that promote or enforce the use of turf and landscaping.

benefits to the residential and street areas. Trees will also help to reduce pollution and the amount of heat radiating off the hot concrete. When doing turf planting or landscaping of any kind, you should be familiar with the local laws and how easements are handled in your community.

The reduction of pollution does not occur just from runoff absorption but also from regular processes such as photosynthesis. Plants create oxygen in the atmosphere while getting rid of some of the more harmful elements at the same time. Carbon dioxide is utilized and changed into oxygen in the presence of light. Pollutants like ozone and other harmful gasses are cleaned out of the air. We all benefit from the natural processes that occur in turfgrass plants.

Some people would argue that grass is a hassle to mow and that it does nothing more than to drive our allergies crazy. Pollen is caused by every plant that blooms for the most part and can be just as much of an irritant whether it comes from trees and weeds or from a grassy area. Most weeds produce a higher level of pollen than turfgrass plants.

There is also the cover crop role that turfgrass plays when mowed on a regular basis. Of course, grass responds to mowing very positively and is in the best shape to keep other weeds and woody trees from competing. I mentioned earlier how grass creates a canopy that keeps other weed seeds from germinating. This canopy also keeps other woody species such as trees or other large, noxious weeds from growing.

Mowing does have a great deal to do with the ability of grass to fend off other competing plants that cannot survive at lower cutting heights, but if grass were left alone, eventually woody species would overtake the area. Grass works as a cover crop and is very good at providing a thin coverage of the soil so other plants can not germinate. In Missouri, if native plants were left to take over an area, the area would resort eventually to primarily large timbers of hickories, oaks, and eastern red cedars.

FIGURE 9.2 Fairways can be mowed in ways that create a more difficult line of play.

Grass is like corn in the sense that breeding has changed the natural reproduction cycle of the plant so that it requires human intervention. Grass does not produce nearly the amount and quality of seed on its own to reproduce and create the same lush cover we are accustomed to. Plants have been bred over the years so that they require human intervention at the harvesting stage. This means that, if left alone, the grass plants that we use today would probably be overrun by more native and aggressive types of grasses. Grass is a plant that can spread and perpetuate itself on its own, but mowing and other cultural practices give it the characteristics that we value.

The point I am trying to make is that we depend on turfgrass to provide us with several beneficial effects. Turfgrass also depends on us to a certain degree when it comes to the healthy-looking grass that we see in our lawns and on golf courses today. Without the mowing frequency, grasses would become more native-looking and not like what we are accustomed to.

Landscaping is an industry driven in large part by America's desire for better looking turf. Mowing was a major part of what started that trend. Landscapers use turfgrass to accent their designs and also to create areas for outdoor enjoyment. Turfgrass has many other aesthetic and practical functions in the landscape. It is used for a large portion of what most homeowners consider to be the yard. Landscaping is then placed around building foundations and out in the lawn so that is it accented by the turf. The trick is to use the grass in a way that it is not overwhelming but allows the turf to create many quality outdoor areas.

Outdoor areas of turfgrass are generally family areas where the kids can play and mom and dad can barbecue. They are also places where the family pet may use the restroom or a place where the car can be washed. The uses for turf are endless and the placement of grass should be carefully considered in any application.

The needs for and placement of turf on a college campus are much different than in a lawn, but the uses are still going to be the same. There is a need for areas for gatherings, sports, and recreation. This brings up the next point in turf placement, which is a consideration of the traffic around and on the turf. College students are a prime

---

TURFTIP

The trick is to use the grass in a way that it is not overwhelming but allows the turf to create many quality outdoor areas.

example of what repeated foot traffic will do to an area of turfgrass. The first thing a college student will learn at a campus is the shortest distance between A and Z. This creates a multitude of worn paths and areas that need constant repair. Since students wear the paths in the shortest routes, it makes sense to place walkways in areas where they will be utilized and use the turfgrass in locations where it will not become unsightly and worn. A good friend of mine who once was in charge of the grounds on a college campus used to wait two months after a building was opened before establishing sod or finishing seeding in order to redirect sidewalks, landscape, and seed based on traffic flow.

Worn areas in turf create problems for maintenance. They also can create problems for the people who have to deal with the mud that gets tracked around from walking through these barren areas. It only makes sense to either eliminate turf in an area with a path or to do rigorous maintenance to keep the integrity of the grass. On a golf course, paths from both spiked shoes and from golf carts traveling over an area numerous times are a constant battle (Figure 9.3, Figure 9.4). There are ways to alleviate the stress to the grass and cultural practices that can keep bare spots to a minimum.

Probably the best way to eliminate this type of damage to turf is by using some sort of aeration to reduce the soil compaction. A core aerator will work wonders for this type of area if used on a regular basis, meaning at least a couple of times a year. A core aerator will reduce compaction by pulling cores out of the soil and redistributing them on top of the ground. This will help to eliminate thatch and level uneven ground over time. Core aeration is one of the more common ways to deal with foot traffic areas and keep the turf looking its best. Overseeding can be done in combination with core aeration to maximize the results to the turf injury areas.

Another way to eliminate compacted stressed areas of grass is to use a power rake to loosen the soil surface. This will provide a temporary fix but is not nearly as effective as aeration in terms of really pulling up and loosening the compacted areas down into the

200 Turfgrass Installation: Management and Maintenance

FIGURE 9.3 Golfers often ignore obvious signs that an area has been driven over too many times.

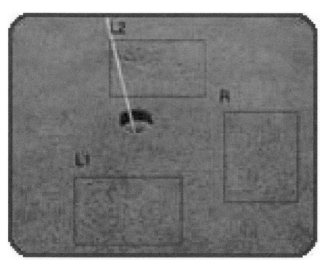

FIGURE 9.4 You can see in the marked areas how much wear is possible around a golf hole in just a short amount of time.

> **TURFTIP**
>
> Core aeration is one of the more common ways to deal with foot traffic areas and keep the turf looking its best.

root zones. If you are planning on repairing the area and preventing further traffic, power raking after core aeration can provide an excellent fix for the problem.

A really good way to solve a compaction problem is by using a deep tine aerator (Figure 9.5). This is a practice that is more common for compacted areas that occur in greens, but if you have this piece of equipment available it will work wonderfully to fix areas damaged or compacted by feet or vehicles. The deep tine aerator can be used to pull plugs out of the ground or simply to drive a deep hole into the ground, which also fractures the soil. Either tine type will provide deep penetration to fix a compaction problem (Figure 9.6). This type of aeration is also really good for curing drainage problems and other wear issues. The problem with deep tine aeration is that it requires a very specialized piece of equipment that can be quite expensive to purchase. You must also have a tractor big enough to power such a machine (Figure 9.7).

Turf in the landscape is a challenge to keep looking good and growing. The establishment of the turf can be a problem if you are fighting outside circumstances such as people, pets, or even automobiles. It is often hard to seed in situations where the grass is being torn up as you are trying to establish it. I find that sodding can be an easier solution so that pedestrians do not sink up to their knees in mud right after you seed. Sod should not be traveled on immediately after being laid, but it will be ready for traffic much sooner

FIGURE 9.5 Deep tine aerators fracture the soil leaving channels for drainage and reduced compaction.

than new seeding. When you are seeding turf, it often seems that no barrier is large enough to keep people off the grass you are trying to get established. When you catch people crossing your project, they are always quick to deny seeing the 4-foot snow fence, orange pylons, and large shrubs you planted to keep the area safe from traffic. They always seem to look puzzled by the fact that they have lost their shoes in what seems to them has become a bad idea. It will be a struggle to convince people that newly established areas are off-limits, but make every effort to do so and do not assume that straw will indicate to them that it might be newly seeded. I can recall pulling many a golf cart out of such areas.

There are many things that can damage turf. I have found that dog urine is perhaps the most common problem with smaller areas of turfgrass injury. I have even seen the occasional human urine injury, but nothing surprises me. Often turf abuse occurs from special events where traffic is much higher than normal. I'm not just talking about commercial and public properties: a special event could be a child's birthday party in the backyard or a concert in the local park. The point is that it is unusual repeating traffic that packs the soil down, causing a restriction to root growth and development.

Cars, golf carts, and other wheeled vehicles can be just as damaging to turf areas. This is includes bicycles, ATVs, and many other

Turf in the Landscape 203

FIGURE 9.6 (a) Golf green before aeration.

FIGURE 9.6 (b) Cores are harvested with an aerator.

FIGURE 9.6 (c) Complete aeration of green is done.

FIGURE 9.7 (a) Cores are dragged out with a drag mat.

FIGURE 9.7 (b) Heavy topdressing of sand is applied which will need to be broomed in.

FIGURE 9.7 (c) Green is finished and will be irrigated to work the sand in.

forms of transportation that you might not be thinking about. What usually happens is that, much like foot traffic, the response of the human operator is to follow the exact same path that the person in front followed. I always joke that when it comes to traffic on turf we are like a bunch of cattle following aimlessly along the same path because the guy in front of us already went that way.

Construction damage is another notorious culprit in turf injury. When there is work to be done, turf injury is not always on the minds of the construction staff. Often trees and existing grass just get in the way. Turf is quickly removed with large-scale pieces of equipment that can get the job done in a hurry. One of the major problems from large equipment is the weight and the long-term compaction that some of this equipment can cause to the soil. Construction equipment is just one of many things that can damage grass. I have seen instances where the soil was compacted so badly that the old soil had to be removed and new less compacted soil put in its place.

When compaction occurs, there are many issues regarding reestablishment of soil in that area. I am not inferring that all compacted soils should be deemed worthless and replaced if construction has occurred, but I want you to be aware that soil compaction can be so extreme that soil replacement will be necessary. If aeration and all other practices do not improve the soil condition, then replacement becomes the only alternative possible.

TURFTIP

If aeration and all other practices do not improve the soil condition, then replacement becomes the only alternative possible.

The problem with compaction is that what often may take an afternoon to tear up could take years to fix or the turf could be ruined permanently. Compaction is always worse when it occurs in wet or saturated soils. What happens is that, as the soil is "smashed" down, the air pockets are more or less forced out of the soil completely. This leaves a hard-packed soil that is not good for much of anything without replenishment. Remember, soil is just a growing medium for plants. Soils can typically be improved over a period of time with practices like aeration, which will help to break up compacted soil profiles.

Turf in the landscape is prone to a wide range of changes over a lifetime, and the environment in which it is growing is more than likely going to change at some point. You should be aware that you will need to adapt the turf to all the conditions that present themselves to that particular grassy area. I mentioned the construction of new buildings as a possible scenario that would cause change to the existing turf and soil conditions, but there are many simple effects that can happen that will create the need for a change or increase in cultural practices.

Let's say that grass is seeded in a newly constructed location. We can look at the a possible series of events that would affect a grassy location in a 10-year time frame for the average home.

Changes to a turfgrass environment over 10 years at an average household:

- *Year one: new seeding planned; soil deemed low in nutrition and pH could take 2 to 3 years to maximize turf's potential; homeowner is dealing with bare spots and weeds in the lawn; neighbors do not treat for weeds,*

- *Year two: grass is looking much better; weeds are 90% in check; boy who mows the yard is cutting too short and crabgrass puts a damper on appearance midsummer when drought occurs,*

- *Year three: lawn looks better; owner just bought a pet and it looks like "Spot" is causing dead areas and damage; overseeding will be necessary and maybe a more durable grass.*

- *Year four: little Taylor just turned 5 and playground is added to the backyard; lots of bare areas from the kids; the dog has worn a permanent path around the fence border.*

- *Year five: another baby on the way; left side of the house will need an addition; owner brings in the skid steers and back hoes to dig the foundation; lawn looks worse than it did on day one.*

- *Year six: construction is done and the late fall seeding job looks good; new fescue grass that replaced what was left of the bluegrass lawn appears to be doing much better; the grass is growing better than ever with the exception of that area under the tree by the property line—that tree looked much smaller when the place was built.*

- *Year seven: aeration and lime applications have improved that area under that big tree; gas leak on the mower has killed a large spot in the yard that will not grow any grass.*

- *Year eight: late storm took out a huge tree on neighbor's property that had been providing some shade; tore up the lawn where it fell and the grass was scorched from the hot sun afterwards.*

- *Year nine: city decided to run new water lines in the subdivision; agreed to seed areas when done but looks like they put back some really clumpy grass.*

- *Year ten: homeowner got a promotion; decided to sod the area in the front that the city tore up last year; backyard looking better since the dog kennel got built and the kids do not use the playground set any more.*

This is just an example of things that can happen in any one household that affect the grass condition in the yard. Changes in sun exposure, traffic, water, and all other environmental conditions affect the condition of the lawn (Figure 9.8). Getting a lawn to grow and stay healthy is a never-ending process. A homeowner might see this as a hassle, but I like to think of it as job security for all the rest of us.

The concept of turf management has been around for many years but the recognition the industry is receiving now changes the profile of a turfgrass manager as well as the responsibilities we have for using turfgrass in our landscapes in the future. I hope that there will be a day when people stop and take notice that turf plays a role in our everyday lives and that, frankly, no one pays attention to it unless it becomes unsightly or is poorly maintained. I hope that if nothing else I have provided you a better understanding of turfgrass. I also hope that you know more about how to establish, grow, and maintain it on a year-to-year basis. I hope that you can learn to appreciate an industry that has provided me with so much enjoyment and that you have taken away something from your reading.

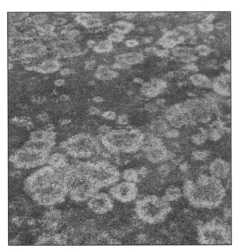

FIGURE 9.8 Grey snow mold.

Chapter 10

Irrigation

All plants need water, but not all plants need the same amount. This idea is not revolutionary, but it defines the challenge that faces those in the landscape management profession. This chapter is meant to shed some light on an issue that is still quite misunderstood by many in the trade: how much water does a landscape need? Further, it seeks to define the proper use and maintenance of the many components that comprise a large, commercial-grade irrigation system. Besides exploring the methods to employ in managing a system, special emphasis is placed on monitoring the landscape to derive useful feedback that will guide the irrigation-management process.

BASIC SYSTEM HARDWARE

Systems have become rather complex, integrating the many components into an efficient and serviceable end product has never been tougher

Systems can take many shapes and forms; many are derivatives of agricultural systems. Some systems still utilize essentially ag com-

FIGURE 10.1 Solid set irrigation system used to establish a crop at a commercial sod farm.

ponents, such as sod production or utility turf areas. Solid set irrigation (individual aluminum pipes that couple together)(Figure 10.1), wheel roll, linear movement, and center pivot systems are employed where terrain is suitable. These typically utilize impact-type emission devices, although gear-driven rotors may also be suited to this application as well. These above-ground or temporary approaches work well if only seasonal irrigation is needed or underground installation is not practical (allowable) for physical or economic reasons.

The next level of sophistication would be a quick coupling system, using either impact heads mounted on the valve key or some form of mobile sprinkler that uses the power of water flow to drive the mechanism across the area to be irrigated. Some golf courses still use the quick coupler approach, employing an irrigator that works moving key-mounted sprinklers throughout the night. This is a low-cost approach at installation; however, the ongoing labor cost mounts through the years.

Each of these approaches offer some unique benefits in a particular situation or niche, but they share the drawback that they are labor-intensive and will typically be less efficient (in terms of uniformity

> **TURFTIP**
>
> A poorly installed and maintained "in ground" system can easily waste vast quantities of water.

of water application) than a permanent system. One note of caution: a poorly installed and maintained "in ground" system can easily waste vast quantities of water. Automated systems do not guarantee improved irrigation performance, although that is clearly a goal of a permanent landscape irrigation system.

Controllers

Beyond these various "manual" irrigation systems are automated systems that feature a control clock (controller). There are four basic categories of controllers associated with landscape systems. While not commonly installed any longer, the hydraulic clock is still in use on older sites.

This system typically uses an external (domestic) water source for the controller and a separate source for the sprinklers. The controller itself is electro-mechanical, using electric motors to power the timer and rotary valve, and copper or polyethylene tubing to control valve function. By releasing or building pressure in the upper chamber of the valve, the hydraulic clock can operate the remote control valve (RCV).

There are a couple of operational hazards associated with this hardware. First, the clock "leaks" while a valve is open. This can be a safety concern, and it can lead to a bog forming adjacent to the unit. Second, if the pressure in the control source line drops due to an interruption in service, then it is impossible to close valves once they've opened. Obviously, there is eventually a wear factor that affects the rotary valve; it cannot provide a truly positive seal indefinitely. When this happens, valves close slower (or not at all) and the controller must be overhauled.

The next advance was the electrically actuated hydraulic valve. This allowed the controller to be fully electrical: no mud puddle at the clock and no seals to replace. Like the hydraulic clock, it is also electro-mechanical in nature. With age, these controllers tend to lose accuracy in terms of keeping time of day as well as valve run time. If a controller of this vintage were still in service, it would be wise to verify the cycle duration periodically because it will tend to truncate runtime as the mechanism wears out.

Solid-state controllers are the norm in commercial systems. They have many features and offer unparalleled scheduling flexibility. Gone are all moving parts: no motors to fail or gears to wear out. Unfortunately, with all the miniaturization and sophistication came a great deal more complication. These controllers can do incredible things, but all the bells and whistles were intimidating to many who tried to use them. Where the electro-mechanical units were bulky and unpredictable, they were undeniably simple, a virtue that leads many to continue to use them to this day.

Manufacturers eventually recognized this weakness in solid-state class controllers and came up with the "hybrid" clock. Hybrids use solid-state control circuits and processors, and some combination of rotary and slide switches. The number of buttons tended to be reduced as well. The result is a more intuitive user interface, while maintaining the dependability of solid-state control. Most modern commercial controllers fall into this category. Some strictly solid-state units are still on the market; they are often meant to be integrated into a computer-controlled system, termed "central irrigation control."

Regardless of the type—hydraulic, electro-mechanical, solid-state or hybrid—each allows the operator to implement an irrigation schedule by initiating irrigation and duration at predetermined times.

Valves

As we already learned, there are two broad categories of automated valves, the hydraulic and electric types. These terms are somewhat misleading, though, as it implies that they are wholly different in construction. In fact, all common remote control valves (RCVs) are hydraulic valves. This designation is born of the fact that water pressure is used in both cases to control valve operation.

All valves used in landscape applications utilize a flexible rubber compound disc or wedge (diaphragm) to create two separate areas within the valve cavity (Figure 10.2). The area above the diaphragm is used to control valve operation. As we already saw, a "hydraulic" valve uses water supplied from the controller to manage water pressure in the upper chamber of the RCV. The now common electric valve operates on the exact same hydraulic principle. Rather than using the clock to directly reduce the upper chamber pressure, causing the valve to open, a solenoid creates enough magnetic force to

TURFTIP

All valves used in landscape applications utilize a flexible rubber compound disc or wedge (diaphragm) to create two separate areas within the valve cavity. The area above the diaphragm is used to control valve operation.

lift a plug off the exhaust port of the valve. Since the exhaust port is larger than the inlet port, upper chamber pressure drops enough to allow the valve to open. In both cases managing upper chamber water pressure is the basis for valve function.

In reality, modern RCVs are correctly termed electrically actuated, hydraulic valves, recognizing that electric force does not drive valve function; it only facilitates control of upper chamber water pressure. In terms of electrically actuated hydraulic valves there are two varieties manufactured. They can either be normally open or normally closed. The descriptor "normally" expresses how the valve will react in a no-voltage condition. Most RCVs are normally closed, since if there is no voltage supplied by the controller, the valve will be shut (closed). By the same token, a normally open valve will be on (open) when no voltage is supplied to the solenoid. The main application of the normally open valve is to serve as a master control valve. Not all controllers are equipped to operate a normally open master valve, so careful planning is required when using this type of valve.

A different use of a standard RCV is to plumb it into the mainline upstream of all the other RCVs. In this configuration it becomes a master control valve. Virtually all modern-era controllers have a master valve/pump relay circuit designed to operate a normally closed valve. It is a very useful addition to any commercial system, as it can prevent emergency calls to shut down a stuck RCV. It also prevents

TURFTIP

Not all controllers are equipped to operate a normally open master valve, so careful planning is required when using this type of valve.

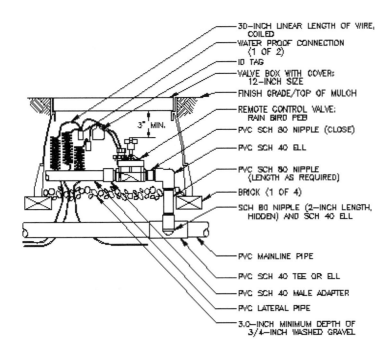

FIGURE 10.2 All valves used in landscape applications utilize a flexible rubber compound disc or wedge (diaphragm) to create two separate areas within the valve cavity. The area above the diaphragm is used to control valve operation.

long-term water waste if a plumbing failure were to occur in the mainline network. Although it adds some cost to the project, it can pay for itself many times over the life of a system.

Beyond RCVs there are several other valves that are key elements of a commercial-grade irrigation system. Isolation valves should be included on any mainline that crosses a paved street or extends more than 500 feet from the point of connection (POC). This allows subsections of the main to remain in service while repairs are completed as a result of damage to the mainline or fitting failure(s). Isolation valves were traditionally brass or bronze gate valves; these tend to gather sediment in the channel where the gate seats and thus are less likely to positively seal after several years. The ball valve is a viable alterna-

> **TURFTIP**
>
> Virtually all modern-era controllers have a master valve/pump relay circuit designed to operate a normally closed valve.

tive in many applications. A ball valve with integral unions allows the unit to be serviced or replaced with minimal "down time" for the POC.

Quick coupling valves (Figure 10.3) are accessories of convenience; they provide a ready source of water for maintenance personnel who need to fill a spray rig or water in some new plants. A specially machined key is inserted into the bore of the valve to engage the source. They are plumbed into the mainline, usually on a swing joint. They should be anchored on a stake or otherwise stabilized so they can't be broken off if someone yanks hard on the hose that is connected to it.

Vacuum relief valves are critical for larger mainlines such as those found on golf courses. Although not all such systems must be so equipped, if the mainline traverses significant elevation changes, then vacuum relief valves are worth their weight in gold. If there is a mainline blowout at the toe of a long, straight run of a large diameter mainline, gravity is going to force as much water as possible out of that break. The result—at the top of the hill (the proper location of a vacuum relief valve) a significant vacuum will be created, a vacuum potentially capable of collapsing the pipe. The relief valve admits air to the pipe, preventing the vacuum from forming to a destructive degree. In drip applications on uneven terrain, relief valves are likewise advisable.

While getting air into the system is critical in certain situations, getting the water out of the mainline and laterals is a necessary part

Irrigation **217**

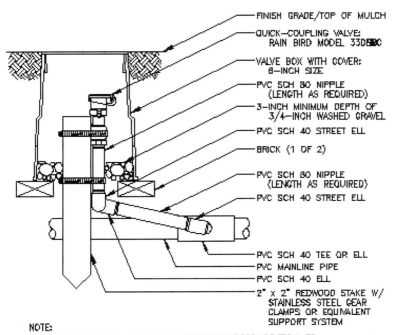

FIGURE 10.3 Quick coupling valve. *(Detail Courtesy of RainBird)*

TURFTIP

Vacuum relief valves are critical for larger mainlines such as those found on golf courses. Although not all such systems must be so equipped, if the mainline traverses significant elevation changes, then vacuum relief valves are worth their weight in gold.

of maintaining a system in cold climates. If water is left in the pipe, it will freeze, possibly leading to pipe damage. Drain valves allow the pipe run to drain off water by virtue of gravity. As such, drain valves should be located at low spots in the pipe run. If a system is to be fully self-draining, then air relief valves must be included in higher areas of the terrain. Automatic drain valves are available, but will drain each time the mainline is depressurized. If a master control valve were a part of the system, then this would lead to a significant waste of water, as the mains would have to be recharged at each irrigation event. In such cases the drain valves would be built using manual ball valves (see Figure 10.4). These valves would be plumbed into mainline tees at low spots in the pipe network. A sump or drain line giving the drained water an avenue of escape must be provided for these to perform in a useful manner.

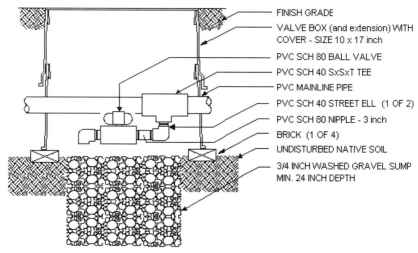

NOTE:
1. OFFSET BALL VALVE SO IT WILL CLEAR THE MAINLINE PIPE WHEN CLOSED.
2. AUGER 12 INCH DIAMETER HOLE TO 24 INCH DEPTH.

FIGURE 10.4 Manual drain valve.

> ## TURFTIP
>
> Drain valves allow the pipe run to drain off water by virtue of gravity. As such, drain valves should be located at low spots in the pipe run.

In some cases, cold climate systems are equipped with a fitting near the POC that allows the mainline to be charged with compressed air. A glad-hand air fitting, such as those found on semi-trailers, offers a convenient method of connecting an industrial air compressor to a mainline. Then either quick coupling valves (with keys inserted) or drain valves allow the water to be forced out of the plumbing network.

Anti-Siphon Protection

It is important to protect water sources from contamination if the supply is associated with a municipal water supply or a domestic well.

> ## TURFTIP
>
> Reverse siphoning or "backflow" occurs when a sudden drop in pressure is sustained (such as the failure of a water main) within the supply system.

In these cases, a cross connection control device is desirable, and likely required by local plumbing code. If a fertilizer/ chemical injection system is employed, a high level of protection is essential. Reverse siphoning or "backflow" occurs when a sudden drop in pressure is sustained (such as the failure of a water main) within the supply system.

There are three types of in-line devices that are commonly used in landscape applications. The simplest is the atmospheric vacuum breaker (AVB), which features a "floating" plunger that moves up to seal the water passage under normal flow conditions. If there is a drop in pressure on the supply side of the device, the plunger drops in response to the vacuum. These are common on residential systems and typically are integrated into the RCV assembly. The device must be set above grade and at a level above the highest sprinkler on the lateral being protected. On slopes an independent AVB is set downstream of the RCV at the top of the slope. Local code will often dictate how far above the sprinklers this type of device must be installed.

The next variety of backflow prevention is the pressure vacuum breaker (PVB); it operates on the same basic principle as the AVB. Only one device is required for a site, as it can be upstream of the RCVs. This device is well suited to small, flat sites, as it must be installed at an elevation above any sprinkler on any lateral. Again, local code will often dictate this minimum elevation relative to sprinklers.

TURFTIP

There are three basic types of sprinklers used in commercial systems: fixed spray, rotor spray (including impact drive), and flood bubblers.

A more common alternative for a commercial system is the reduced pressure backflow device (RP). This device is mounted above grade, but can be at an elevation below the sprinklers and still function. One device is required per site and, in many jurisdictions, must be tested periodically by a certified technician to assure it will function properly. It works by reacting to a drop in upstream pressure and then venting all downstream water to the atmosphere in such cases.

Emission Devices

There are three basic types of sprinklers used in commercial systems: fixed spray, rotor spray (including impact drive), and flood bubblers. Fixed spray heads are typically termed "pop-ups," as the nozzle pops up from 4 to 12 inches depending upon the application. They are commonly used in relatively narrow (16 feet or less) or irregularly shaped areas because of the design flexibility they offer. Rotors feature a rotating nozzle, so they are useful in larger areas. Many varieties are produced to work in areas as narrow as 16 feet. Larger units have a radius of throw that exceeds 80 feet. These often are used in golf applications and can be equipped with an integral valve (valve-in-head). Flood bubblers are usually utilized to provide supplemental tree irrigation or for hedgerows.

With a new emphasis on conservation sweeping the industry, a discussion of components is no longer complete without a look at the variety of "drip/micro" irrigation devices that comprise what has become known as xeric irrigation techniques, or simply drip irrigation. As with much of the equipment used in the landscape trade, drip parts are born of products developed for the agricultural realm. There are three basic forms of drip emitters: non-pressure compensating, pressure compensating, and micro-spray. Each has specific advantages depending upon the application.

Non-pressure compensating emitters use a small orifice or turbulent flow path to restrict flow and generate a very slow discharge

> **TURFTIP**
>
> With a new emphasis on conservation sweeping the industry, a discussion of components is no longer complete without a look at the variety of "drip/micro" irrigation devices that comprise what has become known as xeric irrigation techniques, or simply drip irrigation.

rate. As the name implies, on long runs of hose the discharge rate will vary as the pressure fluctuates, due to friction within the hose and elevation changes. They are the cheapest form of emitter and can be prone to clogging. Hose runs of greater length can be accomplished with the use of pressure compensating emitters. As the name implies, the discharge rate is buffered against minor changes in pressure in the hose. The result is more uniform performance at distal ends of a hose run. Certain forms of this emitter offer some level of resistance to clogging. One form of pressure compensating emitter is the extruded or "in-line" system. A polyethylene hose is manufactured with emitters inside the hose at consistent intervals (such as 12, 18, or 24 inches). This product is used to install "wall to wall" irrigation systems or for hedgerow applications.

Micro-sprays, not surprisingly, are overhead sprays on a smaller scale. The flow rate tends to be higher than that of typical emitters. They work well where vandalism is not a problem, and small or irregularly shaped areas must be overhead-irrigated for cultural reasons.

Commercial irrigation systems have become rather complex, and integrating the many components into an efficient and serviceable end product has never been tougher. A solid understanding of system hydraulics and actual site terrain is essential to engineering a sound irrigation system.

MAINTAINING A COMMERCIAL SYSTEM

Commercial landscape irrigation is the next item on the purveyor's 'to-do' list, as it is a highly visible source of water use (and perceived cause of large volume waste).

A commercial irrigation system is composed of hundreds of parts: sprinklers, nozzles, fittings, pipes, and more. Each can fail due to wear, vandalism, or poor system design. Although landscape maintenance is typically focused on keeping the plant elements in an aesthetically pleasing condition, this is only part of the job. Since the irrigation system must function within the landscape environment, maintenance practices must be geared to account for "the big picture."

A well-maintained landscape looks good month after month, and since it is always changing (growing, hopefully), the performance of the irrigation system must be sustained despite the growth. This fact leads to the need for maintenance personnel to work as a team. Most large maintenance contractors have independent irrigation service technicians, and it makes good economic sense to run crews this way.

A Well-Manicured System

One potential problem is that those who manicure the site never see the irrigation system in operation. Even if the site looks good, there can be significant conflict between the plants and the irrigation system. Spray pattern blockage is common on sites that are maintained this way. Consider a scenario with a hedge adjacent to a large turf area. Low-hanging branches interfere with the discharge of the turf rotors, causing stress to appear. The irrigation technician may notice the obstruction, but his job is fixing heads, not trimming xylosma hedges. And so as the plants grow, the dry spots multiply and enlarge. The different crews must communicate well and work together to keep the irrigation system well-manicured.

Or perhaps after years of thatch build-up and being run over by mowers, the 4-inch pop-up heads don't clear the turf any longer (except perhaps the day after it is mowed). The constraints of time (time is money) usually preclude spending many hours raising and leveling sprinklers that are not properly oriented for maximum performance. The key is to recognize the problem early, fix a few each week, and prevent the situation from deteriorating at the same time. System performance is tied directly to the level of care taken in eliminating conflicts between overhead irrigation devices and the plants they sustain.

Drive-by Maintenance

The marvels of technology certainly have made an impact on how irrigation systems are maintained, particularly in the commercial

---

TURFTIP

Most modern controllers are remote-ready or can be equipped with a radio-based remote transmitter with relative ease. Although they cost $1000 to $1800 for a set, in the competitive business of commercial maintenance they have become almost as essential as a screwdriver.

realm. Most modern controllers are remote-ready or can be equipped with a radio-based remote transmitter with relative ease. Although they cost $1000 to $1800 for a set, in the competitive business of commercial maintenance they have become almost as essential as a screwdriver. The task of visually checking system integrity on a consistent basis is the foundation upon which a successful maintenance program is built. If the system is not being monitored properly, consistent excellence is not even an option.

Unfortunately, remote technology has ushered in the era of "drive-by" maintenance. The urge to simply plug in and take a 40-mph tour as valves pop on for a few seconds is tough to overcome. If this is the normal system test at a property, the subtle problems will go unnoticed for months on end. The result is a slow demise in plant material. Clogged nozzles, leaky wiper seals, and minor lateral leaks all work together to degrade system performance. A walking tour where systems can run for at least 90 seconds each is essential, at least monthly, during the irrigation season. You can tell when something isn't right just by the sound in many cases when careful observation of system operation is practiced.

A common contributor to system degradation is the practice of haphazard nozzle replacement. When a missing or damaged nozzle is replaced, does the technician use the first nozzle insert that can be pulled out of a pocket? Where applicable is the filter screen replaced also? It is not difficult to check the adjacent head to figure out which nozzle insert is really needed. This simple step will make a significant impact upon system uniformity for months to come.

If your main goal is to fix geysers and little else, the efficiency of the systems under your care is dropping slowly but surely. Failing to address these "small" problems eventually leads to one large, costly (perhaps your job) problem.

One inevitable result of drive-by maintenance is an increase in water consumption (see Table 10.1). A poorly maintained system uses extra water for at least two reasons. The more water that comes out places it isn't supposed to, the more demand the system places on the supply. Consider a well-maintained turf rotor system that demands 40 gallons per minute (gpm) and runs 25 minutes four nights per week in July. Assuming the working pressure is at the top end of the recommended range for the sprinklers, this system will continue to operate more or less normally even if leaks push demand up to 50 gpm.

Next, assume that this inadequate maintenance program allows some nozzles to become partially clogged. Some stress areas are sure to appear. What will the drive-by maintenance technician do? Probably pump another 5 minutes onto the run time to clear up that little problem. Initially this band-aid fix may have the desired effect, but at what cost?

So a quick peek at the scoreboard (or the water meter) shows that all is not well. Whereas the system should demand 16,000 gallons for the month, drive-by maintenance yields an actual demand of 24,000 gallons. That means one of every three gallons applied is wasted.

Get on the Conservation Bandwagon

It is no secret that in the new millennium many areas of this country are facing unprecedented water restrictions due to drought, political wrangling, or the demand of a swelling population. The result? The spotlight is on the green industry like never before and it is not just a cause celèbre or passing fad; this time water conservation in

TABLE 10.1 The effect of inadequate maintenance practices on water consumption (July).

Level of Maintenance	System demand (gpm)	Run Time (per start)	Number of Starts / Month	Total Gallons Applied / Month
High	40	25	16	16,000
Drive-by	50	30	16	24,000

the landscape is going to happen. There has been a drought here and there in the arid West that caused local agencies to tout conservation and make threats of dire consequences if things didn't improve. But a wet winter or a few good snowfalls, and the status quo quickly resumed.

Such is not the case any longer. Many parts of the country, not just in the sunbelt, have already implemented landscape-focused legislation that ultimately will require landscapes to be built better and maintained better. An increasing number of water purveyors are in the process of implementing a performance-based rate structure where landscape water use is measured against reference evapotranspiration (ETo). ETo is a prediction of water loss (reported in inches), typically from a crop (fescue turf) being monitored by a weather station. Plugging environmental data generated by the station into a mathematical equation derives ETo. The more water applied above threshold levels (tiers), the more each subsequent drop costs (Figure 10.5). The following is an excerpt from a water purveyor's web site:

- *Customers can accrue penalties on their water bill if their water use exceeds their water allocation. Landscape allocations are based on the actual water needs of turfgrass according to current weather conditions.*

- *Customers who use more water than their allocation enter into the three penalty tiers, which charge progressively more as*

Tier	Rate (per ccf*)	Use (percent of allocation)
Low Volume Discount	$0.48	0-40%
Conservation Base Rate	$0.64	41-100%
Inefficient	$1.28	101-110%
Excessive	$2.56	111-120%
Wasteful	$5.12	120%+

*One ccf =748 gallons[1]

FIGURE 10.5 Tiered rate structure

water waste increases. Penalties are paid only on the amount of water that exceeds the allocation, not on total usage.

- So let's take a look at the cost of using drive-by maintenance practices for the valve evaluated in Table 10.1, using the tiered rate structure.

Table 10.2 illustrates that in one month for only one RCV, there is potentially an increased cost to the owner of almost $20. Multiply that by the number of valves at a site and months in the irrigation season, and you can see that this translates into a significant disadvantage of continuing to use drive-by maintenance techniques at the site.

Rate structures patterned after this tier system are on the horizon in many areas of the country. Why? Because soggy turf and water-stained sidewalks from excessive runoff are all too common in commercial landscapes, a fact that has not gone unnoticed by water purveyors. The difference is that now, at the dawn of a new millennium, many regional water supplies clearly will not suffice to meet long-term demand. Throughout the 1990s, purveyors worked hard at indoor water-use reduction. Now common are low-volume showerheads, 1.6-gallon per flush toilets and the like in homes and commercial buildings. But now, they've opened the door and are looking outside for more ways to conserve.

TABLE 10.2 The potential cost of drive-by maintenance under a tiered rate structure (assume 3,800 irrigated square feet and that turf requires 100% of ETo).

Level of Maint.	July inches applied (I)	July ETo	% ETo appl-ied	$ of water @ tier 1	$ of water @ tier 2	$ of water @ tier 3	$ of water @ tier 4	$ of water @ tier 5	Total $ of water
High	6.75	7.1	95	$4.92	$7.12	0	0	0	$12.04
Drive-by	10.13	7.1	143	$4.92	$9.84	$3.28	$6.57	$6.70	$31.31

That means commercial landscape irrigation is the next item on the purveyor's "to-do" list, as it is a highly visible source of water use (and perceived cause of large volume waste). Those who wish to continue managing commercial landscape infrastructure need to get on the conservation bandwagon, or you may wind up under it.

IRRIGATION MANAGEMENT

The majority of the water savings will come through not applying irrigation when it is not needed.

While it is clearly essential to have a solid system maintenance program that is adhered to carefully, that is only half the battle. A system that performs with good application uniformity does not guarantee anything in terms of system efficiency.

A Matter of Semantics

Two terms that frequently emerge in any discussion of irrigation performance are uniformity and efficiency. The two have become almost synonymous in landscape jargon; however, they have different meanings, and must be viewed differently in terms of how system performance is evaluated. While system uniformity clearly affects system efficiency, they have no clearly defined correlation because of the many factors that affect ultimate efficiency.

System uniformity describes how consistent the amount of water applied to each square foot of landscape is over the area served by a given RCV. As a frame of reference, assume that natural rainfall approaches 100% uniformity. That is, all parts of the ground surface receive the same amount of precipitation when the storm is over. Obviously, even a well-designed and maintained system cannot duplicate that level of uniformity. It can, however, be evaluated to see how it stacks up against rainfall.

One common method to achieve this evaluation is a system audit. The focal point of an audit is a catch can test. The common form this takes is by using catchment devices (Figure 10.6) that are manufactured with an integral measuring scale (either depth or volume) on each device. These catchments typically are placed in a grid or other consistent pattern over a representative area of the landscape. A less exotic option is to use even-sided catchments (tuna cans are popular) to capture the irrigation water and then measure the depth of water in each one.

In either instance the recorded values are analyzed to determine a ratio between the average amount caught in all the cans and the average amount caught in those cans with the lesser amounts of water in them. The number of catchments to include in this "lesser" category is a question that is up for some debate in the industry. One approach uses the lower 25% can average as the denominator (bottom number) of the ratio. This has empirically proven to generate

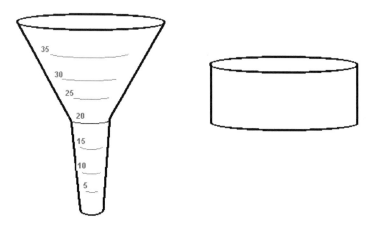

FIGURE 10.6 Comparison of two commonly used catchments: on the left, a specially designed catchment device that can be read directly either in volume (milliliters) or depth (inches); the tuna can, on the right, can be used by measuring the depth of water in each can with a ruler.

rather low uniformity ratings for many landscape situations. The result is that the run time required to overcome this calculated deficiency in uniformity tends to inflate the irrigation schedule to a level that can lead to unnecessary water application.

Another approach is to use the average of the lower half of the cans as the denominator of the ratio, which will tend to yield higher uniformity ratings on many irrigation systems. This approach will usually yield lower calculated valve run time, as there is a smaller uniformity deficiency that the irrigation schedule must overcome. Ultimately, the less uniform water is applied, the longer a valve must run to keep all areas in an aesthetically acceptable condition.

Now that we have a handle on how uniformly our sample system applies water, we must also consider how efficiently it performs. Overall system efficiency boils down to this: of all the water that flows through the valve, what percentage of that water is effectively utilized by the landscape? Except in a laboratory, that is a difficult value to nail down. For the real world, estimating system efficiency comes down to using two pieces of data, the estimated plant need and the applied irrigation.

In the previous discussion of drive-by maintenance, the concept of plant water need (ETo) was introduced as a benchmark of estimated system demand. The same benchmark value can be used to

---

TURFTIP

Overall system efficiency boils down to this: of all the water that flows through the valve, what percentage of that water is effectively utilized by the landscape?

help assess system efficiency. A certain percentage of the applied water will not be applied precisely within the root zone. The timing of irrigation events has a certain level of error compared to exactly when it is needed. These are just two factors that lead to less than perfect efficiency. But there is a point where these unavoidable obstacles no longer account for inefficiency. This unavoidable loss is generally assumed to be 10% (i.e., 90% efficiency, in terms of how scheduling is performed, is considered very good with overhead irrigation). This evaluates how well the manager performs; it does not account for inefficiency associated with irrigation hardware performance. When viewed together, good system uniformity and good system management equate to the right conditions for excellent system efficiency.

Managing a System Wisely

The key to efficiency is nailing down exactly what "the right conditions" are. Each landscape is unique; let's review why this is so. While this concept seems elementary, there are definable factors that when properly considered lead to clues of how to set up these elusive "right conditions."

TURFTIP

A key element of system design is the concept of hydrozoning, which simply means that all landscape elements that are served by a single RCV must have closely matched irrigation requirements.

A key element of system design is the concept of hydrozoning, which simply means that all landscape elements that are served by a single RCV must have closely matched irrigation requirements. It seems pretty simple, but many sites are developed with little thought given to how well system performance matches plant needs. The price of ignoring this key design step is steep and leads to a slow decline in landscape quality in many cases.

Even an identical plant palette on two sites does not guarantee that management practices will work at both sites. Although climate is usually considered a regional characteristic, for two plants the climate can seem much different at the same site. Consider photinia hedges on opposite sides of an eight-foot-tall block wall oriented east to west. The plants on the north side of the wall will need less water, especially in the cooler months. Why? Because the wall will cast a shadow on one hedge, while reflecting "extra" heat on the other one. This is an example of how blanket assumptions about plant water need can lead to inefficiency.

If all the hedges are on a common irrigation schedule, one side will be drowned and the other will likely suffer some degree of stress. This is an example of evaluating what is often termed a landscape microclimate. Reflected heat from a glass and steel office building will dramatically impact how the turf there is irrigated. But the surrounding turf will need significantly less irrigation to sustain healthy growth. Failing to recognize these site-inflicted influences leads to irrigation inefficiency. Other subtle factors that can impact the true needs of a landscape in the urban environment include structural climate control equipment (HVAC discharge, boilers, etc.), which can have a dramatic effect on the climate from the plant's perspective.

Got Dirt?

Another critical consideration is what the plant must contend with below-grade. How good are soil conditions? In urban sites, the soil is often difficult to manage well for two reasons. First, mass grading operations are geared to sustain stable structures, which means soil compaction to roughly 95% density. This is clearly not in the best interest of the plant. Second, by virtue of the process, the soil profile is typically layered in thin "lenses." Unless each lens is of the same parent material, the resulting soil profile (a vertical section of the upper 2 feet of soil) is likely to be difficult to assign to a textural class and tougher still to judge how well water will drain through it.

Depending upon the terrain, soils in a relatively small geographic area may vary greatly. In a valley setting, where the author works currently, there are three very different soil series that come into play as the area develops. In the foothill region the soil is dominated by decomposed granite. On the valley floor, the soil is a sandy loam-based texture. To the east end of the valley the soils are very uniform, deep alluvial silts. Projects developed in each of these regions must be managed differently.

Each soil type has differing characteristics in terms of physical and chemical properties. As a general rule clay soils will contain more organic matter per volume than a sandy soil will. Clay soil also tends to be "richer," which is a subjective measure of chemical properties such as cation exchange capacity (CEC). CEC measures the ability of the soil particle itself to bind elemental particles (typically in dissolved form) to the surface of the particle. High organic content also helps impart the ability of the soil to hold essential nutrients, such as phosphorous and iron, that the plant needs for normal growth. These conditions encourage a healthy soil profile to develop where plants can thrive.

There are two basic soil characteristics that represent the physical makeup of the soil. In terms of physical properties, some predictions can be made as to how applied irrigation water will move into and through the soil (see Table 10.3). First, how fast can applied water be absorbed into the soil profile? Certain generalizations can be made based on soil texture; sand absorbs water more quickly than loam- or clay-dominated soil. In highly disturbed urban soils, these predictive habits of "typical" soils can sometimes be thrown out the window. That is one of the reasons why careful observation of the

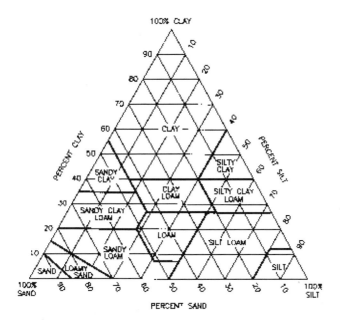

FIGURE 10.7 Soil triangle: this matrix of soil textures allows the three different types of soil particles (sand, silt, and clay) to be organized into classes based upon the percentage of each particle within a sample.

landscape it is critical for success. That said, it is useful to understand the differences and why they occur.

From Table 10.3 the following generalizations can be drawn. Soils dominated by sand will absorb and release water easily, thus they cannot sustain a moisture (or nutrient) reservoir for plants over an extended period of time. Clay soils tend to behave in an inverse man-

TABLE 10.3 Expected characteristics of landscape soils (without amendment) based on textural differences.

Texture	Run off Potential	Expected Rate of Drainage	Moisture Holding Capacity	Potential for Productivity ('richness')
Sand	Low	Rapid	Low	Low
Loam	Moderate	Moderate	Moderate	Moderate
Clay	High	Slow	High	High

> **TURFTIP**
>
> Soils dominated by sand will absorb and release water easily, thus they cannot sustain a moisture (or nutrient) reservoir for plants over an extended period of time.

ner, while loams provide a good balance across the spectrum of characteristics. Not surprisingly, loamy soils are most desirable from a management perspective.

The second result of the physical condition of the soil is how water moves, once in the soil profile. A natural force resulting from the physical properties of fluids is capillary action. This force, combined with gravity, affects how water will move in the soil. Gravity wants to pull water vertically; capillary force adds a horizontal component to this process, meaning water will tend to move sideways as well. In uniform, well-prepared landscape soils this process of water movement is rather predictable. In general, the larger the soil particles, the less lateral movement will be exhibited within the profile.

The bottom line is this: how water moves into and through the soil at every site can be different, especially in new landscape developments. How uniformly the soil profile was prepared and amended (if at all) will have a dramatic effect on how best to apply water to that particular site.

Scheduling Irrigation

Here's where the rubber meets the road in terms of system efficiency. If you make good decisions standing at the clock, the system will likely reach that magic 90% management efficiency plateau. Overall system efficiency is the end result when the maintenance practices and management practices come together. The water meter tells the system efficiency story each month. If your system doesn't have a meter, install one; it is the single most important water management tool available.

You Can't Manage What You Can't Measure

Without feedback, it is impossible to evaluate how well a system is functioning. Recall the fact that "big brother" (and likely your clients who pay the water bill each month) is taking a much more "hands-on" approach to landscape irrigation efficiency. If asked to demonstrate what steps are being taken to control water use, what could be better than a graph that compares actual water use vs. expected plant demand (ETo)? See Figure 10.8.

The beauty of such a feedback-driven approach to management is that when things go awry, you have early warning. There is time to react and correct the problem(s) before the client has a chance to get belligerent.To implement such a feedback system, the water meter must be read consistently, and the actual area being irrigated must be known with a good degree of accuracy. The following equation converts a volume of water (V) to its equivalent amount of irrigation water (I) in inches[3]:

$$I = V \times Cu/A$$

where:

I = Amount of irrigation water (inches)

V = Volume of irrigation water (gallons or ccf (100 cubic feet) of Cu below)

A = Landscape area (sq. ft.)

Cu = Unit conversion (for gallons, use 1.6043; for ccf, use 1200)

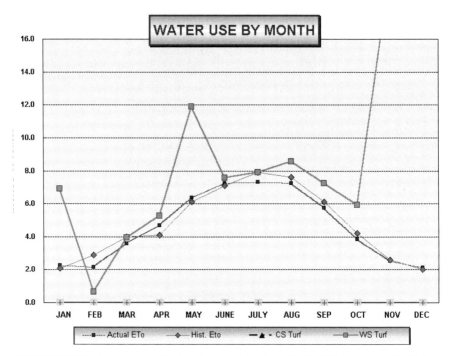

FIGURE 10.8 An example of a tool (a spreadsheet application) that graphically displays historical ETo, actual Eto, and applied irrigation from an actual parkway irrigation system; note that system problems are clearly highlighted, such as schedule tampering, broken equipment, etc.

At the beginning of each month, the actual amount of water applied can be determined by reading the meter. Not every meter reads in the same units of measure; a common unit is 100 cubic feet (ccf). Others read in terms of acre-feet or simple gallons. If the data gathered is in any other form, a conversion must be made to change it to gallons so the above equation will work correctly.

One note regarding calculated "I". This value assumes equal irrigation to all parts of the site, so it is not a means of identifying a particular hydrozone where water is applied in excess. It is a very useful tool to create a benchmark for system management and hardware performance to be judged. If "I" is consistently more than 115%

of ETo for the same period, there is some work to be done. Carefully evaluate the hardware to see if this could account for the waste.

When you are satisfied that maintenance practices are not the cause of the poor performance, the management practices must be reviewed. Are runoff and erosion common occurrences? Is there moss or algae growing on areas with bare ground? When you stick a soil probe in the landscape (if this is a novel concept, understand that a probe is an essential tool for a professional irrigation manager), do you hear a sucking sound upon removal? Do mowers usually leave ruts in the turf? If the answer to any of the above is "yes," management practices are in need of some improvement.

Each microclimate and hydrozone must be evaluated to see where scheduling changes can help enhance how well the system can meet plant needs. The first change to implement where soggy conditions persist is to extend the irrigation interval (cut days from the schedule). With regular observation and use of a soil probe, the fact that more days can pass between irrigation applications will often become obvious. When evaluating planter beds, a dry soil surface reveals nothing about when irrigation is needed. Only a quick check of the soil profile down where the roots are can offer a clue as to when irrigation is really needed.

If the controller at the site does not support multiple program scheduling, replace it. For commercial landscapes, a minimum of three independent programs should be available to place different hydrozones on different irrigation intervals. Remember that these intervals must change throughout the irrigation season. The goal of irrigation is to tailor the irrigation interval to the needs of the plant. Irrigation on a strict calendar basis often leads to excess runoff, deep percolation, and root zone saturation.

Once the correct irrigation interval trend is established, then run-time adjustments can be considered. The idea is to establish the amount of run time needed to hydrate the entire root zone, and then apply that same amount each time irrigation is truly needed. This is the revolutionary part of the process. Traditionally, technicians tweak the run time whenever scheduling adjustments seem necessary. This turns the process into a guessing game, and few will be able to guess correctly with great consistency. By reversing the process and managing the interval with great care, the run-time question need be answered only once. The run time becomes a relatively static value, used only to make minor conservation adjustments. The majority of the savings will come through not applying irrigation when it is not needed.

Time is on Your Side

So how is the correct run time established? All you need is a watch and a soil probe. Select at least one RCV in each hydrozone and conduct a field test to determine the total amount of valve run time needed to adequately irrigate the entire root zone. Begin a few days (less with sandy soils) after the last irrigation by probing the soil and squeezing the soil core. If you can extract free water from the sample, it is safe to assume the plants can also. Check again tomorrow (unless, of course, there is visible stress). When you get to the point that the sample tests "dry," simply turn on the valve and note the time. While the valve runs, watch for signs of ponding and runoff. If these occur, then shut off the valve, wait 15 minutes (or longer in clay soils), and then apply another pulse of irrigation of the same duration, all the while checking with a probe to see how deeply water has penetrated into the soil profile (Table 10.4). If soil compaction or incompatible soil layers prevent water penetration below a relatively shallow depth, consider some form of mechanical operation to improve soil infiltration characteristics.

Each landscape is different; the soil conditions and specific plant needs may require adjustments to these values. When considering non-turf hydrozones with coarser-textured soils supporting plants of a xeric nature, these values should be increased by 50% to take full advantage of site conditions.

TABLE 10.4 General guidelines for determining depth of irrigation penetration in uniform loam soils of good tilth.

Cool Season Turf	Warm Season Turf	Mixed Shrub & Ground Cover	Shrubs	Trees
6	10	10	12	16

Once the total run time required to drive water down to an adequate depth for each hydrozone is achieved, a site-tailored irrigation schedule can be built. Using the notes regarding the required run time, don't forget to use multiple start times where ponding or runoff was noted. A schedule that will properly hydrate the entire root zone is easy to structure. Programming the interval becomes the item to adjust as time passes. Except for turf on a rigid mowing schedule, a skip-day-type schedule is most beneficial for the landscape. In this way the calendar can be ignored, and irrigation applications are made based primarily on expected plant need. A seven-day calendar cycle is often too restrictive. Wednesday is not a good reason to irrigate; instead, irrigate when the soil moisture reserve is nearly exhausted.

A system maintenance program that features drive-by techniques cannot produce excellent results in the long term. Maintenance personnel must work as a team to assure the irrigation system is well manicured also. Finally, irrigation management must be viewed as a feedback-driven process; both water consumption and soil conditions must be evaluated regularly in order to practice site-tailored irrigation. This approach leads to a more expansive, healthier root system, capable of supporting robust, durable plants that are less susceptible to short-term drought. Landscapes managed in this fashion will be more attractive, conservative, and enjoyable to maintain.

Chapter 11

Golf Course Management

Managing a golf course has tremendous rewards but can be a very challenging career choice. A golf course manager must have both scientific and artistic skills. The person must be creative and also have good management and business abilities. I want to discuss some of those skills and look at the various aspects of golf course management.

I mention business management skills as one of the most important characteristics a golf course professional can have. The reason is that, in a sense, running a golf course is no different than running any other business. As a superintendent or even an assistant superintendent you are responsible for as many issues as any small business owner would face, and sometimes you may be under even more pressure due to the demands of a board of directors, membership committee, or other owners. It is business savvy that will set you apart from the people who fail in the golf course industry.

Start first with the biggest problem anyone in business faces, which is good employees. I know from being a past superintendent myself and from currently being an employer that finding quality employees is the biggest problem in the green industry. As a superintendent it is your job to bring in top-quality professionals who will enhance the work you are doing and lighten your headaches. You must also bring in crew leaders that will assist you in the hiring and management process. This is especially true at larger courses where you might not be doing all the hiring yourself. Hiring top-quality employees can make your job a much easier task. The task of hiring and firing never seems to get easier even with years of experience, and I have always been the kind of manager who wants to give everyone the benefit of the doubt. You do, however, need to be able to eliminate excess baggage at your workplace if it is causing conflicts.

The management of most courses goes through various channels, which often include multiple bosses. I am referring to the most common situation, which is a board of directors. This might be in addition to the membership, who may feel that they are your boss because they contribute to your salary or wages. Dealing with the membership and with the ownership of the club requires a great tolerance for people and also a level of public relation skills. A superintendent does get the opportunity to act as the supervisor and head up all golf course operations, but there is often accountability to a board of directors or group of owners. There are a few golf courses with only a single owner, but those are usually smaller 9-hole courses.

It is often the back-and-forth dialogue between a superintendent and the board of directors that establishes the work orders for the course. The line of communication between the board and the superintendent is one of the most important. In many instances I have seen the board of directors ruin a course because they did not allow the superintendent to do the duties that were necessary to improve the course quality. Too many times I have seen boards giving advice to their superintendents on turfgrass management skills when none of the people seated on the board knew the first thing about golf course maintenance in general. This is where proper channels need to be established for communication between the board and the person in charge of the course. Do not allow the board to monopolize the general scope of work that it takes to run the course. The maintenance decisions should be made by the superintendent or greens keeper in charge of the course. This is the person who knows how grass should be maintained. I always hate it when a member is giving a superintendent an education on how the grass should be mowed when he or she probably has not mowed a lawn in 10 years.

Guidelines need to be established for dealing with any and all issues involving the golf course operations. There should be channels of communication established to make sure that people involved with the golf course are not confusing the issues at hand (Figure 11.1). In some instances an unauthorized person—a member perhaps—may be relaying information to workers. Too many times I have had someone give me the agricultural perspective of greens maintenance or

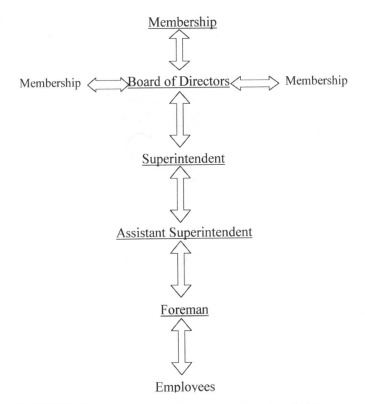

FIGURE 11.1 Dialogue tree—if communication is handled in this fashion there is less confusion for everyone involved.

his personal theories on turfgrass care. It is important that situations like that are handled politely and that the person is reminded that he or she needs to redirect the issues to the entire board, instead of assuming that he has some authority to tell you what to do.

I do want to stress that the lines of communication should not be entirely blocked from the top of command to the bottom. A person should not feel that he or she can only communicate with the immediate supervisor, and the supervisor should be able to address issues with all the employees under his or her supervision. In dealing with day-to-day tasks and maintenance, the flow chart above will alleviate a lot of issues.

All employees have roles in the chain of command, but good employees will let you know when something is important to them. It is important for anyone to listen to those who work above or below to find new ideas in course improvement or to fix existing problems.

A golf course superintendent needs a lot of patience. There are many problems in course maintenance that simply require time to fix, from purchases that need to be made to waiting for newly seeded areas to germinate. You must be patient and realize that perfect turf conditions are not possible overnight. High-quality turf will take several years to get into that condition. All too often a superintendent does not possess the personality skills and traits to deal with people effectively. You need to be willing to learn and change as the industry changes. Like a good doctor, if your skills become outdated, chances are you are out of luck and someone else will be doing your job for you. I mentioned how important it is to attend training courses and to find out what changes are occurring in the golf course industry each year.

You will also need to be patient about the career choices you make within the golf course industry. There are always openings for those who are exceptionally qualified and especially for those who possess a degree or years of experience. What I always jokingly say is that a good assistant must "hurry up and wait" for that jump into the driver's seat as the superintendent. Good superintendents who want to excel within the industry unfortunately have to move around the country to get those higher-paying superintendents' jobs. There are only going to be a limited number of job opportunities within your region, no matter how large it is. There is only so much open space available to build golf courses, and only the more elite clubs are going to pay you the big dollars. There are also few people with a tremendous resume to get those higher-paying jobs, therefore the good ones are paid a premium.

If you are sitting in the wings waiting for a superintendent's position, you are more likely to be promoted by being an assistant at that course first. Being the assistant at a course prepares you for what is to come should the superintendent's position open up. There are very few positions that have been filled by someone who was not an assistant on a course first. Being the assistant does not mean that you are a shoo-in for the job, but if you are patient and build a reputation with your course, it is a great stepping stone to success. Movement through the ranks occurs quickly for those with a good work ethic, so if you are willing to relocate, there will be opportunities for you.

COURSE MAINTENANCE

There are many aspects to a golf course manager's duties and responsibilities that do not deal directly with turfgrass maintenance. A superintendent is responsible for all the areas on the golf course and the overall appearance of the entire grounds. I mentioned earlier about the various areas of turf and the typical mowing heights associated with those areas. I want to mention the other locations that are considered in-play areas that may or may not involve turfgrass. The superintendent may also be somewhat involved with the clubhouse operations at smaller courses, although this is very unusual at a larger golf course.

TURFTIP

There are two primary sand locations that you may encounter when you are on a golf course. These are sand traps and fairway bunkers.

The first areas I want to mention are the sand areas. There are two primary sand locations that you may encounter when you are on a golf course. The first areas are the sand traps, (Figure 11.2) which lie in close proximity to the golf greens. These areas are maintained with a rake or a machine that provides a raked effect. These areas should be raked frequently and in a manner that will cause the golf ball to be slightly buried in the sand. This is a care area that does not have grass, but certainly is a maintenance point on the course.

FIGURE 11.2 Sand trap

The other sand areas found on some golf courses are the fairway bunkers. These areas catch golf balls out in the fairway. The fairway bunkers are typically wide areas that are often more shallow than the sand traps that surround a green. The idea is for these areas to create obstacles in play. They are maintenance points for a superintendent and will need to be raked often. On a larger course they will be maintained by a specialized piece of equipment that is designed to give the trap a freshly raked appearance. There is also a need for both sand traps and bunkers to be edged on a regular basis to keep as clean-cut an appearance as possible.

A type of trap that is becoming more common on courses is known as a grass bunker. It usually consists of a low area where the grass is allowed to grow to a very tall height that makes it difficult to hit the golf balls out. I have seen traps with grass over six inches high. Mowing and maintenance to keep the integrity of the grass bunker area can be difficult. These traps are much cheaper than sand traps and do not require the constant raking or need for specialized equipment if they are designed correctly.

Moguls are an intricate part of course design and play. Moguls can occur in any of the play areas. Moguls are large bermed areas of dirt that add challenges to the course layout for the golfers. They can be used to somewhat hide greens or create obstacles in the middle of a fairway. They do create a challenge for mowing, and

---

TURFTIP

There is a need for both sand traps and bunkers to be edged on a regular basis to keep as clean-cut an appearance as possible.

> **TURFTIP**
>
> Mowing and maintenance to keep the integrity of the grass bunker area can be difficult. These traps are much cheaper than sand traps and do not require the constant raking or need for specialized equipment if they are designed correctly.

not all mowers have the ability to go over a mogul without scalping the grass.

One of the most important resources a superintendent has on the golf course is the water supply. Very few courses get water from municipal water systems or sources due to their astronomical cost. Water is normally taken from a lake or pond that is centrally located on the golf course. This lake is tapped into for irrigation for most of the course. If the superintendent does not manage the lake and pay attention to its needs, there could be catastrophic losses to the turf on the golf course.

> **TURFTIP**
>
> Pollution is one major consideration for the water system on the golf course. If over time your water system becomes contaminated or polluted, you stand the potential for damage to the course turf, flora, and fauna. The biggest potential pollutant is of course from pesticide residues and runoff.

Pollution is one major consideration for the water system on the golf course. If over time your water system becomes contaminated or polluted, you stand the potential for damage to the course turf, flora, and fauna. The biggest potential pollutant is of course from pesticide residues and runoff, which is common to all golf course environments. It is very common to have the pond located in such a fashion that a majority of the runoff water on the course runs into it. Water sources are often designed to collect as much water as possible so there is plenty of supply for the irrigation systems. The problem is that runoff water can collect pesticide and fertilizer residues from the large areas that are treated with chemicals. A golf course uses a very high number of chemicals in its daily maintenance, and seldom is consideration given to the potential damage to the water system.

In addition, make sure that you are not keeping your ponds too clean from aquatic weeds, vegetation, and cover. Plant life in and around the water source is critical to the purification of the water system and for healthy fish life. Many superintendents like to have immaculate-looking pond banks and even have been known to spray repeatedly to remove all vegetation from the lakes and streams on the course. This is not a good practice because you may remove too much vegetation which can affect water pollution and kill fish in the lake as well. It is also possible, however, for a water source to become overgrown with algae. Many aquatic weeds love golf course ponds because of the high levels of nitrates in the water from fer-

TURFTIP

Plant life in and around the water source is critical to the purification of the water system and for healthy fish life.

tilizers. Many of these weeds can become evasive and some can become problems in irrigation systems, so there is a balance that will need to be maintained in the water system.

Do not be afraid to leave some naturalized areas on the course or even some wetland areas. These are beneficial to water quality and provide a natural cleaning system for the ground water supply. Try to not eliminate every last bit of vegetation in the lake and also leave some cover in and around it. Taller grass and shrubs can help to slow run-off and create a more desirable water supply. There are ways to naturalize plantings without looking unsightly or unkempt. Be sure and leave an ample amount of grass along the water's edges because erosion can start in any pond. This is especially true for levees that are built on the west side of a pond.

If you are constructing a lake on a golf course, you should always use precautionary measures to ensure long-term success of your levee system. Riprap or large white stones can be placed on the inside of the levee to prevent water from lapping away from the pond bank, causing a massive erosion problem. This is a very common problem for levees constructed on the west sides of ponds. You can also use a retaining-wall system for the interior face of the levee. This will give

---

TURFTIP

If you are constructing a lake on a golf course, you should always use precautionary measures to ensure long-term success of your levee system. Riprap or large white stones can be placed on the inside of the levee to prevent water from lapping away from the pond bank, causing a massive erosion problem.

you a nice surface for the water to splash against and will not wash the bank away. It will also create a much nicer appearance to the pond in general, though it will not give you a natural look if that is your plan (Figure 11.3).

All of the areas found on a golf course share an overall importance to the way a course is designed and played. These areas are also maintenance points that will need to be kept up as neatly as the turfgrass areas.

Many superintendents express a lack of interest in the beautification of the golf course. The attitude that course features simply make mowing more difficult will not create a very interesting golf course nor will it provide you with a very long career as a greens keeper or worker of any kind. I think that landscape design issues and other key horticultural concepts need to be carried out over the entire golf course. That not only makes the course a challenge to play but beautiful to look at. I think that a combination of tree care, floral designs, and landscaping features should be used throughout the course whenever maintenance of such features is possible. If you find yourself in the underpaid and overworked category, chances are that you are not looking to increase the maintenance points out on the course. You should still consider the small things that will make a difference over time and add aesthetic beauty and additional challenges. The sand traps I mentioned earlier are a feature that many small 9-hole courses do not offer, but they are features that can greatly enhance the layout and design of the course without a lot of extra work. Small flowerbeds can also be added even to the tiniest course to create an overall effect that will keep the people coming back.

One way that beautification can occur on a course with a limited staff is through a membership committee set up specifically to consider improvements and beautification of the course as a whole. Often members will create an auxiliary that participates in planting and maintaining flowerbeds on the course. Such programs should be implemented with the approval and help of the superintendent. The superintendent should be aware and involved with

254 *Turfgrass Installation: Management and Maintenance*

FIGURE 11.3 Island green in a small 9-hole course has a trademark ninth hole that is completely surrounded by water.

all aspects of golf course management and maintenance. This is one reason that projects involving "free" help from members often fail so horribly. If a superintendent is left with additional headaches and maintenance that he or she was not aware of in the first place, the workload just becomes more overwhelming and cumbersome. I love to have support and help from a membership and think that an active membership helps to create a positive golfing experience for everyone.

There are many ways a golf course can raise money to promote course improvements. Extra funds can be created from fundraisers and tournaments. Donations can be sought out from the membership to create memorials or other features that can be both sentimental and attractive at the same time. You can take up collections to raise money for tree plantings or even place a cash box in the clubhouse. Any one of these ideas can create additional funds that can be kept out of budgets and used strictly for the purpose of beautification of the course (Figure 11.4).

Unilateral actions by memberships and boards where a superintendent is left out of the loop create the biggest struggles for the golf course superintendent as a whole. There is no way to avoid conflicts if the decisions of the superintendent are not supported by the board of directors. This is not to say that every time a superintendent needs a piece of equipment the golf course should just buy it.

FIGURE 11.4 Large recycling center—recycling beer and soda cans is a very low-overhead way to create extra money for projects.

I do think that it is important that the person running a course be knowledgeable enough to make good decisions about the care of the golf course. It should not be up the board to tell the superintendent how he or she should maintain a course. If at a later time the goals are not met, then specific suggestions should be made or a better person be hired to run the course. Anyone in charge of a golf course should know about basic turfgrass care and general maintenance.

BUDGETING AND FINANCE

I want to focus on the most overlooked part of golf course management, which is the business part. The expenses that are involved in a golf course budget are quite substantial, and the costs to run a course are only getting higher each year—at the same time that a slight decline in the number of golfers is occurring. This is creating a deficit for many courses and a reduction in the number of smaller courses.

A golf course is a business, and the superintendent is the manger of that business. The owner and the board can be thought of as the money managers of the course, but the superintendent must be involved in every level of the spending process. The budget should be established and voted on by the board or owners but should take into consideration what the superintendent needs to run the course for the year. The superintendent will know more about the chemical needs, equipment concerns, and the areas on the course that need attention and improvement. It is impossible for a budget to be established without the input of the superintendent as to what will be needed for the actual maintenance of the golf course for the upcoming year.

Budgeting is a critical part of your job if you are in the superintendent's position. Many courses rely heavily on the superintendent of the course for the input in the yearly budget. If the superintendent's costs go drastically over budget, then he or she may be searching for work elsewhere. I always made sure that my budgets were figured as accurately as possible in terms of future needs and checked against past performance.

A good manager should not always look at a budget as just a number to beat or match from year to year. A budget is a guideline for doing business and a starting point for the expenses in a fiscal year. There is always room for cost savings in budgets, but growth can be very positive, as it may mean that there are more members or that expansion is planned. It takes an open-minded board of directors and an aggressive superintendent to sell ideas that will be both beneficial for their careers and the overall condition of the course. If positive growth is occurring, the numbers on the budget should naturally be higher from year to year. With profit comes the need for continued growth, and with growth comes the need for added expenses. It is this juggling act that makes your budget work.

If you overspend in your budget, than you can find yourself in trouble in a hurry, but do not set yourself up for failure by providing false numbers or unrealistic figures to the golf course directors and boards. Do not try to slip by with too few chemicals, equipment, or people. I have seen too many cases where superintendents are hired to turn a course around and they only end up hurting the course by lowering the operating costs. It should be expected, as with any business, that there be an allowance for additional expenses each year just to keep up with any inflation that may have taken place.

Chemicals are an often overlooked part of golf course budgeting. As turf improves and perfection of play is demanded, chemical usage tends to go up. This is especially true in bentgrass areas where the need for chemical usage is very high and turf treatments are absolutely necessary. When bentgrass areas are added to the course, the playing conditions may improve but the need for chemical use goes up. I think I have witnessed this mistake more than any other. When courses add in bentgrass tees or irrigation in the fairways, what happens is that everyone assesses the costs of water, seed/sod, and so on but no one understands that this new turf will require a considerable amount of chemicals and fertilizer to maintain a reasonable integrity. It is important to once again understand the needs of turfgrass to properly budget for the chemicals used to treat that turf.

Equipment is still the largest expense on most golf courses. Golf courses require very large, specialized pieces of equipment that cost a lot of money to purchase and maintain. This is especially true of the reel mowers that are commonly found on golf courses. Reel mowers are high-maintenance machines with many moving parts. Not only are the costs of purchasing such equipment very high, but keeping these items running is also very difficult. There are many other specialized pieces of equipment, such as sand trap equipment, aerators for greens, and mowers that are capable of incredibly low mowing heights and that are not used in other turfgrass applications.

When figuring out a budget for such items, you must look at it in several different ways. It is possible to constantly turn over newer equipment with low hours and replace it with as little depreciation as possible (Figure 11.5). It is also possible to buy cheaper older equipment at a lower cost and just repair the equipment as needed. The problem with the second scenario is that it doesn't take into account the value of your time. This is the most overlooked part of equipment usage. Anyone can figure up what a unit will cost them and what they will have to pay a person to operate it. What is hard to figure is how often that machine will break down and what will be the costs in repairs and lost time. This is one reason that the trend of leasing equipment became so popular in the 1990s. Leasing was offered by many companies and started a trend of being able to replace equipment on a more frequent basis.

When the equipment you are using is modern and up to date, the quality of the course is usually much higher and also easier to maintain. You need to assess down time compared to repair costs. Down time should be assessed when purchasing any mechanical items for the course. Another thing to consider when buying equipment

FIGURE 11.5 Old, worn-out reel mower. Many courses wait until the items they are using have no value, which creates more of a price shock when there is nothing worth trading in at the time of a new purchase.

is how many work hours can be eliminated by having a larger or newer piece of equipment. When costs are lowered in one area, for instance labor, that money can be used in some other area of the budget. This might be a hard sale if your board or ownership is very conservative. You must show why if money can be saved in some areas, it can be used more productively somewhere else.

When preparing a golf course budget there are many factors to consider whether you are at a course that is prospering or one that is floundering. Take a look at the expenses in a budget. This list does not include a clubhouse and or other facilities outside the golfing areas.

Typical items budgeted for on a golf course for one year:

- *Employees (labor)*
- *Wages*
- *Benefits*
- *Equipment*
- *New items needed*
- *Replacement of existing equipment*
- *Chemicals*
  - *seed*
  - *fertilizers*
  - *pesticides*
- *Fuel*
- *Equipment repairs*
- *Parts*
- *Labor*
- *Irrigation*
- *Expansions*
- *Water, if applicable*
- *Taxes*
- *Employment*
- *Social Security*
- *FICA*
- *Medicare/Medicaid*
- *Insurance*

- *Property (golf course and equipment)*
- *Liability*
- *Workman's compensation*
- *Buildings (not including club house and cart sheds, etc.)*
- *Utilities*
- *Renovations*
- *Course beautification (landscaping)*
- *Unforseen expenses*
- *Miscellaneous expenses*

It is important to evaluate your specific needs each year and take into consideration any special projects your course may be considering at the time. Adding irrigation is an example of an improvement that often creates a desire for a better turf, which will lead to unexpected expenses. There is also the potential for a catastrophic loss on a golf course. There should always be room in a budget for unexpected disasters or emergency money. A major irrigation problem can cost thousands of dollars to correct. If the problem persists for any length of time, major turf damage could occur or even the loss of a green. Losing a green can be one of the most damaging scenarios to occur on a course from both the perspective of play and the problem of correcting the situation.

Chapter 12

The Putting Green

THE ART AND SCIENCE OF TURFGRASS MANAGEMENT

The putting surface is the "dance floor" of the golf course. You can get by with problems on any other part of the golf course, but players will be most disturbed when there are problems with the putting surface. As defined in the rules of golf, the putting green is an area specially prepared for putting and containing a $4^{1}/_{2}$-inch diameter hole. If budget cuts have to be made, this is the last place where any cuts should happen.

There is both art and science involved in golf course turfgrass management. In my opinion, you cannot teach the art part of the business. You either have it or you don't. Interning, working at nationally known courses, and being mentored by successful superintendents will immensely help students become successful when they get their own facilities. Someone that has worked at a famous facility will be more attractive to the course that is in the hiring process, especially if they have a reputation for outstanding putting green surfaces.

> **TURFTIP**
>
> Talk with as many experienced professionals as you can as you enter your career. There is no substitute for experience. Many elements of caring for a golf course are a matter of personal experience in a particular area of expertise.

Science

High quality greens are free of mat, thatch, and grain. The surface is fast, firm, smooth, true, and consistent throughout the course. Only a properly struck shot should hold the green. If the green holds shots from the rough, it's too soft and will not provide the smoothness and ball roll that is demanded. Maintenance practices that will influence putting green quality are mowing, irrigation, nutrition, cultivation, pest control, and topdressing. In order to properly manage greens, you must know the physical and chemical nature of the soil in the rootzone and the basin below the growing medium.

Three things to keep out of golf greens:

- *Mat*
- *Thatch*
- *Grain*

Mowing

To achieve top quality conditions the putting green must be mowed daily. Assuming the turf is healthy, never skip a day of mowing during the growing season because you will lose ball roll quality. You must determine the height of cut (HOC) that will work on your greens. Generally, top quality greens are mowed at a HOC of <0.140 inch. In some situations mowers with a groomer attachment will help to encourage an upright habit of growth. Vertical mowing, spiking, and brushing are tools that will help to incorporate topdressing into the turf canopy and keep the surface free of puffiness.

Steps to quality greens:

- *Mowing*
- *Irrigation*
- *Nutrition*
- *Cultivation*
- *Pest control*
- *Topdressing*

Irrigation

Irrigation should only be done as needed. Wet, sloppy, or soft greens are not conducive to top playing conditions. The surface should be maintained as dry as possible. Only enough irrigation water should be applied to keep the turf alive. A light and frequent irrigation regime can be used effectively on push-up greens. The perched water USGA

TURFTIP

Do not irrigate excessively. Too much water is bad for playing conditions on a golf course.

TURFTIP

Fertilizer may not be your friend. Do not use rule-of-thumb methods for fertilizing your course. Have the soil tested and apply fertilizer based on the results of the soil test.

(United States Golf Association) greens should be deep-watered infrequently until the water runs out of the tile line or the root zone will remain too wet.

Nutrition

Nutrition is an important cultural practice for maintaining top quality playing conditions. When I started in this business it was not uncommon to apply 8-10 pounds of nitrogen per M ft^2 per year in the north central United States. In the late 1970's we started drastically reducing the amount of nitrogen by 60-80%, to 2-3 pounds per M ft^2. Jim Latham once told me, "You're going to run out of gas (nitrogen) one of these days," but what else would the sales manager of Milorganite® fertilizer say? The results were less puffiness, less mower scalping, fewer disease problems especially Brown Patch, and much better playing conditions. My opinion is that if you locked the fertilizer storage shed and threw away the key, the putting quality of greens at many golf courses would significantly improve. Our greens were once called phosphorous mines because of the many years that Milorganite fertilizer was used. Phosphorous was blamed for Poa annua encroachment on the bentgrass putting surface and fairways. In the future, we are probably going to be regulated by government legislation on how much phosphorus can be used. Potassium must be applied frequently to keep sufficient levels in sand-based rootzones.

It is imperative that a reputable laboratory does a chemical soil analysis every two to three years.

Cultivation

Cultivation or aerification will prepare or pre-condition the putting green turf for environmental stress. Tom Mascaro, the inventor of the aerifier, said, "The wrong time to aerify is when the turf is weak." No truer words could have ever been spoken. Cultivation should be done on an "as needed" basis. Spiking or "fine tining" on a regular basis is an excellent way to allow oxygen into the rootzone. On push-up greens, deep-tine cultivation is very beneficial to encourage deeper rooting. Water injection or the Hydro-Jet can be a valuable tool to promote root growth during periods of environmental stress.

Pest Control

Pest control is most important on putting greens because there is a zero tolerance threshold level for blemishes. With the low amount of nitrogen fertilizer that we use on greens today, there is not much recuperative ability if spots of disease occur. The putting green is not the place to use curative disease control practices. With a good preventive disease control program, fewer chemicals may be used than with the high curative rates of application needed after disease has been detected.

> **TURFTIP**
>
> Pest control is most important on putting greens because there is a zero tolerance threshold level for blemishes.

TURFTIP

It is sad, but ball roll or the speed of the green gets the most attention from the player.

Topdressing

Topdressing is an integral part of producing a high quality putting green surface, if applied at light frequent rates that match the growth rate. See Figure 12.1. It will help create quality putting surfaces that are firm, putt smooth, putt true, and are consistent and fast enough. It is sad, but ball roll or the speed of the green gets the most attention from the player. So in order to satisfy players that insist that the

FIGURE 12.1 Topdressing by hand

greens are too slow, you must constantly work on providing enough ball roll or speed that is acceptable.

ABOUT GREEN SPEED

For putting green management you should develop a regular maintenance program that will be used most of the year, and also a tournament maintenance program that will be used only for special circumstances. Practicing good agronomics and effective communication will allow both your putting greens and your golfers to be champions.

CULTURAL PRACTICES AND CIRCUMSTANCES THAT INFLUENCE GREEN SPEED

Green speed is influenced by the following factors:

- *Mowing*
- *Rolling*
- *Moisture*
- *Nutrition levels*
- *Topdressing*
- *Weather conditions*
- *Smoothness & trueness*
- *Cultivar selection*
- *Plant Growth Regulators*

Mowing

Mowing has less impact on speed than you think. I have seen greens that are mowed above 0.125" that easily rolled 10 ft. on the stimpmeter. It's when you need to have green speeds of 11 ft. that the mowing height has to be below $1/8"$. Even then you will have to mow more frequently than once a day. During special events I have mowed some greens up to 5 times in one day and double rolled to reach green speeds of 12.0. This is inflicting a considerable amount of stress on turf and you better know when enough is enough or there can be rapid deterioration of the putting surface. Remember, speed kills.

One thing is for sure: Once you produce these fast conditions, there will be discussion among members demanding that they should have these conditions everyday. This is where the trouble begins. Expectations go up and the bar is raised. It is sad but true; golfers judge quality by how fast the greens are rolling. It gets down to communication, communication, and more communication. You better be able to tell them why and get them to believe it.

The use of groomers and vertical mowing are valuable tools to keep the turf growing upright rather than vertically. See Figure 12.2.

FIGURE 12.2 Turf grooming

Many creeping bentgrass cultivars have a tendency to grow more vertically than others, especially Penncross. Grooming and/or vertical mowing can promote a more upright habit of growth of bentgrass. Along with topdressing, both of these vertically cutting blades will help produce a smooth putting surface.

The quality of cut is not only important to gain speed, but a clean cut heals quicker than a ragged cut. On closely cropped putting surfaces, a thin bedknife must be used, and the bedbar must have the proper attitude. Toro® has a "deep penetrating bedbar" that is tilted forward to keep the bedknife from scuffing. This is very helpful on low HOC greens, but because of the aggressiveness of the deep penetrating bedbar, it may scalp the turf. First time users should install it first thing in the spring when the mowing HOC is higher than it is for the prime time of the season. The HOC should be gradually lowered in increments of 0.010-0.020 inch every 7-14 days until you reach your target green speed.

Everyone knows that the direction of mowing is changed every time the green is mowed, unless you are double cutting or vertical mowing, and you turn around and come back on the same pass. Usually greens are mowed in four different directions (from 6 to 12, from 3 to 9, from 5 to 11, and 2 to 8 o'clock). To avoid getting a walk-mower ring (or a triplex ring) around the outer perimeter we only mow the cleanup three times per week. For those that use a triplex mower exclusively for mowing greens, you may want to consider using a walking greens mower on the cleanup pass.

Rolling

Rolling of putting greens became very popular around 1990. The better players came back from a tournament that they played in and saw the greens being rolled. They told everyone how nice the putting conditions were after the green had been rolled. Now, the club or green chairman was recommending that a roller be purchased, and it was an easy sell.

272 *Turfgrass Installation: Management and Maintenance*

FIGURE 12.3 Rolling the greens

Clubs in the northeastern US were rolling putting surfaces up to five times a week. Refer to Figure 12.3. Dr. James Beard's research showed that rolling more than two to three times per week caused thinning of the putting surface. There is a half-life or carryover of rolling speed enhancement to the day after the green is rolled. During normal play we would roll putting greens one to two times per week. For special events we would double-roll by turning around and going back on the same pass. This would increase ball roll by 1.0 ft.

At Ozaukee Country Club, Mequon, Wisconsin, we had a triplex mower with vibrating rollers and a Salsco (sidewinder) greens roller. Considerably more training was needed with the sidewinder than the triplex roller, but after a good operator had some experience, the greens could be rolled almost as fast as the triplex. I would strongly recommend using the sidewinder instead of the triplex roller because of less damage to the putting green surrounds and there are no wheel tracks behind the rolled surface.

I learned that the green mowed with the triplex mower set at 0.100 inch was not as fast as the green mowed with the walking greens mower set at 0.115 inch. In my opinion, this is because the drum on the walking greens mower is also rolling the putting surface as it is being mowed. Rolling helps create a smoother putting surface. The reason the ball roll is increased is because the ball does not bounce; it just simply rolls.

Soil Moisture

Soil moisture content is a very important cultural practice. Good putting surfaces cannot be good landing areas. A soft or spongy putting green surface will footprint, which is not conducive to smooth ball roll. High quality golf green putting surfaces must be firm, not soft, and have a deep root system. Only enough irrigation water should be applied to keep the turf alive and not to soften the surface to accept shots. On a good green there is a certain amount of resiliency that will accept a properly hit iron shot from the fairway, and not a 3-wood hit from the rough.

Nutrition

To produce high quality playing conditions, nutrition is very important and the soil must be chemically balanced. Putting greens should

---

TURFTIP

Rolling helps create a smoother putting surface. The reason the ball roll is increased is because the ball does not bounce; it just simply rolls.

have a chemical soil analysis done at least every three years; this is something that you cannot guess at. Phosphorous does not leach out of a soil that has a higher cation exchange capacity, but it does leach on new sand-based greens. Potassium must be applied frequently throughout the growing season. Calcium, magnesium, sulfur, and trace elements must be available to the plant and in proper balance.

Nitrogen has more influence on putting green speed than any other cultural practice. Turf managers that like to have dark green turf on putting green surfaces are not likely to have good ball roll. Nitrogen response is usually more green color and increased density. Every turf manager must determine how to manage nitrogen for good quality putting conditions. Very little or no nitrogen should be applied to putting green surfaces during periods of warm/humid weather in the summer when microbes are very active. There really is no good soil test for nitrogen, so you must ask the grass. I prefer using exclusively the sprayable products to supplement nitrogen to putting greens. There is no worry of mower pickup or granules moving to create problems.

Changing from a topdressing mixture of sand, soil, and peat to light, frequent applications of straight 100% sand topdressing has significantly improved the putting quality of greens. Green speeds have significantly increased since we initiated the sand topdressing program at Ozaukee CC in 1976. The putting surface is firmer, truer, and more consistent. The ball rolls smoother, does not bounce and, therefore, ball roll is increased.

TURFTIP

Green speeds are difficult to maintain when the weather is warm and the relative humidity is high.

> ## TURFTIP
>
> Contact the National Turfgrass Evaluation Program (NTEP) to find which cultivars perform best in your geographical area.

Green speeds are difficult to maintain when the weather is warm and the relative humidity is high. On days when it is windy and the humidity is low, the ball will roll very well. In the spring and especially in fall when the nights are cool, it is very easy to attain high green speeds. During warm/humid weather conditions, more mowing and rolling is needed to produce desired green speeds.

Putting surfaces must provide smooth conditions that allow the ball to roll smoothly. Uphill putts that are properly struck should not bounce. Aggressive vertical mowing (with the blades set at least $1/8$ inch deep), in conjunction with topdressing, will help eliminate grain and/or puffiness. This should be done when the turf is actively growing, and not during periods of high temperatures. The lines from the vertical mowing will be seen for a couple of weeks, but they will not affect ball roll.

On a new project, or even on existing courses, you should consider using one of the "new generation" Penn A or G cultivars if the members or players are going to demand fast greens. These grasses need more intensive management, so you must know if you will have the resources to properly maintain these tightly knit bentgrasses. Quadra-tining, spiking, and/or vertical mowing is required to incorporate topdressing into the turf canopy.

Contact the National Turfgrass Evaluation Program (NTEP) to find which cultivars perform best in your geographical area. There are many bentgrass selections now available that can produce high quality putting surfaces. Some have excellent heat tolerance for areas in the transition zone and beyond.

Plant Growth Regulators

PGR's are a very useful tool in putting green management for enhancement of turf density and also for suppression of seedheads.

There is nothing better than melfluidide (Embark® 2S) for suppressing seedheads. Timing of the application is very critical; it must be applied before the seedheads form. What I would do is carefully watch for the first seedhead to form and then run.

For maintaining putting greens that have a mixture of Poa annua and bentgrass, I like to use the PGR Primo Maxx®. It tank mixes well with contact fungicides, is foliar absorbed so it does not need to be irrigated or watered in, and there is no widening of the leaf blade with Primo. Most PGR's do increase turf density (which creates friction), so I do not see increased green speed with growth regulators. I have heard for several years that our greens used to be faster than they are now. My conclusion is that, in the late 1970's, our greens were not as tight knit as they are now, so maybe they were faster. I do not believe that they putted any better than now.

It is important that this craze or demand for more speed on the putting green surface does not get you into trouble. You must know when know when to back off! After you see the symptoms of thinning or deterioration of turf, it is too late. Damage, which may be caused by environmental stress, anthracnose, or Poa annua decline, is usually fatal, as Dr. Gayle Worf, Retired Professor of Pathology at the University of Wisconsin, used to say.

Determining when to "back off" is much easier said than done. Keeping your eyes and ears open is very important. Listen to what others have experienced from trying to provide championship conditions for too long, or during periods of high environmental stress. These same players who told you how great the golf course was when you had the greens rolling at 12 feet, will dump you like a bad habit after the putting surface turf deteriorates.

> **TURF TIP**
>
> Greens that are covered with the geo-textile fabric may be injured more than non-covered areas during extended periods of ice sheeting.

WINTER PLAY?

Play on golf courses in the northern part of the US is a very touchy subject. When there are only a few players on completely frozen turf, there will not be any significant damage. It is those nice days during the winter when the top inch of frozen ground thaws that you need to be concerned about. There will be many players out and that will cause foot printing of the playing surface. This is where the trouble lies; when the weather is bad, it's ok for me to play... but when it gets nice and warm, you won't let me play. The best procedure is to have a policy set and approved that when the golf course is closed, it is closed for the season. I have seen golfers walk right by the "course closed" sign on their way to tee it up. Then governing powers can take action.

Play should not begin in the spring of the year until after the frozen ground has thawed and everything has firmed up. Someone will ask how much it will cost to repair the damage. You cannot put a price tag on disrupting the surface from foot printing that occurs on partially thawed ground. It may take a year or more before the signs of rutting damage are no longer noticeable.

Geo-textile greens covers can be used during the winter. The best benefit of this practice is keeping the golfer off of greens. Other benefits include better seed germination in cool weather and helping to prevent desiccation, especially during periods of low temperature without snow cover. Greens that are covered with the geo-textile fabric may be injured more than non-covered areas during extended periods of ice sheeting. There is no easy way to deal with winter, but use good agronomic practices and common sense to provide top quality playing conditions for the majority of the clientele.

Chapter 13

Fairways

THE IMPORTANCE OF FAIRWAYS

When I attended college in 1962, a question was asked about the most important area on the golf course. My answer then was fairways; my answer was wrong. The correct answer is greens. The measure of a prestigious golf course back in 1962 was bentgrass fairways or fairways that were mowed so low (below 1¼ inches) that the bluegrass just disappeared and Poa Annua took its place. Golfers and superintendents alike tagged these fairways as bentgrass when in reality they were 90% Poa Annua. As superintendent at Milwaukee Country Club from 1973–1989, I felt that I had the best bentgrass fairways in the country. The reason is because they were originally planted to bentgrass. I am sure you are asking what variety and I really can't tell you. When these fairways were planted in the early 1920's, the only varieties available were sea side and colonial bentgrass. I can only speculate that many varieties were in the seed lots that were purchased in that era.

Situations to avoid for healthy fairways:

- *Low mowing*
- *Overwatering*

- *Poor drainage*
- *Compaction from mowing equipment*
- *Shade*
- *Golf carts*

What evolved were many low mowed fairways that were predominantly Poa Annua. Due to low mowing, overwatering, poor drainage, compaction from mowing equipment, shade, and golf carts, fairway survival was always an issue. Many superintendents tried to overseed bentgrass into Poa Annua with little success. Until you make a better environment for the bentgrass and eliminate many of the before mentioned problems, good bentgrass fairways will never to be a reality. We have learned, through sponsored turf grass research and superintendents that have managed Poa Annua fairways over many years, that maintenance programs had to change. I believe that the biggest reason today for the sustainable success of bentgrass fairways is the change to lightweight mowing and clipping removal. When you factor in better fungicides, growth regulators, and state-of-the-art irrigation systems, the rate of success in establishing and maintaining bentgrass fairways has dramatically improved. We can now enjoy the low mow bluegrass fairway along with lower maintenance costs. Low mow bluegrass has great playability and is a viable option to bentgrass in the north.

FAIRWAY CONVERSION

The biggest obstacle to fairway conversion is the time of year when it needs to be done. Rather than give up their golf course for a few weeks, members, owners, and golfers would opt for what they have as good enough. This is like saying that that black and white television is good enough. The cost of fairway conversion is always a consideration in both the cost of the project and the lost revenue. An important question to ask is whether a better golf course will increase revenue? If your golf course experiences turf loss in fairways, will your golfers go somewhere else to play? Will they come back?

Reasons for Fairway Conversion

Cost and playability figure heavily in planning fairway conversions. These reasons are outlined as follows:

1. The number one reason is playability. Whether you choose bentgrass or low mow bluegrass, you will have consistent playing conditions throughout the golfing season. The fear of losing your fairway will diminish. Although both bentgrass and Poa Annua are cool season grasses, Poa Annua is less tolerant of heat and more susceptible to a variety of diseases.

2. The second reason for fairway conversion is cost. Primarily, most cost savings will come from applying less pesticide and fertilizers. Poa Annua is very susceptible to Summer Patch and anthracnose disease; bentgrass and low mow blue grass is not. I believe that where Summer Patch is a problem year-in and year-out, fairway renovation becomes a much stronger case. I also believe that the farther south you are in the northern regions, the greater success you will experience. Lowering the risk of anthracnose and Summer Patch represents a large saving in your pesticide budget. Factoring in less hand watering and fewer fungicide spray applications means a reduction in labor. In many instances, your cost of the fairway conversion can be recouped in just a few seasons.

Benefits of Fairway Conversion

The benefits of fairway conversion to the membership and owners are three-fold. Specifically as follows:

1. A variety of turf grass for better and consistent playing conditions

> **TURFTIP**
>
> Ideal grow-in weather is warm days and cool nights, weather generally occurring in late August or early September in northern regions.

2. Long-term savings to the members and owners
3. Peace of mind; not worrying about losing turf in mid to late summer

To have a successful fairway conversion program, timing is critical. In general, the best time for fairway renovation is the 2nd or 3rd week in August. (This may vary a little depending on your location in the northern regions.) Starting your renovation too early, you'll face the problem of competition of crabgrass and diseases. The main concern here is damping off (Pythium). Ideal grow-in weather is warm days and cool nights, weather generally occurring in late August or early September in northern regions. Starting your renovation too late, you'll face slower germination and a more desirable grow-in period which means a less than perfect fairway when the golf course opens the following spring. If your fairway renovation is timed right, you can experience bentgrass seed germination in 7 days and be back playing on the new fairway in a month. It is always a good idea to keep golf carts off the newly renovated fairways until the following spring.

It is possible to finish the renovation of an 18-hole golf course in a week with enough equipment, but a safer time frame would be ten days. Golf play needs to be kept off the renovated fairways for the period of 24 hours after Round Up® is sprayed and when your crew is working on an individual fairway. Communication to the golfers is important; a letter to the membership is in order. Have an open forum so concerned golfers have the opportunity ask questions about the renovation.

Before getting started, remember the list I pointed out about Poa Annua: over watering, poor drainage, heavy mowing equipment, and shade. Correct as many of these factors as possible before embarking on a fairway renovation program. To get started you'll need to have all equipment, seed, and fertilizer on site before you begin.

Equipment and Supplies

Make sure you have collected all the materials need for fairway conversion before you proceed too far in the process. The equipment and supplies needed for fairway conversion are:

- *Sprayer with boom*
- *Walking spray boom for edges*
- *Safety equipment, disposable shoe protectors*
- *Plywood for drift protection*
- *Utility vehicles*
- *Ropes and stakes*
- *Marking paint*
- *Mower for scalping (a flail mower works nicely)*
- *Tractor, push or knapsack blowers*
- *Turf vac*
- *Slit seeder hand and tractor mounted*
- *Drop seeder hand and tractor pulled*
- *Roller*
- *Seed*
- *Fertilizer*

Table Steps

Be sure to follow an understandable procedure during the conversion. You will need the cooperation of staff, management, and members. Following are the table steps for fairway conversion:

1. Close golf course to play. (Monday is usually the best day.) Be sure to put a notice in the club newsletter marking the beginning date. Place informational flyers around the club, in the pro shop, and locker rooms.

2. Put ropes up around the entire fairway. Leave an opening as a gate to get equipment in and out. Make sure that you include enough room to seed rough areas next to the fairways. It is a good idea to seed the intermediate in bluegrass after seeding the fairway in bent or low mow bluegrass.

3. Use plywood to protect the green front from any drift spray of Round Up. It is also recommended to use plywood at the entrance gate into the fairway.

4. Fill sprayers making sure that the walking spray boom is calibrated the same as the large boom. Read and follow recommended rates on the Round Up® label along with all safety instructions.

5. Use marking paint to outline the edges of the fairway. This gives you a way to make sure the fairway is the same size and shape as when you began. This also gives you a chance to establish new edges if so desired. Make sure to repaint the edges throughout the renovation.

6. Using the walk-behind spray boom, spray the fairway and apron near the green. Make sure to hand spray out leaving enough room for the larger sprayer to make turns. Mark the side of the spray area with flags where you stop. This will indicate to the operator of the large sprayer where to shut off and restart the spray. Make certain to have two crew members doing this task; one spraying and one pulling the hose.

It is a good idea to have disposable shoe protectors that can be removed before leaving the fairway.

7. Spray the fairway making several passes across the tee end of the fairway and mark along the edge where you stop. This will give the operator a mark where to turn the sprayer off and restart at the turn. Start spraying at the green, on one side of the fairway spraying in rows working toward the opposite side of the fairway. Never spray in a circle. It is a good idea to use a tracker when spraying. Realizing that foam indicators can be used at each end of the boom, but without a tracker you must be certain you have no clogged nozzles. See Figure 13.1.

8. Be sure to wash the equipment before leaving the fairway. If your entrance gate onto the fairway can be placed close to a bunker, use the bunker as a wash station.

9. Wait until the next day (after dew has gone) before moving to the next step.

FIGURE 13.1 Spraying the fairway

FIGURE 13.2   Cleaning debris from the fairway

10. Using a reel mower or, if available, a flail mower, set as low as possible, mow over the area several times, scalping the turf.

11. Use a blower to remove debris to the sides of the fairway, where it can be vacuumed or swept. If you feel there is still too much grass remaining on the fairway, step 10 may be repeated.

12. The next step is aerating. There are many aerifiers on the market available to you. Core aerating is the most critical step in the renovation process. Aerify the fairway at least three times, bringing up as much soil as possible. The aerifier holes, along with the soil from the cores, will enhance seed-to-soil contact. The aerifier holes serve as a good seed bed, an avenue for new roots when the turf begins its reestablishment.

13. After the fairway has been aerated, the plugs need to be pulverized. This can be done by using a flail mower or vertical type mower.

14. Seeding is the next step. It is critical to research available seeds through your local university. There has been research done on

disease resistance and growth habits of available varieties. As a rule, it is not wise to use an aggressive variety as thatch buildup will become an expensive problem. Bentgrass seed rate should be 2# per 1000 square feet. It is recommended to slit seed ½# in two directions and ½# spread over the top of the fairway with a drop spreader, applying in two directions. When seeding bentgrass, it is also recommended to mix the seed with an organic carrier. I always used Milorganite®. Due to the small size of the seed, the carrier makes the seeders easier to calibrate and also helps with minor wind disturbances when they occur. The ratio for mixing the seed is 2 to 1 ratio, 2# of carrier to 1# of seed. Calibrate your seeders with the carrier at two pounds per 1000 square feet When that has been completed, mix your seed with the carrier at a 2 to 1 ratio. You are now ready to proceed. When slit seeding, be sure the seed has made soil contact but do not bury the seed too deep, no more than ¼ inch. If low mow bluegrass is your choice, the seed rate would be 2½ to 3# per 1000 sq ft. Germination will take two weeks, usually twice as long as bentgrass. Before you start to seed, check your irrigation system to see that all the irrigation heads are working properly.

15. With a steel mat, drag the fairway and then, with the tractor blower, go over the fairway quickly. If a small amount of debris is left, it will help maintain moisture for seed germination

16. Roll the fairway with a drum roller to insure better seed to soil contact.

17. Apply 1# of nitrogen per 1000 sq ft using an organic-based fertilizer or high-grade controlled release, a product that will give a steady release of nutrient well into the fall season.

18. Set your irrigation system so that the new seedlings are watered lightly every two to three hours throughout the day. The seed should be moist, not drowned. Do not irrigate throughout the nighttime hours. If the weather is extremely hot and humid, it is wise to have fungicide protection on hand.

FIGURE 13.3 Dragging renovated fairway after seeding

Within seven days you should have germination. It is conceivable to be mowing your fairway in less than two weeks. Be sure to mow the fairway with a lightweight fairway unit and make certain that the mower is sharp. The height of cut should be the height you are going to maintain on the fairway. For bentgrass, I would recommend the height of cut be at $1/2$ inch.

MAINTENANCE

How do we keep these fairways bentgrass or low mow blue grass? The best way is a sound maintenance program, proper fertility, and a good integrated pest management program. The safest way, chemically, is to use a preemergent herbicide; my choice is Dimension in the spring. Another option is Prograss. This product is sprayed in late fall and it is an herbicide for Poa Annua. When using Prograss, extreme care needs to be observed in calibrating and applying.

FIGURE 13.4 A finished fairway renovation

Overspraying can spell disaster. Before going forward, research is a must. Another option is the use of a growth regulator. Paclobutrazol, sold under the name TGR or Trimmitt, has shown good results when used properly in a good maintenance regime.

Bentgrass Fairway Care

Bentgrass fairway care is the same as the described care of Poa Annua in the following section except for the few differences:

1. Watering-Bentgrass needs less water. The fairways can be maintained firm and fast. They can be deeply watered and left to dry out. If the fairways wilt, they will respond to irrigation.

2. Disease-Bentgrass shares most of the disease problems that Poa Annua does except for Anthracnose and Summer Patch. The closer your golf course is to the transition zone, the bigger

the reason for a fairway conversion program. That's just because of the increased problems that Summer Patch and anthracnose cause in respect to turf loss and the large expense of treatment, usually I might add unsuccessfully. The one disease that is characteristic to new bentgrass planting is take-all patch. Take-all patch appears at first as a patch of bronzed or bleached turf. The grass in the center dies and is replaced by broad leaf weed or, many times, Poa Annua. The black runner hyphae of the fungus can be seen under the base of the leaf sheath, crown, and on the roots. The disease is favored by high pH. Many times the severity of the disease can be minimized with the use of acidic fertilizer such as ammonium sulfate. To preventively treat take-all patch, use azoxystrobin 30 days after first mowing and even equally or more importantly, use azoxystrobin in the fall when nighttime temperatures drop into the 50's.

Bluegrass Fairways

With the introduction of low mow bluegrass, you can still have tight lies and produce excellent fairways at a lower expense. Fertilize the bluegrass with about 3# of N per 1000 square feet with a least a pound and a half in the fall; use a balanced fertilizer with a good residual. Treat for leaf spot in the spring and dollar spot a couple of times in summer; use a broad spectrum fungicide. If you want great playing bluegrass fairways, mow them every day and use Primo® plant

---

TURFTIP

With the introduction of low mow bluegrass, you can still have tight lies and produce excellent fairways at a lower expense.

growth regulator. If you do not have low mow blue grass, mow at a height greater than one inch. Again, mow every day and use Primo®. If you apply 3# of N a season, you will still experience great playing surfaces even if your bluegrass fairways are not low mow.

Maintenance of Poa Annua Fairways at ½ inch

How do you maintain a Poa Annua fairway at ½ inch? There is a simple answer. Just maintain a healthy Poa Annua grass plant. If only this chore was easy, we as superintendents would have it made. Let's start down this slippery slope beginning in the fall. It is important to aerify the fairways in the fall; sometime in mid-September is perfect. This is generally the best time of the year for root development. Follow immediately with a fertilizer that has an extended release. This could be an organic base or a synthetic fertilizer with these characteristics. It is always a good idea to use a fertilizer that has an analysis of 2-1 potash to nitrogen. This high potassium fertilizer will strengthen the cell wall preparing the Poa Annua plant for winter. Some superintendents like to apply a dormant fertilizer also. If this is your plan, use a slow release fertilizer such as an all organic and be sure the grass is dormant.

Winter Damage

Any surge of growth before winter can lead to winter damage. The next step before winter is to water the fairways heavily before drain-

---

TURFTIP

Do not let the fairway go into the winter dry. It is always a good policy to treat a Poa Annua fairway for snow mold disease.

ing the irrigation system. Do not let the fairway go into the winter dry. It is always a good policy to treat a Poa Annua fairway for snow mold disease. There are two types of snow mold to be concerned about. Grey Snow Mold (Typhula Blight) and Pink Snow Mold (Microdochium Patch). Grey Snow Mold occurs at temperatures between 30 and 55 degrees F. As the snow melts, circular grayish or straw-colored to dark-brown infected centers appear in the turf. As the snow recedes in the spring, a grey mycelium can readily be seen. Grey Snow Mold is more of a problem when snow falls on unfrozen ground. Pink Snow Mold can occur without snow. It appears as a reddish-brown spot in the turf ranging from one inch to eight inches after the snow has melted. The pink mycelium of the fungus can be seen. I can name many fungicides that can take care of one or the other of these snow molds, but it is to your financial advantage to use one product that prevents both. PCNB is a fungicide labeled for both Grey and Pink Snow mold. PCNB is the most economical for use on fairways, but does have some down sides. PCNB

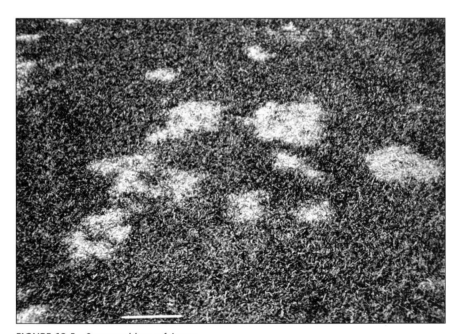

FIGURE 13.5 Snow mold on a fairway

will slow down the green up in the spring, especially when a winter without snow occurs. Do not reapply PCNB once you have put it down in the late fall. A second application may have a root pruning effect on the Poa Annua. There is a combination of products on the market with Chlorothalonil and Thiophanate that are safer to use than PCNB, but are more costly.

Ice cover can occur. This can happen on fairways, tees, and greens. In many cases, attempting to remove the ice does more harm than the ice itself. The rule of thumb for fairways, tees, and greens is that Poa Annua will lose turf if under solid ice for more than 45 days; 90 days for bent grass. If you remove a portion of the ice and it has an odor like that of rotten eggs, in most cases you will lose grass. If you fear solid ice for more than 45 days, crack the ice, if you can, to let in oxygen. That should be sufficient to save the Poa Annua.

Spring

When spring arrives, your next challenge with the Poa Annua is seed heads. From the prospect of playability, seed heads on Poa Annua are not as big an issue as they are on greens. From the standpoint of the health of the plant, it is to your advantage to control the formation of seed heads. The use of growth regulators has been the choice of most. A combination of Proxy™ and Primo® seems to be the first choice as of this writing. Embark® and Embark Lite® also have been used for many years. I have heard of superintendents using large amounts of Aquatrol wetting agents with success. If you want to control seed heads on Poa Annua, it is important to research the topic before making your final decision. You're the one that has to live with the decision, not the superintendent next door. Do your homework.

Mowing

Without question, when mowing Poa Annua fairways, it is to your advantage to use lightweight mowing equipment on fairways and

> **TURFTIP**
>
> When spring arrives, the challenge with Poa Annua is seed heads. From the prospect of playability, seed heads on Poa Annua are not as big an issue as they are on greens. From the standpoint of the health of the plant, it is to your advantage to control the formation of seed heads.

remove the grass clippings. Leaving grass clippings creates an environment that encourages turf disease. Grass clippings draw heat, and this is only going to make your job more difficult. Removing clippings will, throughout the year, give the golfers a better playing surface and really is this not "Job One"? If you start out in the spring removing clippings, you will be removing seed heads as well. What do we do with the clippings? In early years, removing grass clippings from the golf course was a problem, I will admit. However, with the wide spread use of growth regulators, all you need to do is instruct your fairway mower operators to dispose of the clippings in the primary rough. I have heard of concerns about Poa Annua seeds in the rough; you should be mowing the rough at a height that Poa Annua

> **TURFTIP**
>
> Without question, when mowing Poa Annua fairways, it is to your advantage to use lightweight mowing equipment on fairways and remove the grass clippings. Leaving grass clippings creates an environment that encourages turf disease.

FIGURE 13.6 Burned-in fairway striping

invasion should not be a concern. What is the practical definition of a lightweight mower? If you have all the money in the world, you could walk mow fairways. In fact, that was done in a major championship. The five unit mowers, manufactured by Jacobsen, Toro, John Deere, and Ransom, are a practical solution to lightweight mowing. These units give superintendents the ability to collect clippings. These mowers allow the operator to change mowing direction, which helps against grain formation and gives the operator the ability to mow patterns. Wherever the fairway begins to narrow at the green, usually at the first fairway irrigation head, you should use a triplex greens mower with mowing height set the same height as the fairways. I know superintendents who mow these areas lower than the fairway and claim that it is easer to keep the turf alive. Many times this is not the wish of the golfer and many of the medium to high handicappers have a difficult time playing the tight lie. Before lowering the approach mower lower than the fairway, it best to discuss the change with the green committee or owners. If you have par three fairways, also mow the fairway with the triplex mower. What are collars at the green, fairway, or an extension of the greens? Collars are the most difficult area to keep alive. Collars are usually 30 inches wide and always heavily compacted from both the turning of greens mowers and the apron mower itself. The apron mower usually is a walking greens mower. These areas should be on a greens program, except they require aerating at least once a month and always removing the cores.

Spring Aerification of Poa Annua Fairways

My experience over the years is to eliminate aeration in the spring except for the approaches to the green and collars starting at the first irrigation head out. Use a walk-behind aerifier with $1/2$ inch to $3/4$ inch coring spoons and remove the cores, do not grind the cores. Grinding cores will only put added stress on the Poa Annua.

Herbicide Use on Poa Annua Fairways

If you have broad leaf weeds, spray them in the fall before aerating. Herbicides on Poa Annua put undo stress on the plants and, when the hot humid weather comes, it is a larger factor than you realize. If you need to spray for clover, use 6 ounces per acre of Dicamba (Banvel D®). You can usually add this to your fairway fungicide spray, jar test,

---

TURFTIP

If you have broad leaf weeds, spray them in the fall before aerating. Herbicides on Poa Annua put undo stress on the plants and, when the hot humid weather comes, it is a larger factor than you realize.

> **TURFTIP**
>
> My number one choice for controlling pre-emergent crabgrass is Dimension on both Poa Annua and bentgrass fairways.

and spray a test area before proceeding. Repeat application if necessary in two weeks, though I highly doubt that will be necessary.

My number one choice for controlling pre-emergent crabgrass is Dimension on both Poa Annua and bentgrass fairways. There is a relatively new herbicide for crabgrass (Drive®) that has performed well. Be sure to read the label and it is always wise to spray herbicides by themselves unless otherwise stated on the label.

Watering of Poa Annua Fairways

It has always been my experience to maintain a consistent moisture level. You can water deeply and let the fairway dry out on bentgrass and bluegrass. But, this is not an option with Poa Annua. By no means am I advocating overwatering. Overwatering only leads to short roots and disease pressure. Just simply put back the amount of water you lose each night. Many of the newer irrigation systems have evapo-transpiration monitoring capabilities. If your irrigation system does not, you can get this information off the Internet at www.weather.com. The use of wetting agents has become more commonplace on fairways. If your course has heavy soils and you find yourself hand watering and fighting isolated dry spots continually, it may be in your best interest to consider the use of an extended life wetting agent that will be effective for 90 days or more. You may very well offset the cost of the wetting agent application in saved labor cost. Injecting wetting agents into the water system is also an option. The results hinge on how well your water system operates and how evenly water is distributed. The use

> **TURFTIP**
>
> Golf carts are the number one reason for loss of Poa Annua around the green surfaces. You need to keep the carts away from the green approaches.

of a wetting agent will enable you to conserve water by utilizing water more efficiently.

Damage to Poa Annua

Golf carts are the number one reason for loss of Poa Annua around the green surfaces. You need to keep the carts away from the green approaches. Superintendents need to change traffic controls on a regular basis and, in hot periods, daily. I knew of a superintendent in Chicago who took carts off the golf course when the daily high was forecasted at 105 degrees. He was one of the very few superintendents who did not lose turf that day. As superintendent, you need to communicate the importance in following cart regulations to your members in each club newsletter.

Top four diseases associated with Poa Annua:

- *Anthracnose*
- *Dollar Spot*
- *Pythium blight*
- *Summer Patch*

Diseases that Poa Annua are susceptible to are many, but the diseases to be most concerned about are four in number: Anthracnose, Dollar Spot, Pythium blight and Summer Patch. Good integrated turf management can control the disease with the least amount of chemical being used and using as many cultural tools as available. Usually if you control these four diseases, the broad spectrum properties of the fungicide recommended will prevent the balance of the remaining disease affecting Poa Annua. A description of the four diseases follows.

Anthracnose (Colletotrichum graminicola)

Anthracnose is characterized by irregularly shaped patches of yellow-bronze turf ranging in size from a few inches to several feet. Leaf lesion initially appears as elongated reddish brown spots on the leaves. Numerous black fruiting bodies (acervuli) of the fungus can be seen on the foliage as the disease progresses. The acervuli have black spines protruding from them. According to Dr. Joe Vargas, retired entomologist at Michigan State University, Anthracnose can be managed with light applications of nitrogen ($1/2$ lbs N per m) every three weeks in cool weather. In the summer, a light application of Nitrogen can help, but fungicide treatment will also be needed to prevent severe turf loss. Anthracnose can be managed with systemic fungicides like benomyl, fenarimol, propiconazole, thiophanate methyl, azoxystrobin, and triadimefon, every two to three weeks, or with a contact fungicide like chlorothalonil, mancozeb, or maneb plus zinc sulfate, applied every seven to ten days.

Control Measures For Anthracnose

- *Benomyl*
- *Fenarimol*
- *Propiconazole*
- *Thiophanate methyl*
- *Azoxystrobin*
- *Triadimefon*

- *Chlorothalonil*
- *Mancozeb*
- *Maneb plus zinc sulfate*

Dollar Spot (Sclerotinia homoeocarpa)

Dollar Spot is characterized by round, bleached-out or straw-colored spots ranging in size of a quarter to the size of a silver dollar. The spots appear as sunken areas in the turf, especially in turf mowed at $1/2$ inch or shorter. Individual spots may coalesce and destroy the turf in large undefined areas. If the fresh spots are observed in the morning while the grass is still moist, the grayish-white mycelium of the fungus can be seen. Dollar spot can be controlled with vinclozolin, propiconazole, chlorothalonil, or piprodione.

Pythium Blight

The first symptoms of Pythium Blight are circular reddish-brown spots on turf ranging in diameter from one to six inches. In the morning dew, infected leaf blades appear water-soaked, dark, and may

---

TURFTIP

Dollar spot can be controlled with:

- Vinclozolin
- Propiconazole
- Chlorothalonil
- Piprodione

feel slimy as the blades dry, shrivel, and turn reddish brown. In the morning, you can easily see the active mycelium of the fungus, a purplish gray or white cottony appearance on the outer margins of the spots. Because pythium can readily be spread by mowers, it is important to instruct your crew, particularly your mower operator, of the symptoms. When nighttime temperatures remain above 70 degrees F, it is important to check the golf course for pythium. Start in the low and poorly drained areas. If you spot pythium, do not mow; apply fungicide. Use a contact in conjunction with a systemic. A good contact to use is Chloroneb. There are many good systemics available such as mefenoxam, Banol™, and Aliette.

The safest approach is to monitor the weather and spray the systemic on a preventive basis. In some climatic areas, just treating the high risk areas will be enough. Many golf courses closer to the transition zone may require spraying of all the fairways. It may take time for the systemic to enter into the plant because this may vary with the overall health of the plant. That's the primary reason to apply systemic fungicides as a preventive.

Summer Patch (Magnaporthe poae)

Summer Patch is the most difficult of all the diseases to control on Poa Annua. The disease first appears in the warm weather of summer as an irregularly shaped patch, yellow to bronze in color, and ranging in size from 6 inches to 3 feet in diameter. Examining of roots with a dissecting scope will reveal dark-colored ectotrophic called hyphae. Summer Patch first appears in the warm weather

---

TURFTIP

If you spot pythium, do not mow; apply fungicide.

> ## TURFTIP
>
> Summer Patch is the most difficult of all the diseases to control on Poa Annua. The disease first appears in the warm weather of summer as an irregularly shaped patch, yellow to bronze in color, and ranging in size from 6 inches to 3 feet in diameter.

of the summer usually following a rainy period or heavy downpour. Even though symptoms occur during warm weather, the initial infestation occurs in the spring when soil temperature reaches 65 degrees F at a 2-inch depth for at least 48 hours. Only the outer cortical tissue of the root is infected at this time. Low fertility only adds to the severity of Summer Patch. A slow release nitrogen source $1/2$ lb N every three weeks will aid in the battle against this disease. When soil temperatures reaches 65 degrees F, spray the fairways with DMI fungicide or azoxystrobins. When the summer heat lasts for more than three months, a second spraying on fairways within 30 days of the initial application is necessary. Once the disease symptoms are present, thiophanate methyl manages Summer Patch the best. A second application three weeks later is usually required. The thiophanate methyl needs to be drenched into the soil to be effective. According to Dr. Joe Vargas, "Unless adequate levels of nitrogen are also applied, fungicide treatment will not be effective." It may be to your benefit to include thiophanate methyl in your preventive fungicide fairway program when managing Poa Annua fairways.

Insect Problems

Insect problems on fairways need to be monitored closely. People in the east, with Poa Annua fairways, face the added problem of the annual bluegrass weevil (Hyperodes weevil). Turf damage has been localized in the eastern United States as far west as Pennsylvania and south into West Virginia. Adults are small 1/8 to 3/16 inches long. Black weevils with wing covers are coated with fine yellow hair, yellowish scale, and scattered spots of grayish-white scales. Newly emerged adults are orange-brown in color and require several days before coming fully black-pigmented. The larva is crescent-shaped, legless, and has a creamy white body. According to Dr Harry Niemczyk, "On golf courses where grubs and annual bluegrass weevil are major targets, a combination of imidacloprid (Merit) plus a pyrethroid insecticide applied from mid to late April prevents damage from the first and second generation annual bluegrass weevil larvae. This treatment should also prevent larval infestations of bill bug, black turfgrass ataenius beetle, Japanese beetle, masked chafer, European chafer, and first generation cut worm. On golf courses where the grub species are not major targets, single application of the labeled pyrethroid insecticide during the third week of April should prevent damage from annual bluegrass weevil larvae."

Black turfgrass ataenius (BTA) beetle infestations have occurred in most of the states where cool-season grasses are grown. The adult BTA is a shiny black beetle 3/16 inch long and 1/8 inch wide. The thorax has small pits scattered over the top surface and wing covers have distinct longitudinal grooves. BTA are known to feed in the roots of annual bluegrass, bentgrass, and Kentucky blue grass. First symptoms of injury appear in mid-June to mid-July; turf shows sign of wilt similar to isolated dry spots. The C-shape white grub is very small, but the third instars can be separated from the other grubs using a 10X hand lens. The tip of the BTA abdomen has two distinct white anal pads and the few raster bristles are scattered randomly. The biggest mistake I see with using systemic insecticides is using them too soon. The time the adults are laying their eggs is when the Vanhoutte spirea is into full bloom. Wait two to three weeks after the Vanhoutte spirea is into full bloom, then apply your insecticide. You will not only kill the BTA, you should have season-long protection against Japanese beetle grubs.

Chapter 14

The Teeing Ground

THE TEEING GROUND SURFACE

The first thing that you see at the beginning of the round on any golf course is the first tee. If is all "beat up" with bare spots, divots, not level, and just poorly presented, you are not going to have a very favorable first impression. I do not believe that it has to cost a lot of money to have the teeing ground presentable. The teeing ground should be unobtrusive, have good turf, be level and smooth, and point you in the right direction. Things that are very costly and not needed are signposts with the hole's data, flower and/or shrub beds, steps, rock walls or railroad ties, and expensive furniture or accessories. The teeing ground surface needs to be large enough to facilitate the production of a high quality playing surface for the amount of rounds of golf that will be played. The tee site should be unobtrusive and blend into the surrounding landscape.

MAINTENANCE AND CULTURAL PRACTICES

The teeing ground is maintained today much like putting surfaces in the 1950's and 1960's. Tees at medium to high-end public golf

> **TURFTIP**
>
> Mowing in the same direction every time or "burning in stripes" is not an acceptable practice.

courses and most private clubs in the northern half of the US have bentgrass or a mixture of bentgrass and Poa annua turf. Like the putting surface, the teeing ground surface needs to be firm and not soft and spongy.

Mowing

In 2001, I discovered that the turf on Poa/bentgrass tees (and collars) performed significantly better when mowed at 0.25 inch, than it did at 0.33 inch. We used the Toro®-1600, which is 26 inches wide and equipped with an aluminum-grooved roller. Ideally, Poa/bentgrass tees should be mowed at least every other day and daily during special events like the member-guest invitational. Like greens and fairways, the direction of cut should be altered in at least three different directions. Mowing in the same direction every time or "burning in stripes" is not an acceptable practice.

> **TURFTIP**
>
> Sprinkler heads should be located off the teeing ground surface.

Irrigation

Just like the putting surface, the teeing ground surface must be firm and not soft. A firm surface will hold up to traffic, wear, and tear much better than a soft playing surface. Only enough irrigation water should be applied to keep the turf alive. Infrequent deep watering is the preferred irrigation practice that allows the surface to be dry most of the time. If possible, sprinkler heads should be located off the teeing ground surface. This will eliminate problems of sand getting into the drive assembly, constantly leveling heads, and breaking mower blades from hitting the sprinkler head.

Nutrition

To produce high quality turf, the chemistry should be balanced according to the results of soil tests. To maintain bentgrass, a 4:1:4 to 4:1:8 ratio of NPK (N: nitrogen, P: phosphorus, K: potassium) should be maintained. In order to get divots to recover quickly, around four pounds (#) of nitrogen should be applied per year to maintain a mixture of bentgrass and annual bluegrass in the upper Midwestern US. A dormant application of 1.5# N is applied in late November to early December of which every ton was formulated with 1,400# sulfate of ammonia (21-0-0) + 150# mono-ammonium phosphate (11-52-0) + 600# sulfate of potash (0-0-50). The analysis of this formulation is 14-4-15 and it is applied at the rate of 470# per acre to furnish 1.5# N, 0.4#P, and 1.6#K.

The buffering of the irrigation water with pHairway® (a mixture of sulfuric acid and urea) supplied 1.5 to 1.8# of Nitrogen for the year. This will vary from year to year depending on irrigation needs and how low you set the pH of the treated water.

White Gold (0-0-24 + calcium) is applied monthly at 100# per acre. Sulfate of ammonia (sprayable) and liquid potassium thiosulphate (0-0-25) are included in the fungicide tank mixes for spoon-feeding 2 to 4# of N & K per acre. Manganese sulfate and iron sulfate are also added to the tank mix at 5# per acre. Mono-phosphoric potassium (MPK 0-50-30) is included in the tank mix at 5# per acre for added disease suppression.

Cultivation and Topdressing.

Core aerification is done in the fall of the year. Until 2001, we used ³/₄ inch hollow tines, broke up the aerifier plugs, dragged the soil back into the turf, and removed the debris that remained on the surface. This method worked well for many years.

In 2001, we began topdressing tees and approaches with straight 100% sand. This came about by observing how much firmer the approach was in the area adjacent to the green where sand was regularly applied. Several top golf courses have initiated sand topdressing programs on tees, approaches, and even fairways. It definitely will help provide a firmer playing surface. Aerifier cores must be removed so that the underlying heavier soil does not seal off the surface. The basic rule of physical chemistry is never put a finer textured material over the top of a coarser material. In other words, do not cover sand with soil.

The golf course superintendents that I have talked with are applying sand topdressing on tees at four to eight week intervals. Rates

TURFTIP

The basic rule of physical chemistry is never put a finer textured material over the top of a coarser material. In other words, do not cover sand with soil.

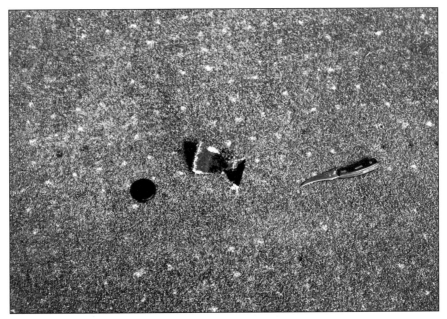
FIGURE 14.1 Aerifying and topdressing

vary from 0.10-to 0.25-inch per application. It is my guess that the day will come where tees will be topdressed at the same frequency as greens.

It is debatable whether divot mix or green sand in bottles or buckets that are located on the teeing ground for golfers is actually an asset or a liability. I believe that medium to higher end courses might be far better off to send personnel out to fill divots than let the players do it. I have witnessed on a Monday morning following a busy Sunday, that there were about one-half of the divots filled with green sand. Many of them were over-filled, which caused damage to the mower blades. The same thing is true when players fill divots in fairways from the sand or mix that is carried on the golf car. Divots that are filled by the golf course maintenance staff are more likely to be done correctly, if they were trained properly.

Pest Control

Disease is the main pest problem on tees. It is my belief that if you want to manage bentgrass, it is best to use contact fungicides, not the sterol-inhibitor materials. If you have a significant amount of Poa annua in the turf stand, it may serve you best to use the sterol-inhibitors.

My disease control program included chlorothalonil (ChloroStar or Daconil-Ultrex) + vinclozolin (Curalan) + thiophanate (SysTec or 3336) + propiconazole (Banner®) @ 2.0 + 0.4 + 0.4 + 0.2 oz per M ft2. This 4-way mixture applied every 14 days provided excellent control. Thiophanate was included in the tank-mix to help suppress earthworm activity.

The PGR, Primo®, was also included in the tank-mix at rates from 0.10 oz per M ft2 in the early part of the season, then we increase the rate each spray application by 0.05 oz until we reach 0.5. Top growth rate is kept manageable and Primo® does not reduce the recuperative ability of divots. It may even speed up the recovery.

We have had very few insect problems on both tees and fairways. On a few occasions we had to apply an insecticide for cutworm activity because the birds were creating an unsightly mess from digging up the larvae. We have been doing everything we can to address

TURFTIP

Practice tees at golf courses receive the most complaints after bunkers.

member complaints about earthworm activity using methods such as sand topdressing and fungicides that contain thiophanate.

Other than being level, why do we need to maintain the teeing ground so intensely? Some tees at certain courses are mowed lower today than putting surfaces were 30 years ago. How high is the bar going to go?

PRACTICE TEE

Practice tees at golf courses receive the most complaints after bunkers. In new course design, I am sure a good golf course architect takes into consideration the location of the practice range. It should be located close enough to the clubhouse and pro shop to be convenient but far enough not to interfere with routine activities. When planning a location for the range, make sure that the range tee faces either north or south so the golfer is not facing into the morning or afternoon sun. I have seen a practice tee under the dining room, not a pretty sight in August. I have also seen practice areas located ten minutes drive by golf cart to the first tee. I can safely say that in 95% of the golfing facilities across the country, the practice tee is not adequate in size. Many times the ranges are not deep enough and/or not wide enough; some are located on water and use floating golf balls. Some ranges use Caymen balls that only travel half the distance of a regular golf ball.

The superintendent knows that the practice range needs more attention than other areas on the golf course, but is given little time to work because the golfers are using the range from early morning to dusk. Whether the range tee is bluegrass, bentgrass, rye grass, or Bermuda, these species are never adequate for the wear the tee receives. In the northern climates, most practice tees are invaded by poa annua that germinates in the void left by the divot. To use a preemergent herbicide would only prevent over seeding from germinating and the golfers' club head striking the ground would continually disturb the chemical barrier.

Grasses suitable for practice tees:

- *Bluegrass*
- *Bentgrass*
- *Rye grass*
- *Bermuda grass*

The answer in most cases is to put in place some artificial playing surface to be used during spring and fall days when there is extreme moisture. It is important to build drainage into the practice tee and playing field to correct any drainage problem that may occur on established facilities. By having good drainage, you will not only have an easier time managing the grass, but the range can be used sooner after a heavy rain. Most golfers want the practice tee to mirror the fairway turf on the golf course, so if you have bentgrass fairways, you should have a bentgrass practice tee, bluegrass fairways-a bluegrass tee, and Bermuda-a Bermuda tee. If using bluegrass, over seeding divots with rye grass is a sensible option.

The practice field can be as elaborate as the amount of money you have spend on it. Many practice ranges have target greens. The size of the green should be size proportioned to the distance from the center of the practice tee; in other words, the 150-yard target green can be 3000 square feet and a 250-yard target green can be 5000 square feet. Many target greens are built with bunkers. Bunker care adds extra expense and if you can do without them, you will save money and many labor hours. I have seen some bunkers lined with white road pack to give the appearance of sand. I believe aerification is a must once a year to relieve the compaction created by the ball picker and at least 1# of N annually on the practice field is a must. Most practice fields are surrounded by a fence to keep the golf balls inside. This fence needs to be maintained on a regular basis. Weeds and grasses need to be kept away from the base of the fence. This can be accomplished by using a non-selective herbicide.

FIGURE 14.2 Fertilizing the tee

The practice tee should be fertilized on the same program as your playing tees. The divots need to be filled every morning with a seed-sand divot mix that contains an amendment that raises the cation exchange and water-holding capacity of the mix. The hitting station should begin in the front of the tee, working towards the back throughout the season.

Many golf facilities have a short game area where a golfer can practice every shot that may face them on the course. If it is at all possible, a golf course architect should be consulted in the design. When a good short game area is incorporated in your practice facility regimen, it is a great asset to the golfing program.

Chapter 15
Topdressing

INTRODUCTION

Like the old saying goes, "Topdressing is the turf industry's castor oil, it is awful to take but it certainly does a lot of good things for us." (Figure 15.1). It is the heart and soul of high quality putting surfaces. In the mid 1970's, the use of straight 100% sand became popular in Wisconsin. In 1974, we had a Golf Turf Symposium in Milwaukee where the keynote speaker was Dr. John Madison, University of California at Davis, who presented an alternate method of maintaining putting green surfaces that had been researched for several years (Figure 15.2). This method was the use of frequent and light applications (3 cubic feet/1000 square feet) of straight sand for topdressing on putting greens. The results were faster, truer, more consistent, and smoother putting surfaces. We were called "mavericks" for jumping on this new method of maintenance back then, and it took many more years before other areas of the country accepted the use of straight 100% sand for topdressing. Now, it is the standard on most golf courses in the US.

FIGURE 15.1 25 years of topdressing common Bermuda grass.

FIGURE 15.2 Dr. John Madison—this is where it began.

EFFECTIVE SAND TOPDRESSING PROGRAMS

Requirements for an effective sand topdressing program:

1. Selecting the proper sand

2. Determining the appropriate rate of application

3. Adjusting the frequency of topdressing to match the rate of growth

The sand that was selected in the Milwaukee area was called Lakeshore sand. My friend, Joe Cannestra, superintendent of Brown Deer Park Golf Course, Milwaukee, WI, found a pile of this sand in the parking lot. We were collecting sands from various pits and giving the samples to Jim Latham, an agronomist with Milorganite, who took them to the laboratory where a physical analysis was done. This Lakeshore sand was very uniform and on the fine side (90% of the particle distribution was between 0.14mm-0.40 mm.) Most of the experts told us that this was too fine and we should find coarser sand. The rest is history. We went with this fine dune sand from western Michigan for the next twenty years.

In 1975, per Dr. John Madison's research in California, we aerified the 5th green at Ozaukee CC with $^5/_8$ inch hollow-tines, removed the cores, and backfilled the holes with straight 100% sand. Seven applications of sand topdressing were made at three-week intervals for the rest of the year. This test green was faster, firmer, smoother, and truer than any green on the golf course. In 1976, we began the sand topdressing program on all the greens at Ozaukee.

In 1983, at the annual spring meeting of the Chicago District Golf Association at Butterfield Country Club, I each gave a presentation on sand topdressing of greens. There must have been at least 300 people in the audience that included club officials, golf professionals, and superintendents. I talked about the positive things that we had experienced with sand topdressing. One participant insisted that he wanted to have the closing statement about the subject. His opinion was that there are too many locker room agronomists and they should leave the driving to us. He also believed that sand topdressing on greens was just a quick fix that may turn into a time bomb down the road. At that time, most of the Chicago area golf courses were still using a mixture of sand, soil, and peat called "silt dressing".

Originally the rate of application was 3 cu.ft. per mft^2 per application every three weeks. Today, because the putting surface is so tight knit and there is only room for a small amount of material, we are only applying <0.5 cu. ft./mft^2 per application at two-week intervals in the spring and three-week intervals in summer.

BENEFITS OF TOPDRESSING PUTTING GREEN SURFACES

Putting surface smoothness is a result of topdressing, no matter what material is used. Before the mid-1970's, topdressing was not a regular practice. In fact, it was done only in conjunction with aerification. Specifically, the benefits include:

- *Smoothes the putting surface*
- *Helps decompose thatch*
- *Reduces graininess*
- *Produces a denser and finer textured turf*
- *Promotes recovery from injury*

TURFTIP

Putting surface smoothness is a result of topdressing, no matter what material is used.

> **TURFTIP**
>
> On new establishment of putting surfaces, it is particularly important to apply topdressing in order to obtain a smooth surface. A tighter and denser turf is produced with frequent applications of topdressing. Greens that are not regularly top dressed have turf that is looser with less density.

- *Modifies the surface soil*
- *Aids in overseeding and renovation*
- *Protects crown of the plant from traffic and winter injury*

On new establishment of putting surfaces, it is particularly important to apply topdressing in order to obtain a smooth surface. A tighter and denser turf is produced with frequent applications of topdressing. Greens that are not regularly top dressed have turf that is looser with less density. Biological or mechanical turf injury recovers more rapidly when topdressing is applied.

> **TURFTIP**
>
> For over seeding and/or renovation of existing greens, topdressing will help provide contact between the seed and the growing medium.

> **TURFTIP**
>
> Topdressing will protect the crowns of turfgrass putting surfaces from severe weather conditions.

The intermingling of topdressing sand into the biomass enhances the decomposition of thatch. This process is best accomplished by frequent applications that will not allow thatch to accumulate. Along with frequent topdressing applications, brushing and/or vertical mowing will help to prevent graininess. Again, frequency is the key.

As an example, the area between rolls of sod will heal more quickly when topdressing is used to fill the joints. Push-up greens that are soil-based cannot tolerate traffic and wear and tear like sand-based greens can. Modifying with straight sand topdressing after incorporating it into the underlying soil has helped produce a playing surface that holds up better to traffic, especially during periods of wet weather. For over seeding and/or renovation of existing greens, topdressing will help provide contact between the seed and the growing medium. This can be best obtained either in holes and/or grooves created with vertical blades.

Topdressing will protect the crowns of turfgrass putting surfaces from severe weather conditions. In 1999, we changed from the Lakeshore dune sand to a material that Waupaca Materials-Greens mix put together for us called "fines free." In 2001, they screened out a fraction of the coarse material and created a material called "coarse free and fines free" which is dry and easier to incorporate into the turf canopy. Although it is not as fine as the dune sand was, we now have a topdressing material that is working. See Figure 15.3.

FIGURE 15.3 Third year of sand topdressing

INCORPORATING SAND INTO THE TURF CANOPY

There are many more advantages than disadvantages of using straight sand for topdressing putting surfaces. One disadvantage is that the sand is very abrasive and care must be taken not to drag too much when it's hot because it will dull the mower blades.

TURFTIP

There are many more advantages than disadvantages of using straight sand for topdressing putting surfaces. One disadvantage is that the sand is very abrasive and care must be taken not to drag too much when it's hot because it will dull the mower blades.

The vertical mower is an excellent tool to use to incorporate the sand into a very dense, tight turf canopy. We have tried every kind of brush that can be imagined, but there is no place for the topdressing sand to go, so it lays on top of the putting surface waiting for the mower to pick it up and dull the blades. After 30 years or more of not using a vertical mower, we are doing it again set at $1/8$ inch deep, double vertical mowed on the same pass. When you make a pass across the green, you turn around and come back on that same pass in the opposite direction. We have also done this with the mower groomers and with greens rollers.

TURFTIP

Intervals between topdressings may vary from two to five weeks depending on the rate of growth. I see turf managers avoid the application of topdressing because it causes mowers to become dull, therefore creating slower greens.

Over 40 years ago, Dr. Joseph Duich advocated double vertical mowing by turning around and coming back on that same path. Grass usually grows down hill and you will not be cutting any grain when you travel down hill. In order to cut any grain, you must travel uphill or into the grain. Vertical mowing, quadra-tining, spiking, or "fine tining" with the AerWay can provide some place for the topdressing to intermingle into the putting surface. This will also help to break up any paper-thin layers that have been created from the heavy topdressing applications made to backfill aerifier holes.

Intervals between topdressings may vary from two to five weeks depending on the rate of growth. I see turf managers avoid the application of topdressing because it causes mowers to become dull, therefore creating slower greens. It is a "Catch 22" situation where "you're damned if you do and damned if don't." The best greens that I've ever seen are frequently top dressed.

To reduce the problem of dulling the mower blades, topdressing sand should be brushed into the putting surface after vertical mowing and/or spiking and, if possible, irrigated that evening. If long extended wet periods prevail, sand should be applied at lighter rates and the frequency should be increased to one- to two-week intervals. More brushing and vertical mowing will be needed when intervals are

TURFTIP

To reduce the problem of dulling the mower blades, topdressing sand should be brushed into the putting surface after vertical mowing and/or spiking and, if possible, irrigated that evening.

less frequent in order to get the topdressing incorporated into the putting surface turf. At one-week intervals, there are a number of superintendents applying dry topdressing sand with push fertilizer spreaders or small pull-behind spreaders. Some are applying sand even lighter at two times per week to reduce mower dulling. Turf managers must experiment and find the method that will work best at their facility.

Chapter 16
Soil and Water Testing

SOIL

The function of soil is to provide a secure anchorage for plants and to provide all the essential nutrients necessary to promote and sustain plant and microbial life. From productive soil, there comes self-supporting energy in the form of living green plants. Plant life alone has the ability to accumulate and store chemical energy that becomes the basic energy and food supply for sustenance of all other forms of life. Carbon, the main elemental substance of plants, comes from carbon dioxide, which plants breathe through their leaves from the air. The plants draw in water and nutrients through their roots from the soil and circulate them upward through their structure. Much of the water taken evaporates while keeping the leaves and interior surfaces moist for gaseous exchange with the atmosphere.

To produce high quality turf you must maintain a balanced soil, have a supply of nutrients that are available to the plants, and have good structure that will allow deep rooting. The chemical and physical nature of the content of the rootzone (soil) is the basis of what the quality of the end product is going to be.

Soil as a Source of Plant Food

Soils differ in their total content of nutrients and in the quantity available for use by plants. Seldom do they contain enough available nutrients to meet growth requirements of turfgrasses. Additional amounts must be supplemented by the application of fertilizer. Intelligent selection and use of fertilizer requires an understanding of the soil as a source of plant-food materials.

Deficiencies of plant nutrients may be due to shortages in the total quantity present in the soil. Deficiencies also may be due to the fact that nutrients are in forms so insoluble that the plant cannot obtain enough to supply its growth requirements. Some soils have a high fixing power. When liberal quantities of fertilizer are added to such soils, very little of it may be available for use by plants in a relatively short time after the application. The nutrients have been changed to insoluble compounds or are held so tightly by the soil particles that their rate of solution is too slow to supply adequate plant needs. Good management practices will adjust fertilizer applications and soil conditions to maximize availability of nutrients.

Nutrient deficiencies may develop when excessive amounts of one or more elements are present in a soil and the quantity of others is limited. Under such conditions the balance which nature attempts to maintain in the available supplies of various materials is upset. Those present in limited quantities are thrown out of the solution entirely because of the excess of the others. For example, in highly alkaline soils having an excess of basic calcium and magnesium, elements such as potassium, manganese, iron, and boron may be unavailable even though the soil contains sufficient quantities to supply the needs of grass under normal conditions. There are several practical remedies. Frequent light applications of the deficient materials dissolved in water can be made, so the grass has a chance to get them before they become unavailable.

The four objectives of soil testing are:
- *Determine the available nutrient status of the soil*
- *A basis for fertilizer needs*

> **TURFTIP**
>
> When turf shows evidence of different growth characteristics in certain areas, separate samples should be taken from those areas.

- *Determine the seriousness of deficiency or excess*
- *Economic evaluation of fertilizer recommendations*

Soil Sampling

The most accurate test is of little value if the sample is not representative of all the soil in the area. Soil on greens and tees is usually uniform. One composite sample consisting of 10 to 12 plugs taken at many locations is enough. When turf shows evidence of different growth characteristics in certain areas, separate samples should be taken from those areas. Sampling depth is very important. Plugs should be taken to an exact and uniform depth from below the surface as recommended by the laboratory. Thatch or turf should not be included in the sample.

Use only a reputable laboratory. Results are of practical value only when the person interpreting them knows the soil, has had fertilizer experience with turf, and thoroughly understands the problems of its maintenance.

The soil analysis audit and inventory report should include:

- *The sample location*
- *Total exchange capacity*
- *Soil pH*

- *Organic matter content (humus) %*
- *Estimated nitrogen release—lb/acre*
- *Anions—soluble sulfur and phosphorous*
- *Exchangeable cations—calcium, magnesium, potassium & sodium*
- *Extractable minor elements—boron, iron, manganese, copper, zinc, and aluminum*

Besides a soil analysis, plant-tissue tests determine the quantity of raw plant-food materials in the conducting tissues of the plant. Tissue tests are designed to determine whether nutrients are actually getting into the plant. A tissue testing method should take into account the balance of nutrients within the plant, calculating the ratios of all nutrient levels within the plant, then comparing ratios between healthy and unhealthy plants.

IRRIGATION WATER QUALITY

Again, like soil sampling, a representative sample is a must. Samples from ponds should be collected when they are in use for irrigation. It may be necessary to collect several samples during the season to correlate evaporation and dilution. Well water should be collected after the well has been pumped for a period of one hour. See Table 16.1.

Three principal methods are used to express the concentrations of constituents in water. They are:

1. Grains per gallon: the English system of units. It is used to report hardness of water. To change grains-per-gallon to parts per million (ppm), multiply grain by 17.2.

2. Parts per million (ppm): defined as 1 part of a salt to a million parts of water or 1 milligram of salt per kilogram of solution. Water analysis is usually reported on a volume basis. 1 liter (1.057 qt.) of water weighs 1 kilogram (1,000 grams or 2.2

TABLE 16.1 A standard irrigation water test package

Parameter	Desirable Range	
Hardness	grains/gal	<10 gr/gal
Electrical Conductivity	EC	0–1.5 mmhos
Sodium	Na	0–50 mg/L
Salt Concentration	TDS	<200 ppm
Sodium Absorption Ratio	SAR or	0–4
Adjusted SAR	SAR Adj	0–7
pHc	8.4	
Residual Sodium Carbonate	RSC	0–1.25 meq/L
Calcium	CA	40–120 mg/L
Magnesium	Mg	6–24 mg/L
Potassium	K	0.5–10 mg/L
Sodium	Na	–50 mg/L
Iron	Fe	2–5 mg/L
Total Alkalinity	as $CaCO_3$	1–100 mg/L
Bicarbonate	HCO_3	0–120 mg/L
Chloride	Cl	0–140 mg/L
Sulfur as	SO_4	0–414 mg/L
Boron	B	<0.33 mg/L
Manganese	Mn	<0.2 mg/L
Copper	Cu	0–0.2 mg/L
Zinc	Zn	1–5 mg/L
Aluminum	Al	0–5 mg/L
Nitrate (ppm)	NO_3	0–5 mg/L
Total Phosphorous (ppm)	P	0.005–5 mg/L

lb.). Therefore, 1 milligram (one-thousandth of a gram) per liter is 1 ppm. Parts per million is used to report boron, nitrates, nitrate-nitrogen, and a few other materials found in relatively small amounts. See Table 16.2.

3. Milliequivalents per liter (me/L): the most meaningful method

TABLE 16.2 Monthly and annual summaries for chemical transport in fairway runoff

	J	F	M	A	M	J	J	A	S	O	N	D	Ann.
Fairway													
Irrig. (mm)	8.9	16.4	62.3	134.0	97.8	136.3	170.1	166.0	98.1	74.7	28.0	15.8	1008.4
Runoff (mm)	10.5	9.3	3.9	14.2	24.0	22.0	10.9	10.7	18.2	20.4	9.2	12.6	165.9
ET (mm)	38.8	58.0	124.2	172.1	183.5	197.7	208.4	204.4	156.3	119.8	72.0	43.5	1578.7
Perc. (mm)	16.6	12.0	4.5	1.3	5.8	5.0	1.5	1.2	4.0	8.2	8.2	14.3	82.6
NO$_3$-N (ppm) Runoff	2.6	1.7	0.8	5.1	7.0	1.1	1.0	0.9	2.0	2.2	13.3	3.3	3.5
NO$_3$-N (ppm) Percolate	0.43	0.13	0.03	0.12	0.03	---	---	---	---	0.02	0.09	0.06	0.13
Pendimethalin Runoff (ppb)	14.1	8.2	9.0	12.9	12.0	16.8	14.1	10.3	15.2	14.2	11.2	16.3	13.5
Green													
Irrig. (mm)	51.7	66.5	115.0	118.3	175.5	206.5	215.4	42.8	51.4	102.2	51.0	56.8	1253.1
Runoff (mm)	2.9	2.3	1.0	4.6	7.3	8.2	4.1	5.0	7.1	6.8	3.4	4.0	56.7
ET (mm)	73.5	91.9	136.8	146.4	216.8	237.6	234.3	58.3	80.2	126.1	71.5	75.4	1548.8
Perc. (mm)	30.8	35.3	26.4	45.8	62.8	47.1	25.5	31.4	50.2	46.7	32.9	36.4	471.3
NO$_3$-N (ppm) Runoff	4.5	2.8	7.5	6.6	2.2	4.2	2.7	19.6	27.9	6.9	7.1	6.6	9.0
NO$_3$-N (ppm) Percolate	83.9	4.4	46.3	46.8	1.1	14.6	3.4	44.3	97.0	26.7	52.2	61.5	39.1
Bensulide Runoff (ppb)	1.4	1.3	15.0	21.7	13.0	11.8	8.8	7.2	5.6	4.4	3.8	3.0	8.5
Bensulide Perc. (ppb)	0.7	0.5	0.5	0.5	0.5	0.6	0.9	0.5	0.7	0.5	0.4	0.6	0.6

of reporting the major constituents of water. This is a measure of the chemical equivalence of an ion. Me/L is a unit of measure that puts all ions on an equal weight basis.

If necessary, a water test may include other heavy metals such as:

- Arsenic
- Beryllium
- Cadmium
- Chromium
- Cobalt
- Fluoride
- Lead
- Lithium
- Molybdenum
- Nickel
- Selenium
- Vanadium

There are tests for bacterial count and for detection levels of chemicals and pesticides.

Salt-Related Problems

Adverse aspects of soil salts, as high soluble salts or excessive Na, on turfgrass growth may be manifested as direct injuries to plant roots and shoot tissues as salts are taken into the plant, or as indirect effects on soil physical properties. Direct stresses include inducement of water deficits by high salt concentrations, ion toxicity, and nutrient imbalances. High Na can destroy soil structure causing problems with water

> **TURFTIP**
>
> Salt-affected soils are divided into three classes: saline, sodic, or saline sodic. Soils within each class share common characteristics in terms of problems and field symptoms.

infiltration, poor drainage, and low soil oxygen. Turfgrass species and cultivars within a species vary in tolerance to different salt injuries. Salt-related problems occurring from existing salts in the soil or from salts added by irrigation water are:

- *Water deficits induced by salts*
- *Soil permeability problems ensuing from degradation of the soil structure from excess Na*
- *Ion toxicities from certain specific ions*
- *Ion imbalances that lead to nutritional problems*

These primary problems often lead to secondary turfgrass management problems such as weak turfgrass stands that are less competitive to pathogens, weed encroachment, and physiological drought. See Figure 16.1.

Characteristics of Salt-Affected Soils

Salt-affected soils are divided into three classes: saline, sodic, or saline sodic. Soils within each class share common characteristics in terms of problems and field symptoms.

Saline soil is a nonsodic soil containing sufficient high total soluble salts to adversely affect the growth of most plants. A high soil salt content attracts water and reduces water availability for plant uptake. Ion toxicity to shoot and root tissues may occur. Ion imbalance can occur in some soils leading to deficiencies of Ca, K, NO_3, Mg, Mn, or P.

Sodic soil contains sufficient exchangeable Na to adversely affect growth of most crops and soil structure of most soils. It causes soil structure breakdown that inhibits water movement into and through the soil. Ion toxicities, especially Na and Cl, can occur in some soils. Ca, Mg, and potassium deficiencies are common under high Na. Primary cultural practices are:

- *Use salt-tolerant plants.*
- *Add gypsum to generate readily soluble Ca to replace Na.*
- *Leach excess Na from the rootzone.*
- *Deep aerification.*
- *Fertilize to alleviate nutrient imbalances.*

FIGURE 16.1 Natural native material helps resist soil and water problems.

Saline-sodic soil contains high salts that adversely affect the growth of many plants. The primary problems of a saline-sodic soil are the same as a saline soil, especially reduced water uptake due to high soil osmotic potential. If soluble salts decline without replacement of exchangeable Na by Ca, the physical problems associated with a sodic condition start to become apparent. Field and plant symptoms vary but symptoms observed for a saline soil are most common. Important cultural practices are:

- *Using high salinity resistant plants*
- *Leaching to remove salts*
- *Providing Ca*
- *An enhanced cultivation program if Na dominates and starts to deteriorate soil structure*
- *Applying fertilizer to correct nutrient imbalances or deficiencies*

Water quality can present many turfgrass management problems in attempting to produce high quality turfgrass that will provide the top quality playing conditions your clientele is demanding.

High pH, Carbonates and Bicarbonates

When irrigation water is high in bicarbonates (HCO_3) or carbonates (CO_3), these ions react with any Ca and Mg in the water to form $CaCO_3$ and $MgCO_3$, which have low solubility. This leaves Na to dominate while reducing Ca and Mg as competitive or counter ions needed to replace Na. When the irrigation water is already low in Ca and Mg, soil HCO_3 and CO_3 levels increase and Ca and Mg are then precipitated out as carbonates in the soil. This also reduces the effectiveness of gypsum treatment or addition of elemental S (to react with lime and form gypsum) since the insoluble carbonate forms are favored. Amending irrigation water with sulfuric acid (H_2SO_4) can counteract the adverse HCO_3 and CO_3 levels. If the Na accumulation in the soil is high, additional gypsum or sulfur + lime amendments would be required.

Sodium Absorption Ratio (SAR)

SAR indicates the relative activity of sodium ions as they react with clay. From the SAR, the proportion of sodium on the clay fraction of soil can be estimated when irrigation water has been used for a long period of time with reasonable irrigation practice. SAR was one of the first methods of assessing the potential hazard from sodium buildup.

Adjusted SAR

A refinement of the SAR called the "Adjusted SAR" (SAR Adj) is now commonly used. SAR Adj includes the added effects of precipitation and solution of calcium in soils as related to $CO_3 + HCO_3$ concentrations. SAR Adj can be reduced by increasing the Ca content of the water or by acidifying the water to reduce HCO_3.

Electrical Conductivity (EC)

EC is usually reported in millimhos per centimeter (mmhos/cm) and measures the total salt content of water. Multiplying EC by 640 gives a good approximation of the total ppm salt or total dissolved solids (TDS). Multiplying EC by 10 gives a good approximation of the total cation or anion concentration in meq/L. Understanding the complexities of the salt-affected problems is the key to making the best management decisions.

Salt-affected sites are becoming more common due to:

- *The use of wastewater that may contain salts*
- *Coastal sites that are subject to saltwater intrusion*
- *Flooding*
- *Salt spray*
- *The use of sands that are easily salinized*
- *Water conservation pressures that limit salt leaching*

Ion Toxicity in Water

The ions that most often cause toxicity problems are Na, Cl, B, HCO_3, and pH >9.0. Toxicity problems can be from both soil accumulations over time as well as from immediate direct foliage injury on sensitive plants from irrigation.

Sources of irrigation water from lakes, rivers, or effluent water can contain constituents that are of concern:

- *Biological material (algae and fungal spores)*
- *Suspended solids (sand, silt, or clay)*
- *Biodegradable organics (carbohydrates, fats, and proteins)*
- *Stable organics (phenols, pesticides, or chlorinated hydrocarbons)*
- *Pathogens (bacteria, viruses, or parasites)*
- *Residual chlorine (toxic chlorinated organics)*

Nutrients in irrigation water are of concern because of human health hazards from nitrates, promotion of algae growth from phosphates, and nutrient content such as N, P, K, and S that can influence growth.

LONG-TERM EFFECTS OF IRRIGATION WATER ON SOILS

It is written, "As the water is, so then shall be the soil." If a given soil is irrigated with the same water over a long period of time, the soil will assume the characteristics of the water. Much of the water used for irrigation in the Midwestern US has a pH of >7.5, and a soil pH of >7.5. The pH of a highly buffered soil cannot be significantly lowered, no matter how much elemental sulfur is applied.

TURFTIP

A working knowledge of irrigation water quality is indispensable for the production of fine turfgrass. The quality of the irrigation water that is available may significantly influence your turfgrass management practices.

When water quality is impaired because it has a potential permeability hazard from sodium buildup, the quality of the water can be improved by the addition of amendments. A reduction in permeability hazard will occur by increasing the Ca content or by reducing the HCO_3. Addition of gypsum ($CaSO_4$) is a common way of improving Ca content of waters. Addition of sulfuric acid will reduce the HCO_3 content. Amendments do not have to be added directly to the water, but can be applied to the soil on which the water is to be used. Other good practices are the leaching or flushing of salts and/or applying Ca and/or K, which will lower Na.

A working knowledge of irrigation water quality is indispensable for the production of fine turfgrass. The quality of the irrigation water that is available may significantly influence your turfgrass management practices. In diagnosing field problems, especially during a drought, water quality must be evaluated. The turfgrass manager should be able to quickly scan a water analysis report and pick out pertinent information.

Chapter 17

Pesticide Application

One of the more dangerous aspects of turfgrass care is the use of pesticides. This portion of the lawn care industry is watched very closely by government agencies and local and state agricultural offices. The state agricultural departments and the Environmental Protection Agency (EPA) have jurisdiction to monitor and prosecute individuals for any misuse of pesticides and chemicals. This is why you should learn how to apply pesticides correctly and be aware of the laws that govern their usage. Without the enforcement of pesticide use, the potential for losing the ability to use chemicals becomes more of a threat for the rest of us each day.

The technology behind the use of pesticides has evolved considerably over the last few years, and many new facts are learned about chemicals every day. The use of biotechnology is one point of change that is limiting and changing our thinking about chemical usage. In some cases the fear of chemical usage is being overshadowed by the fear of genetically engineered plants. Many people who once worried about the pesticides in their foods now worry about the genetic changes within the food they are now consuming. The use of biotechnology is not new to the plant production industry, but the rapid proliferation of these types of plants will eventually change the general use of pesticides due to the built-in resistance to pests. I only mention

TURFTIP

Without the enforcement of pesticide use, the potential for losing the ability to use chemicals becomes more of a threat for the rest of us each day.

this because it is an alternative to pesticide usage in many parts of the agriculture industry, and soon it will affect daily maintenance routines for turfgrass as well.

The most widespread examples of plants that were influenced by biotechnology are Roundup™ Ready soybeans and corn. These seeds were limited in use when first introduced into the agricultural market but soon snowballed when the response from farmers was so positive. The costs of chemical usage and applications were greatly reduced. This was the beginning of a partial acceptance of biotechnology in the plant industry. Currently the same type of technology is being used to develop new turfgrass strains with more desirable characteristics. These changes range from

TURFTIP

The use of biotechnology is not new to the plant production industry, but the rapid proliferation of these types of plants will eventually change the general use of pesticides due to the built-in resistance to pests.

FIGURE 17.1 Pesticide turf and ornamental book for Missouri (it is recommended suggested that you purchase such a study guide to help you in the study and testing process).

disease-resistant grasses to the same Roundup™ type products as are used in the agricultural market. The idea is the to spray over the top of your bluegrass, for instance, and eradicate 100% of the foreign weeds that are present in the lawn. This may sound crazy to some of you, but I assure you that this type of breakthrough is just around the corner.

Technology continues to change, and the use of pesticides will also change with each new development. There is always the unanswered question of what these changes mean to turfgrass management for the future, but for the time being we should look at what is at hand and how we can control pest problems with the least amount of damage to our environment. The reason there are tests and licenses for the use of pesticides is only to protect the people, flora, and fauna that could be affected by the use of chemicals. Pesticides can cause problems in our environment by contaminating drinking water and by other forms of off-site movement in the form of drift. By using pesticides in a safe manner according to the label we are helping to save out environment and yet eliminate target pest problems.

TESTING AND LICENSING

If you are planning on using pesticides in your operation, you must be aware of all aspects of their application. You will need to pass tests in order to get your license and the insurance that is necessary in case of an accident. Every state has different guidelines, but most require a test for lawn care professionals and golf course people with the designation "ornamental and turf category." This is a test that is usually given in addition to the standardized test that must be taken by anyone who sprays for pests, which is commonly known as the "core" test for basic pesticide safety. By passing these two tests you will earn your pesticide applicator's license. You must still provide proof of liability insurance and also pay a fee for your license, but the testing is the most difficult part for most. Most states do require you to have liability insurance for pesticides, and the costs for this type of insurance are often quite high.

> ## TURFTIP
>
> If you are planning on using pesticides in your operation, you must be aware of all aspects of their application. You will need to pass tests in order to get your license and the insurance that is necessary in case of an accident.

Many states offer a more standardized test that does not have a specific category for each of the licenses offered, but category tests are more common. Most of the tests given are a standardized 50-question multiple-choice test. Many tests are offered in combination with the agriculture department and the university extension offices, which helps to promote the educational aspect of the testing. There are courses that can be taken and training given prior to taking the exams to ensure a successful outcome and provide some sort of training with the actual pesticide applications. It is still an ongoing problem for state agencies to catch the numerous people who are spraying pesticides illegally.

There may be required training prior to taking your pesticide exam, but this is not true for all states. This training is only for your benefit, and it will help inexperienced operators who are trying to get started in the lawn care business. Remember that everyone needs to start somewhere and that training will only eliminate problems for all pesticide users. The more experienced people there are who are applying pesticides correctly, the better off the entire lawn care and golf course industry will be. This will mean fewer accidents, fewer problems, and lower insurance costs for everyone involved.

Another part of the training process is the retraining or recertification process. This part of the license process will vary the most from state to state. In many states a 1-day training program must be attended every 2 to 3 years. In other states a point system is used

> **TURFTIP**
>
> Another part of the training process is the retraining or recertification process. This part of the license process will vary the most from state to state. In many states a 1-day training program must be attended every 2 to 3 years.

in which training programs and seminars must be attended throughout the license holder's career and points are awarded. Each person must maintain a certain point count of certified training per year.

SAFETY

Pesticide application should not be taken lightly, and those who pursue a career in turfgrass should learn as much about pesticide safety as possible. You should consider not only the health risks of exposing yourself to pesticides as well as anyone who may come into contact with the chemicals or their residues. This includes the people in your own house who may be exposed to your clothing or foot traffic. I am not trying to scare people, and certainly I am a firm believer in pesticide usage. I do, however, feel that many who pass a 50-question test underestimate the importance of pesticide safety and also overlook the risks involved when using chemicals. You need to understand and respect the products that are being applied to turfgrass and also know why, how, and when chemicals need to be applied.

I always have believed that chemical usage should be used in moderation and that there are often many alternatives to chemical controls. I find that many of the misuses or overuses of chemicals come about from improper diagnosis of diseases and pest problems. When misdiagnosis of a pest problem occurs, overuse of chemicals can result from treating a problem that does not exist. For instance, if a turfgrass plot has leaf discoloration, it may be suffering from a lack of nutrition and not a disease of some sort. Many would see this as an opportunity to spray the turf without first checking to see if a disease problem was actually present. This is why pest diagnosis is as important as the application of the products used to control the disease or pest. There is something to be said for a preventive application of a chemical to avoid disease occurrence, but such treatments can still be performed based on environmental factors that lead you to believe there was going to be a disease problem.

FIGURE 17.2 Nutrient-deficient turf.

346 Turfgrass Installation: Management and Maintenance

The most common types of pesticide applications with turfgrass are weed treatments. These treatments are almost always done in a program that calls for several chemicals to be applied throughout the growing season. There are some who fear the use of pesticides in the lawn, but for the most part there are few risks to those who apply these pesticides safely and according to the law. When using a weed control program, you are treating for some weeds on a preventive basis but this is because they cannot be controlled in any other fashion. There sometimes cannot be the same considerations for threshold levels as there are when dealing with other pests in the lawn care industry. This is still not to say that a lawn cannot look very nice without the use of chemicals or even with some sort of organic approach. It is just more difficult to achieve that plush carpet look that has become so expected within the lawn care industry.

Pesticide safety starts with the handling of the chemicals and the strategies used for pest control. One thing I like to do is to apply

FIGURE 17.3   Patch of clover (it is easier to cover this with a liquid product).

TURFTIP

One thing I like to do is to apply granular products just because of their ease of application and the control I have when spreading the product. In most cases granular products can provide just as good or better results than what you can get by using liquids.

granular products just because of their ease of application and the control I have when spreading the product. In most cases granular products can provide just as good or better results than what you can get by using liquids. However, there are some applications that still give more desirable results when liquids are used. It is through a combination of granular and liquid applications along with some reasonable caution that a beautiful looking lawn and landscape are created. Certain broadleaf plants are harder to get rid of when using a granular product because the weeds may be in patches, and complete coverage is more difficult.

Hopefully, you have picked up some of my lawn care control programs from the previous chapters and understand how an effective program will work. I try to provide sufficient control and not necessarily the overkill that many would believe to be necessary. The next safety issues occur with chemical preparation. There is just as large a risk to a chemical applicator when you are mixing chemicals as there is when you are actually applying them. Start thinking about safety before you ever apply any product at all. Transportation, storage, and mixing of the product can be just as big a safety hazard.

If you are handling a granular product, the safety concerns involved with transport and application are a little less severe. In most cases there is no mixing of the product at all, and the chances for drift when using dry chemicals are also very low. You should always consider off-site movement of any chemical, but with a granular product you can usually see where the product is being applied, with the exception of some dust. The granular product is much easier to transport and certainly easier to dispose of the empty containers or packing. There is a higher risk for small spills of product, because damage often occurs from tears or punctures in the bags. Since most chemicals are not applied in straight, active form, there is often a carrier such as corn cob, clay, or fertilizer so the product can be easily spread. Granular products do clean up much more easily than liquid products and can often be swept.

Liquid chemicals should be handled with a totally different approach. They are more likely to be absorbed into the ground, clothes, or skin. The dust is a concern with a dry product but drifting is not as likely as with liquids. A liquid chemical spill may need to have some sort of absorbent product applied to it to make it easier to clean up, whereas a dry product spill would be easy to sweep up. Liquids also need to be mixed in a tank, which means that pouring is necessary unless the chemical is in a water-soluble package. Many spills occur during mixing, and often these spills are not taken care of as they should be. There are very specific mixing instructions that should be followed, which can be found on every pesticide label.

Spraying or applying pesticides always requires some sort of personal protective equipment, which is also known as PPE. PPE will vary based on the requirements of the pesticide label. The pesticide label will provide you with all the information pertinent to the application of the product you intend to spread or spray. You will find mixing instructions, PPE requirements, and all the other important information about the chemical you are applying.

With granular applications the required PPE may be as simple as a long-sleeve shirt, long pants, and protective shoes. With liquid applications the PPE requirements may include a spray suit, mask, gloves, and even goggles in some instances. There may also be a requirement for wearing an apron when mixing up certain liquid chemicals. The PPE is required for the applicator's protection and should always be carefully followed. I like to use more PPE than suggested instead of

Pesticide Application 349

FIGURE 17.4 A worker pouring pesticides into a spray tank (although this person is not spraying, he is actually at a higher risk for injury because he is dealing with the concentrated product).

less. It is always good to limit your exposure to any pesticide as much as possible. For instance, I will wear rubber gloves at when applying any type of chemical and sometimes even when I am transporting chemicals or storing them to limit my exposure to product residues.

As I mentioned, most of the information that pertains to the application of a product can be found on the pesticide label. The pesticide label holds almost all the critical information not only about using

FIGURE 17.5 Personal protective clothing.

the product but also about what to do if misuse of the product were to occur. The label will also have an EPA registration number for each product. This number means that a record exists for this product so that information is readily available and on file in case of an emergency or an inquiry. The EPA number should be recorded by commercial applicators for each product they are applying every time they apply that product.

The pesticide label also contains safety information, which will include signal words or warnings about the type of product contained in the package or jug.

TURFTIP

The pesticide label holds almost all the critical information not only about using the product but also about what to do if misuse of the product were to occur.

There will also be a child hazard warning, which will read "keep out of reach of children." This should be present on all pesticide labels and is there for your protection. Children are often the innocent victims of pesticide poisoning from accidental ingestion or exposure. This is one of the primary reasons why chemicals should always be stored in their original containers. The EPA numbers, safety warnings, emergency numbers, and other information will be readily visible.

Another piece of information that can be found in the safety portion of the label will be treatment procedures. They will give specific instructions as to what to do should accidental exposure occur. There are usually treatment suggestions in case of ingestion or exposure. A warning might read: "If swallowed, call a physician or the poison control center immediately." If the pesticide label contains the word "warning," then an 800 number must be provided on the label for you or a physician to be able to contact someone immediately.

FIGURE 17.6 Interpreting warnings on pesticide containers.

Signal Word	Toxicity	Approximate amount needed to kill the average person
DANGER	Highly Toxic	A taste to a teaspoon
WARNING	Moderately Toxic	A teaspoon to a tablespoon
CAUTION	Slightly Toxic	An ounce to more than a pint

There should also be a warning that will state how toxic the product is to humans and animals. This part of the label should provide specific information about possible allergic reactions from coming into contact with product. It will list possible symptoms of a reaction to the product and should have suggestions of what to do and not to do with the chemical. This may be valuable if you are not sure whether you are suffering from chemical exposure.

The next section of a label that should be closely read is a description of the result of misuse of the product to the environment. This section will have many environmental warnings and give specific information of the threats the product poses to ground water, plants, and animals in the area. Some labels might even mention additional threats to endangered species.

There are many smaller parts on a pesticide label that many do not understand and others simply ignore. There is a reason for every component of the pesticide label, and all the information in the label provides a "short story," so to speak, about the particular product. By not reading the label thoroughly or reading only the mixing instructions you do pose a risk to yourself and others.

The use classification is another required part of the label, which classifies the pesticide for either general or restricted use. Restricted-use pesticides are more toxic to people and the environment, and they have many additional restrictions for use and handling. A restricted-use product may cause injury to non-target plants. The retail purchase and any use of these products should only be performed by certified applicators or people who are being supervised by those who either have a license or are covered by a certified applicator's certificate.

You are starting to see just how much information is on a pesticide label and why it is important that the label be read very thoroughly before the chemical is applied. One of the most important parts of labels is the statement of the brand name as well as the active ingredients. This is an area that many people confuse because chemicals are often

Pesticide Application 353

Super TRIMEC®
BROADLEAF HERBICIDE

Controls dandelion, chickweed, knotweed, plantain, henbit, spurge, and many other species of broadleaf weeds.

ACTIVE INGREDIENTS:
Isooctyl (2-ethylhexyl) ester of
  2,4-dichlorophenoxyacetic acid 32.45%
Isooctyl ester of 2-(2,4-dichlorophenoxy)
  propionic acid 31.80%
Dicamba: 3,6-dichloro-o-anisic
  acid 5.38%
INERT INGREDIENTS: 30.37%
                                TOTAL 100.00%
THIS PRODUCT CONTAINS:
2.0 lbs. 2,4-dichlorophenoxyacetic acid equivalent per gallon or 21.54%.
2.0 lbs. 2-(2,4-dichlorophenoxy) propionic acid equivalent per gallon or 21.54%.
0.5 lb. 3,6-dichloro-o-anisic acid equivalent per gallon or 5.38%.
Contains Petroleum Distillates.
Isomer Specific by AOAC Methods.
TRIMEC® is a registered trademark of PBI/Gordon Corporation.

785/11-2000 AP081800
EPA REG. NO. 2217-758
EPA EST. NO. 2217-KS-1

Manufactured By
pbi/Gordon corporation
An Employee-Owned Company
1217 West 12th Street
Kansas City, Missouri 64101

KEEP OUT OF REACH OF CHILDREN
WARNING– AVISO

Si Usted no entiende la etiqueta, busque a alguien para que se la explique a Usted en detalle. (If you do not understand the label, find someone to explain it to you in detail.)

Statement of Practical Treatment
IF IN EYES: Hold eyelids open and flush with a steady, gentle stream of water for 15 minutes. Get medical attention.
IF SWALLOWED: Call a physician or Poison Control Center immediately. Contains petroleum solvent. Do not induce vomiting because of danger of aspirating liquid into lungs, causing serious damage and chemical pneumonitis. If spontaneous vomiting occurs, keep head below hips to prevent aspiration, and monitor for breathing difficulty.
IF ON SKIN: Wash with plenty of soap and water. Get medical attention.
IF INHALED: Remove victim to fresh air. If not breathing, give artificial respiration, preferably mouth-to-mouth. Get medical attention.

See below for additional Precautionary Statements.

READ THE ENTIRE LABEL FIRST. OBSERVE ALL PRECAUTIONS AND FOLLOW DIRECTIONS CAREFULLY.

PRECAUTIONARY STATEMENTS
Hazards to Humans and Domestic Animals
WARNING: Causes substantial but temporary eye injury. Do not get in eyes or on clothing. Wear protective eyewear. Harmful if swallowed. Harmful if absorbed through skin. Avoid contact with skin. Harmful if inhaled. Avoid breathing vapor or spray mist.

Personal Protective Equipment (PPE):
Some materials that are chemical-resistant to this product are listed below. If you want more options, follow the instructions for category E on an EPA chemical resistance category chart.

Applicators and other handlers must wear: • Long-sleeved shirt and long pants • Chemical-resistant gloves such as barrier laminate, nitrile rubber, neoprene rubber, or viton • Shoes plus socks • Protective eyewear • Chemical-resistant apron when cleaning equipment, mixing or loading.

Discard clothing and other absorbent materials that have been drenched or heavily contaminated with this product's concentrate. Do not reuse them.

Follow manufacturer's instructions for cleaning/maintaining PPE. If no such instructions for washables, use detergent and hot water. Keep and wash PPE separately from other laundry.

Engineering Control Statements for WPS Uses:
Containers over 1 gallon and less than 5 gallons in capacity: Mixers and loaders who do not use a mechanical system (probe and pump) to transfer the contents of this container must wear coveralls or a chemical-resistant apron in addition to the other required PPE.

Containers of 5 gallons or more in capacity: Do not open-pour from this container. A mechanical system (such as a probe and pump or spigot) must be used for transferring the contents of this container. If the contents of a nonrefillable pesticide container are emptied, the probe must be rinsed before removal. If the mechanical system is used in a manner that meets the requirements listed in the Worker Protection Standard (WPS) for agricultural pesticides [40 CFR 170.240(d)(4)], the handler PPE requirements may be reduced or modified as specified in the WPS.

When handlers use closed systems or enclosed cabs in a manner that meets the requirements listed in the Worker Protection Standard (WPS) for agricultural pesticides [40 CFR 170.240(d)(4-6)], the handler PPE requirements may be reduced or modified as specified in the WPS.

USER SAFETY RECOMMENDATIONS:
Users should:
• Wash hands before eating, drinking, chewing gum, using tobacco or using the toilet.
• Remove clothing immediately if pesticide gets inside. Then wash thoroughly and put on clean clothing.
• Remove PPE immediately after handling this product. Wash the outside of gloves before removing. As soon as possible, wash thoroughly and change into clean clothing.

ENVIRONMENTAL HAZARDS:
This product is toxic to aquatic invertebrates. Drift or runoff may adversely affect aquatic invertebrates and nontarget plants. For terrestrial uses, do not apply directly to water, or to areas where surface water is present, or to intertidal areas below the mean high water mark. Do not contaminate water when disposing of equipment washwater. When cleaning equipment, do not pour the washwater on the ground; spray or drain over a large area away from wells and other water sources.

Most cases of groundwater contamination involving phenoxy herbicides such as 2,4-D and 2,4-DP have been associated with mixing/loading and disposal sites. Caution should be exercised when handling 2,4-D and 2,4-DP pesticides at such sites to prevent contamination of groundwater supplies. Use of closed systems for mixing or transferring this pesticide will reduce the probability of spills. Placement of the mixing/loading equipment on an impervious pad to contain spills will help prevent groundwater contamination.

Physical or Chemical Hazard
Do not use or store near heat or open flame.

DIRECTIONS FOR USE
It is a violation of Federal law to use this product in a manner inconsistent with its labeling.

Do not apply this product in a way that will contact workers or other persons, either directly or through drift. Only protected handlers may be in the area during application.

For any requirements specific to your State or Tribe, consult the agency responsible for pesticide regulation.

AGRICULTURAL USE REQUIREMENTS
Use this product only in accordance with its labeling and with the Worker Protection Standard, 40 CFR part 170. This standard contains requirements for the protection of agricultural workers on farms, forests, nurseries, and greenhouses, and handlers of agricultural pesticides. It contains requirements for training, decontamination, notification, and emergency assistance. It also contains specific instructions and exceptions pertaining to the statements on this label about personal protective equipment (PPE) and restricted-entry interval. The requirements in this box only apply to uses of this product that are covered by the Worker Protection Standard.

DO NOT enter or allow worker entry into treated areas during the restricted-entry interval (REI) of 48 hours.

PPE required for early entry to treated areas that is permitted under the Worker Protection Standard and that involves contact with anything that has been treated, such as plants, soil, or water, is: • Coveralls • Chemical-resistant gloves such as barrier laminate, nitrile rubber, neoprene rubber, or viton • Shoes plus socks • Protective eyewear.

FIGURE 17.7 A pesticide label.

354 Turfgrass Installation: Management and Maintenance

NON-AGRICULTURAL USE REQUIREMENTS
The requirements in this box apply to uses of this product that are NOT within the scope of the Worker Protection Standard for agricultural pesticides (40 CFR Part 170). The WPS applies when this product is used to produce agricultural plants on farms, forests, nurseries, or greenhouses.

Do not allow people (other than applicator) or pets on treatment area during application. Do not enter treatment area until spray has dried.

FOR USE ON RESIDENTIAL AND ORNAMENTAL TURFGRASS SITES AND SOD FARMS (COOL SEASON GRASSES OTHER THAN BENTGRASS)

Do not apply this product through any type of irrigation system. Avoid drift of spray mist to vegetables, flowers, ornamental plants, shrubs, trees and other desirable plants. Do not pour spray solutions near desirable plants. Do not use on carpetgrass, dichondra, St. Augustinegrass, bentgrass, nor on lawns or turf where desirable clovers are present. Use only lawn type sprayers. Avoid fine sprays; coarse sprays are less likely to drift. Do not spray roots of ornamentals and trees. Do not exceed specified dosages for any area; be particularly careful within the dripline of trees and other ornamental species. Do not apply to newly seeded grasses until well established.

Do not spray when air temperatures exceed 85°F. Seed can be sown 3 to 4 weeks after application. Care should be taken not to make applications where runoff could carry the chemical to food crops or grazing lands where cattle, sheep, goats, swine or poultry would be exposed.

SPRAY PREPARATION
Add one-half the required amount of water to the spray tank, than add this product slowly with agitation, and complete filling the tank with water. To prevent separation of the emulsion, mix thoroughly and continue agitation while spraying.

INSTRUCTIONS
Maximum control of weeds will be obtained from spring or early fall applications when weeds are actively growing. Avoid spraying during long, excessively dry or hot periods unless adequate irrigation is available. Do not irrigate within 24 hours after application.

APPLICATION RATES: Apply 2 to 3 pints of product in 20 to 260 gallons of water per acre (0.75 to 1.1 fl. oz. of product in 0.5 to 6 gallons of water per 1,000 square feet). Use higher rates when using the higher volume of water per acre.

The maximum application rate to turf is 0.8 pounds 2,4-D acid equivalent per acre per application per site. The maximum number of broadcast applications per treatment site is 2 per year.

CONTROLLED DROPLET APPLICATOR (CDA): Add 1½ pints of product to the HERBI container and fill with water. Spray contents over 33,000 square feet. Avoid overlapping between spray patterns.

SMALL AREA APPLICATIONS
Not recommended for hose-end sprayers.

Spray anytime during the growing season when weeds are actively growing. On new lawns — wait until the grass has hardened off — usually after it has been mowed at least three times. Poor weed control may result if spray is applied during drought or just before rain. Do not water within 24 hours after treatment.

SPRAY PREPARATIONS FOR HAND OPERATED SPRAYERS

AMOUNT OF SUPER TRIMEC® TBS.	FL. OZ.	AMOUNT OF WATER IN SPRAYER, GALLONS	AREA TO BE SPRAYED, SQ. FT.
1½ Tbs.	¾ fl. oz.	1 Gallon	1000 Sq. Ft.
3 Tbs.	1½ fl. oz.	2 Gallons	2000 Sq. Ft.
4½ Tbs.	2¼ fl. oz.	3 Gallons	3000 Sq. Ft.

WEEDS CONTROLLED

Aster, white heath & white prairie	Black medic	Buttercup, creeping
Bedstraw	Broadleaf plantain	Carpetweed
Beggarweed, creeping	Buckhorn plantain	Chickweed, common
Bindweed	Bull thistle	Chicory
	Burclover	Cinquefoil
	Burdock, common	Clover
Cocklebur	Lespedeza, common	Shepherdspurse
Compassplant	Mallow, common	Spotted spurge
Curly dock	Matchweed	Spurge
Dandelion	Mouseear chickweed	Sunflower
Dayflower	Mustard	Thistle
Deadnettle	Nettle	Velvetleaf
Dock	Oxalis (*yellow woodsorrel & creeping woodsorrel)	(*pie marker, buttonweed, Indian mallow, butter print)
Dogfennel		
English daisy		
False dandelion		
(*spotted catsear & common catsear)	Parsley-piert	Veronica
	Pennsylvania smartweed	(*corn speedwell) Virginia buttonweed
Field bindweed (*morningglory & creeping jenny)	(*smartweed) Pennywort (*dollarweed)	White clover (*Dutch clover, honeysuckle
Field oxeye-daisy (*creeping oxeye)	Pepperweed Pigweed	clover, white trefoil & purplewort)
Filaree, whitestem & redstem	Pineappleweed Plantain	Wild carrot Wild garlic
Florida pusley	Poison ivy	Wild geranium
Ground ivy	Poison oak	Wild lettuce
Groundsel	Prostrate knotweed	Wild mustard
Hawkweed	(*knotweed)	Wild onion
Healall	Puncturevine	Wild strawberry
Henbit	Purslane	Wild violet
Jimsonweed	Ragweed	Yarrow
Kochia	Red sorrel	Yellow rocket
Lambsquarters	(*sheep sorrel)	
Lawn burweed		

*Synonyms

STORAGE & DISPOSAL
Do not contaminate water, food, or feed by storage or disposal.

STORAGE: Keep from freezing. Store in original container in a locked storage area inaccessible to children and pets.

PESTICIDE DISPOSAL: Pesticide wastes are toxic. Improper disposal of excess pesticide, spray mixture, or rinsate is a violation of Federal law and may contaminate groundwater. If these wastes cannot be disposed of by use according to label instructions, contact your State Pesticide or Environmental Control Agency, or the Hazardous Waste representative at the nearest EPA Regional Office for guidance.

CONTAINER DISPOSAL: Plastic Containers: Triple rinse (or equivalent). Then offer for recycling or reconditioning, or puncture and dispose of in a sanitary landfill, or incineration, or, if allowed by state and local authorities, by burning. If burned stay out of smoke. Metal Containers: Triple rinse (or equivalent). Then offer for recycling or reconditioning, or puncture and dispose of in a sanitary landfill, or by other procedures approved by state and local authorities.

LIMITED WARRANTY AND DISCLAIMER
The manufacturer warrants only that the chemical composition of this product conforms to the ingredient statement given on the label, and that the product is reasonably suited for the labeled use when applied according to the Directions for Use.

THE MANUFACTURER NEITHER MAKES NOR INTENDS ANY OTHER EXPRESS OR IMPLIED WARRANTIES, INCLUDING ANY WARRANTY OF MERCHANTABILITY OR FITNESS FOR A PARTICULAR PURPOSE, WHICH ARE EXPRESSLY DISCLAIMED. This limited warranty does not extend to the use of the product inconsistent with label instructions, warnings or cautions, or to use of the product under abnormal conditions such as drought, excessive rainfall, tornadoes, hurricanes, etc. These factors are beyond the control of the manufacturer or the seller. Thus, any damages arising from a breach of the manufacturer's warranty shall be limited to direct damages, and shall not include indirect or consequential damages such as loss of profits or values, except as otherwise provided by law.

The terms of this Limited Warranty and Disclaimer cannot be varied by any written or verbal statements or agreements. No employee or agent of the seller is authorized to vary or exceed the terms of this Limited Warranty and Disclaimer in any manner.

© 1983, PBI/GORDON CORPORATION

FIGURE 17.7 *(continued)* A pesticide label.

referred to by their brand name instead of their chemical name. I will give a common example, the chemical known as Roundup™. This chemical has name recognition world-wide and is known to be a nonselective chemical that will kill most any nonwoody vegetation that it comes into contact with. However, he actual chemical name of this product is glyphosate and is sold under many other brands. Roundup™ is what you will hear more often than not when the chemical glyphosate is referred to. There are many other products that have been marketed so well that they are commonly known by their trade names. I am simply trying to point out that there are many brands of the same products, and you should be sure that you are buying and applying the product you think you are. You can save considerable amounts of money in some cases if one brand of the same material is cheaper than a more recognized counterpart.

I briefly touched on the EPA registration information, which is a must on every label. It is a state and EPA requirement that a product must be acknowledged and registered in each state and not just nationwide. I have had a couple of instances where I was able to purchase a product that the state did not recognize or had no record of at the time and I had to call the company to advise that there was no registration in my particular area. Both times that chemical was recognized and I was able to go ahead and apply and sell the product, but when the inspectors cannot find the product in their books of registered products it can be a pain to get that product recognized. You are also in a sense liable for proving how and why you have the product until the registration is confirmed.

The next part of the label is another one that I think is quite important: the directions for use. This is where the type of product is described in detail, and the way it should be applied is laid out very specifically. There will be information provided as to which pests the product will control, mixing information, rates, and limitations and restrictions.

The last element I want to mention that can be found on the pesticide label is information pertaining to the storage and disposal of the product. I stated earlier that accidents can happen when handling or even storing a pesticide product. Any material that is discarded after use will contain pesticide residues and can be harmful if it is not taken care of correctly. Make sure that when you dispose of pesticide containers you are following federal, state, and local laws regulate the disposal of pesticides and containers.

TYPES OF PESTICIDES

I want to outline the different types of pesticides that can be used and the modes of action that are possible for them. There are two ways a pesticide can work: through contact modes of action or through systemic modes of action. Almost every pesticide group has both systemic and contact action products available. Systemic modes of action mean that the product will spread through the entire specimen, giving a response in the entire plant even though the product may have been absorbed in an isolated area. Contact products can still move throughout the plant, but they provide a more limited means of absorption. Contacts may need to be used to saturate the plant to provide a desirable result in some cases.

The first group of products that are used are known as herbicides. Herbicides kill weeds in most situations by impairing the plants' meta-

---

TURFTIP

Systemic modes of action mean that the product will spread through the entire specimen, giving a response in the entire plant even though the product may have been absorbed in an isolated area.

> **TURFTIP**
>
> Contact products can still move throughout the plant, but they provide a more limited means of absorption.

bolic process in some way. Some herbicides have multiple ways of doing this, which means that resistance is less likely, while other herbicides have only one mode of action. Some herbicides are actually derivatives of plant hormones and cause a forced response that results in death. There are nonselective herbicides, which will kill most vegetation, and selective herbicides, which will only target a specific type of plant.

A herbicide can also be a pre-emergent or post-emergent product. Pre-emergent products control weeds before they actually germinate and work on the premise of providing a soil barrier. A post-emergent product controls weeds that have already germinated and sometimes can keep additional weeds from germinating.

The insecticide family is more complex than many other groups of pesticides and is currently under a lot of scrutiny. The first group of insecticides is the organophosphates, which are the most widely used type. Over 40% of registered insecticides fall into this category. This group is losing registration and use privileges. The overall safety of this group is now being questioned, and many medical studies have shown that these pesticides cause illness. It is uncertain whether it will be completely banned, but several products have already been banned and others are on a list to be phased out over the next several years. Long-term exposure can be a serious risk to people, but the effectiveness of these products has been very good over the years when targeting insect pest control issues. This will mean that replacement products might be quite difficult and pest control may need to be addressed differently in the future.

FIGURE 17.8 Weed-out (a post-emergent-selective herbicide that kills only germinated broadleaf weeds).

The next group of insecticides is the carbamates, which are widely used in and around homes. As with organophosphates, there are health issues, but the risk at this point is much lower than the organophosphate group. Carbamates are an older group of insecticides and once again have been used for many years.

Boric acids and borates are used primarily in and around livestock areas. The use of these products is not as common as the other insecticide groups, and exposure is unlikely for most people unless they are involved in farming operations. These products are used to kill cock roaches and larvae in livestock confinements. There is a concern about the animals that are exposed to this group of insecticides as well as people confined in livestock areas.

The last group of insecticides I want to mention is the pyrethroids. These are some of the newest available and are actually synthetic products. It is believed that this group is safer than the organophosphates, and I feel that many new synthetics will be developed over the next few years. This will be necessary to offset some of the other insecticides that are being taken off the market. There will also be more organic types of products created in the future and others that will use insect pheromones as an ingredient.

The next group of pesticides that you might encounter in the turf industry is the fungicides. Fungicides can be either systemic or contact. There have been many strides in fungicide development in the last few years. Systemic availability is one of the biggest improvements, as well as the length of time that some of the products will last. A lot of the older fungicides were composed of organic mercury and hexachlorobenzene. These two components can cause health problems and only provide limited protection to turfgrass. There was also a problem of limited suspension in the spray tank. Fungicides now have the capacity to be used for preventive treatments, control treatments, or both. Product that can do both are a recent trend. Fungicide use is much more common in the golf course compared to commercial lawn care business. Unless there is a catastrophic disease in a lawn, it is fairly uncommon to use fungicides. This is largely due to their cost.

Fumigants are another group of pesticides that might be encountered in the turfgrass industry. They are highly dangerous and are often used only in the golf course segment of the industry or perhaps in nursery and greenhouse production.

The last group of chemicals is the rodenticides, which is another high-risk category. The reason this group is so high-risk is because the target pests are usually mammals, and what will harm them can also harm humans and pets. Many of the rodenticides that are available are restricted-use pesticides because of their deadly nature and should only be purchased and used by licensed commercial

> **TURFTIP**
>
> Any restricted-use pesticides will require a license to purchase and to use to any degree.

applicators. Many of these products contain such chemicals as warfarin, zinc phosphate, aluminum phosphate, and even strychnine. If you are familiar with any of these components, you will quickly realize the deadly nature of this pesticide group.

Although most pesticides can be purchased by American homeowners for use on their own property, not all chemicals can be bought by just anyone. Any restricted-use pesticides will require a license to purchase and to use to any degree. Many chemicals can be purchase at large shopping centers and are fine for the general public for personal use as long it is not for hire and the labels carry no restrictions. The important fact to remember is that a chemical license will allow a person to treat other properties for hire. Even products bought at a farm, home store, or large shopping center cannot be applied by nonlicensed people for profit. "For hire" is the key phrase that requires the applicator's license. It is illegal to use even a "weed and feed" type of product for hire, but it is fine in most cases for you to apply it to your own yard. Once again, it is important to research all the local and state laws for your particular area before making any decisions on pesticide usage.

APPLICATION EQUIPMENT

There are numerous types of equipment that are used for the application of pesticides. Some of these pieces of equipment are designed for dry products and others are used for spraying. All spraying equipment needs to be accurate and to be able to be calibrated as well. Good-quality application equipment will always be easy to calibrate and should be very accurate with several adjustments. Equipment used for pesticide applications can be very large for farming operations or fit in one hand when spreading very small areas. There is a type of equipment specially designed for application to any terrain and area, and you should know which ones you'll need to use.

The smallest type of application equipment is handheld equipment. Handheld equipment consists of small spreaders and sprayers. Backpack sprayers are one of my favorite pieces of handheld application equipment because they are inexpensive and very accurate to use. A backpack sprayer is easy to carry and gives you very good control of the liquids you apply. You can get backpack sprayers and other handheld sprayers in either a piston pump or a diaphragm pump. The piston pump will usually give you higher pressure for spraying, but the diaphragm sprayer will provide less clogging and is less likely to leak over time. All small sprayers can even be used for alternative applications like liquid ice melts in the winter months. There are also handheld rotary spreaders, which can be used to spread dry product accurately in small quantities. There are several brands of both backpack sprayers and handheld spreaders that you can purchase; buy one that is adjustable, accurate, and very comfortable to use.

The next step up would be an electric pump sprayer, which usually will mount on some type of equipment with a 12-volt battery.

TURFTIP

Backpack sprayers are one of my favorite pieces of handheld application equipment because they are inexpensive and very accurate to use.

FIGURE 17.9 Small handheld sprayers are versatile enough to do small tasks but do not have enough pressure to do larger jobs.

These sprayers most often hold less than 50 gallons and sometimes do not have a boom. Booms are more commonly found on engine-mount sprayers. There is often a wand that usually attaches to these types of sprayers so they can be used as spot sprayers. Spot spraying is the most common use of electric sprayers, but booms can easily be attached to many models.

There are now a few electric rotary spreaders on the market, which are becoming seemingly more popular every day. The idea behind them is to have a "ride on" type of broadcast spreader that eliminates the walking. The only complaints I have heard about these units is a difficulty in calibration and an inconsistency of ground speed while applying.

The most common type of sprayer on the market for turfgrass maintenance is the gas-powered sprayer. This sprayer can be set up a couple of different ways. The first way is to have a tank with a boom mounted in some sort of transport vehicle. Some of these units are self-contained, run off the vehicle engine, and are designed specifically for spraying. They often have features such as a cab and filters for a safer application of product. The problem with these units is that they are very expensive and limited to golf course use in most cases. Other more commonly used sprayers may mount in the back of a utility vehicle. These are usually skid-mounted so they can be moved in and out of the vehicle very easily. They are very common on golf courses, athletic fields, and some lawn care applications. The larger counterparts of skid-mounted sprayers are found in the agriculture industry. Agricultural sprayers are designed to fit in the back of a pick-up truck and are not recommended for any turf use. Most turf sprayers will have tanks that range from 25 to 250 gallons. Anything over 250 gallons is probably more suited for crop and agricultural spraying.

The other type of skid-mounted sprayers is used more for lawn care but also works off an engine-driven system. These systems seldom have booms but instead use a long hose reel, which contains 100 to 200 feet of garden-type hose. The applicator will have a hose-end sprayer that disperses the pesticide while he or she walks over an area. The problem with this type of system is the high amount of exposure that results from dragging around the hose and picking up residues as you are applying the product. Then the hose is

TURFTIP

The larger counterparts of skid-mounted sprayers are found in the agriculture industry. Agricultural sprayers are designed to fit in the back of a pick-up truck and are not recommended for any turf use.

FIGURE 17.10 Skid type sprayers are typically used for commercial lawn care operations).

handled while reeling and unreeling it at each job site. You also tend to walk directly through the areas you are spraying as the pesticides are applied. It is important to wear personal protective equipment and be aware of the exposure you are receiving.

DISPOSAL OF CHEMICALS

It is not uncommon to have leftover pesticides in your tank when a spraying project is finished. Their disposal can cause problems to our drinking water and the environment. Unused pesticides that are disposed of incorrectly cause as many or more problems than applied pesticides. Many people who use pesticides do not know how to correctly dispose of containers or the leftover pesticides in the application equipment.

It is pretty easy to accidentally mix up too much product or to miscalculate the square footage of an area. After all, none of us is perfect, and sometimes plans can change after the product is mixed up. The first rule of thumb is to NEVER put a pesticide product in anything other than its original container. This includes any extra pesticide that you might have left over in a tank and that you are considering saving. This is the fastest way to expose someone who has no idea that there is a pesticide present in that particular container. The best way and safest way to get rid of extra product after it is mixed up is to apply it. I am not suggesting that an applicator should ever double up coverage in an area to simply "use up" what is mixed. What I am suggesting is to have a designated area where extra product can be applied until the tank or spreader is completely empty. By applying the product you are complying with the label and at the same time you are keeping this extra product from being dumped in a ditch or area where it will cause off-site exposure.

Pesticides should never be poured out in ditches or sprayed out in one spot until the sprayer is empty. This is a very fast way to cause the pesticide to end up in the ground water or even in sewer systems. It could also move off-site to a pond, lake, or stream, where it could end up in someone else's drinking water. Make sure that extra product is applied to an area where it will not cause any problems. You should also avoid spraying the extra chemical out over a concrete parking lot, where it might be more likely to move off-site in runoff water.

If a dry product is used, you can just pour the excess back into the original bag. I have heard stories of people dumping it out in

TURFTIP

The first rule of thumb is to NEVER put a pesticide product in anything other than its original container.

the street instead. This is another bad idea and once again will just cause problems as the product is carried away in runoff water.

Rinsing your equipment or spray tank can be just as important as any part of the cleanup and disposal process. The runoff from washing out a spray tank or rinsing the sprayer off can contain residues that can be very harmful. It is important, when a sprayer is rinsed off, that the water be collected whenever possible and excess runoff avoided. It is possible to put in a collection dam around your rinse area, but you should check with your local laws on pesticide disposal and first and foremost always follow the label. Rinsing a tank should never be done around a well or water source. That is a direct way to get chemicals into a drinking-water system.

If you find that you are using a sprayer for multiple uses, you might need to neutralize your tank during the rinse process before changing chemicals in the tank. A triple rinse should be performed and bleach added to the tank on the third rinse cycle. When you do this, make sure that the rinses have made it through the entire system including the pump. Pesticides can remain in the lines and the pump if you do not do a thorough job of rinsing the system out. This process applies to all sprayers, but is much easier to do with smaller sprayers.

I do want to mention that certain pesticides can actually adhere to some spray-tank compositions, and complete removal of the prior chemical can be difficult to accomplish. I do suggest when using smaller sprayers especially that you try to keep a different sprayer

TURFTIP

Rinsing a tank should never be done around a well or water source. That is a direct way to get chemicals into a drinking-water system.

for each type of chemical you are using. I know that this practice is cost-prohibitive on a larger scale, so make sure that you do a through rinsing each time you change chemical types.

The next issue to consider after you are finished spraying is what to do with an empty chemical container. Disposal of the container is another critical part of the pesticide application process. You should only dispose of chemical containers in approved locations. This will ensure that they are properly disposed at a safe landfill area. There is disposal information on the label, which can tell you specifics about any particular product you might be applying.

When an applicator has finished a treatment, a record needs to be made of that application. All of the following information should be recorded each time an application is made and immediately after you are finished. It is important to have all the information, because in many states an inspector will stop by your facility to check the accuracy of your records from time to time.

Items that should be recorded by a licensed applicator for each application of product:

- *Name of noncertified applicator or technician (if applicable)*
- *License number of applicator or technician*
- *Date of application*
- *Name of person requesting the application*
- *Address of requested person*
- *Address or location of application site*
- *Target pests to be controlled*
- *Pesticide trade name*
- *EPA registration number or documentation*
- *Reasonable estimate of amount of pesticide used*
- *Time, air temperature, wind speed, and wind direction*

A chemical inspector will visit you to make sure you are recording the appropriate information, usually once a year. This inspection is given to all commercial applicators at some point during the calendar year. You should be aware that the inspector may want to follow you for the day and do an on-site application inspection. This is done at random by the inspectors and is studied to provide statistical information and other data for the state. The inspector will follow you throughout the day while you are doing your applications to make sure you are doing everything correctly. He or she may ask you many questions and even photograph the mixing and application process. You will not get in trouble for doing something wrong; it is more for your education and for documentation of what and how applications are being made. This is also a good time for you to ask questions so you can find out anything you may have been doing wrong. It is unfortunate that people view the state inspectors as the enemy when in fact they are there to educate you as a chemical applicator and to improve the general use of pesticides.

No matter what role chemicals play in your day-to-day operation, the concepts behind pesticide applications are fairly simple. If we all follow rules and guidelines set up by our state and local governments, we are making an effort to improve the pesticide application process. We must view the pesticide label as an instruction manual and realize that all the information that is printed on those labels is there for an important reason. Lastly, we must make good

TURFTIP

A chemical inspector will visit you to make sure you are recording the appropriate information, usually once a year. This inspection is given to all commercial applicators at some point during the calendar year.

judgment calls as applicators and realize that safety is the most important issue when using and handling pesticides. The pesticides that are being used by the public are relatively safe when used properly, and the control they provide will be appreciated if everything else is handled in the right way.

Chapter 18

Pest Management for Turfgrass

Pest management of turf can be viewed in many different ways. There are some people who would view any pest in the lawn to be one too many. Pests can be weeds, insects, diseases, animals, or anything that is considered to interfere with the appearance of or causes economic injury to the lawn and landscape. Methods of pest control depend on the degree of control you are looking for. The method explained here takes integrated pest management decisions into consideration instead of just treating for 100% control of all pests that are encountered. It is usually preferable to use thresholds for controlling pests instead of eradication to remove all pests.

There are different levels and categories of pest management. Some evasive species can require a management program outside normal control strategies. An evasive species is a species that possesses the ability to reproduce at a rapid rate or is not controllable with traditional measures. In other words, an evasive species could fit any of the following situations.

Examples of an evasive pest species:

- *New insect is discovered that is not controllable with traditional chemicals that are on the market.*

- *A weed present in the landscape or lawn has a tremendous ability to reproduce, such as stolons, rhizomes, and seed.*
- *An insect develops resistance to traditional insecticides and does irreparable damage before new control measure is discovered.*
- *New non-native plant is introduced to an area and reproduces at a rapid rate.*
- *A disease occurs that at first is not even diagnosable and progresses into a serious problem before a new fungicide is developed.*

There are numerous examples of how any pest can become an evasive species. Evasive species are not limited to any one type of pest. Evasive species introduced intentionally or accidentally can create a potential problem to an ecosystem or an environment. Imported species have been an especially challenging problem for decades if not centuries.

The situation of non-native species having their way with new environments has been the subject of some nightmarish stories and tragic consequences that are notorious in creating ecological havoc to this day. The list of problem species that have found their way to non-native shores and the destructive path they have followed makes for some tragic outcomes in forests, meadows, waterways, agronomic fields, and landscapes that are well documented and still unfolding at present.

The newspaper, television, and radio accounts of these interlopers make for stories that read right out of science-fiction scenarios, but they are unfortunately all too real. Perhaps the worst and most potentially catastrophic development when an introduced species meets a new ecosystem is the potential for virtual extinction and/or displacement of a native population with little chance of recovery once the damage has been done.

For as long as there has been commerce, trade, and exchanges among cultures there has been both the potential as well as the realized situation of inadvertent, if not intentional, introduction of species into ranges not prepared to resist them. This occurs when new plants are discovered in areas outside of their native area. People then feel a desire to introduce these plants into a home or lawn that is maybe located hundreds if not thousands of miles from their native habitat. This is one reason that commerce into and out of the United States is monitored so closely. There needs to be a constant monitoring for potential evasive species, whether it is at the checkpoints at our borders or the inspections of freighters that enter our ocean ports.

The list is long when it comes to plants, animals, and diseases that have gained footholds in ecological communities for which no natural enemies or biotic controls can limit populations or adaptations to new environments. The very selectivity and delicate balance that characterize a particular biosystem make the vulnerability to new species especially sensitive. The degree of external control in the form of disease, virus, predators, and abiotic factors plays a crucial role in whether or not an organism will find a new home in strange surroundings. Situations become serious if there are no unnatural or natural controls for the spread of any species. The mere existence of one natural enemy may be the one limiting factor determining whether or not an organism thrives, merely exists, or barely hangs on. If humans have no ability to limit the spread of a species, the consequences can be magnified tenfold.

WEED CONTROL

Most people would like to see a situation in which there are no weeds in the lawn at all, but this type of mentality can lead to excessive pesticide usage in some cases. Weed control success can depend on the situation in the areas around the given location. The best weed control program in the world may not be that successful if none of the adjoining properties are controlling weeds at all. Weed control needs to be done with a program that will use fertilizers to

> **TURFTIP**
>
> The best weed control program in the world may not be that successful if none of the adjoining properties are controlling weeds at all.

encourage vigorous turf growth in combination with pre-emergents and contact herbicides in the yearly program. The method of control depends on the type of turf area in question. 100% control would be the expectation for a golf green but would be an unlikely goal for a 10-acre property. This is not to say that 100% control is not desired; however, the cost of such weed control is often impractical in most cases.

A weed in the lawn can be defined simply as a plant out of place; it all depends on the plant, where it is growing, and whether it matches its surroundings. A fescue plant would not be considered a weed if

> **TURFTIP**
>
> Weed control needs to be done with a program that will use fertilizers to encourage vigorous turf growth in combination with pre-emergents and contact herbicides in the yearly program.

it was in with thousands of other fescue plants in a lawn, but if that same fescue plant was in a ryegrass fairway, it would appear horribly out of place. Most people recognize what they consider to be weeds in their lawn, but everyone's view of what a weed might be is different. An unusual example would be if bentgrass were to spread too far out from a green or tee. This might not be a very desirable appearance, so a superintendent might want to limit the spread of the grass to the intended area only. To a superintendent this problem might be as annoying as that of a homeowner with a half a dozen dandelions in the lawn.

For the most part when a person thinks of weeds they think more along the lines of controlling crabgrass, dandelions, or even ground ivy in the lawn (Figure 18.1). A golf course superintendent might have thoughts of weed control of plants such as poa annua. No matter what the definition of a weed, the most important issue is how and when to control them.

Weeds can be classified into a couple of categories for all practical identification purposes. One is dicot weeds, which are those that come from a plant with two cotyledons. Examples of this type of weed would include dandelion, plantain, purple violet, and clover. The other type is moncot weeds, which come from a single cotyledon. The monocots include the grass family, but there are many undesirable grasses that can occur in turfgrass. Examples of some grassy weeds would be crabgrass, goosegrass, and Bermuda grass.

Grassy weeds are controlled in different ways than are dicot broadleaf weeds. Grasses are usually controlled with a pre-emergent herbicide, which puts down a barrier to prevent the weeds from germinating. There are herbicides that allow post-emergence control of grassy weeds, but they often only work on the younger stages of growth. This means that a late application of a product to control a grassy weed problem is not going to be very successful. An example of a pre-emergent chemical would be pendemethalin, which has been successful for controlling crabgrass for many years. An example of a post-emergent grass control would be dithiopyr, which

FIGURE 18.1 Crabgrass plant (while this crabgrass plant may seem insignificant it is capable of producing thousands of viable seeds in just one season).

TURFTIP

Grassy weeds are controlled in different ways than are dicot broadleaf weeds. Grasses are usually controlled with a pre-emergent herbicide, which puts down a barrier to prevent the weeds from germinating.

controls crabgrass as long as it is not in the tillering stage. (Dithiopyr also provides pre-emergent control of crabgrass but its most reliable feature is post-emergent control.)

Many turfgrass professionals who treat lawns for annual grasses use a double or split application of pre-emergent chemical. I prefer an application of pre-emergent chemical early in the spring followed by a post-emergent treatment later in the spring. Either method will work for controlling annual grasses in turfgrass. It is not possible to wait until late in the season to think about crabgrass control. Without some sort of early chemical treatment in the Midwestern states, crabgrass is almost a guarantee. Since one crabgrass plant can produce somewhere around 10,000 seeds per season, even a handful of plants can produce 1 million viable seeds for the following season.

Dicot weeds in turfgrass are referred to as broadleaves, and the methods of control are quite different. Broadleaf weeds need to be controlled after they germinate instead of before. Broadleaves are much easier to control in the seedling stage, so early control of broadleaf weeds does have its benefits. By controlling broadleaves early in the year you also prevent the plants from seeding out and creating more viable seeds to germinate in the lawn. Broadleaves can become more resistant to chemical control as they mature through late spring and into the summer months; however, the fall

TURFTIP

Without some sort of early chemical treatment in the Midwestern states, crabgrass is almost a guarantee.

is also a good time to control broadleaf weeds. The reason fall is a good time is because weeds are in the process of gathering energy and creating food for survival during the long winter months. At this time they are also more vulnerable to chemical control.

The most common control for broadleaf weeds is the chemical known as 2,4-D. Typically the chemical is used in combination with MCPP and Dicamba. These three chemicals are available in a product called Trimec™, which gives you the best balance of the three components. Any of these ingredients used individually will provide you with reasonable control of broadleaf weeds.

DISEASE CONTROL

Disease control is usually more applicable to golf courses or athletic fields. It is often impractical to treat for diseases in lawn care unless there is an unusual situation where severe economic injury to the lawn is possible. Economic thresholds are a common measure for the use or not of chemical treatments. There are numerous diseases that can occur in turfgrass, but most of them are only treated on golf courses, athletic fields, or sports complexes.

TURFTIP

Broadleaf weeds need to be controlled after they germinate instead of before. Broadleaves are much easier to control in the seedling stage, so early control of broadleaf weeds does have its benefits.

TURFTIP

It is often impractical to treat for diseases in lawn care unless there is an unusual situation where severe economic injury to the lawn is possible.

The golf green is the most common area for treatment of disease in turfgrass. I have covered the more common diseases that occur in turfgrass in the golf course chapter. I want to focus instead on some new diseases that are occurring in many parts of the transition zone and the available control methods.

One type of disease in turfgrass is rhizoctonia. Brown patch (Figure 18.2) is a very problematic and common disease that occurs

TURFTIP

One type of disease in turfgrass is rhizoctonia. Brown patch is a very problematic and common disease that occurs in bentgrass greens. Brown patch is just one type of rhizoctonia disease that you might encounter, but now I am seeing other types of rhizoctonia problems, the worst being Rhizoctonia zeae.

in bentgrass greens. Brown patch is just one type of rhizoctonia disease that you might encounter, but now I am seeing other types of rhizoctonia problems, the worst being Rhizoctonia zeae. These newer diseases are known as yellow patch, sheath spot, and sheath blight, and they are primarily showing up in cool season turf. Cultural control, as with brown patch, would start with a reduction in the amount of water that is applied to an area if you are using irrigation in that location. As with most rhizoctonias, water can increase the progression of the disease as well as the signs of turf injury. Daconil™ (chlorothalonil) seems to be one of the most effective chemicals in controlling many of the rhizoctonias, although there are others that will work.

Another disease that was at first thought of as a turf condition is known as take-all patch. This disease first was thought to be a troublesome symptom of unhealthy turf and not actually caused by a pathogen. There still are no defined symptoms for early detection of this disease, but the pathogen that causes take-all has been

FIGURE 18.2 This is a disease known as brown patch. Although this is more commonly found in cool season grasses there are now problems developing in warm season turf as well.

diagnosed as Gaeumanomyces graminis avenae. This pathogen will cause take-all patch to occur in bentgrass but only in newer varieties. It has been discovered that it usually only occurs in greens four year of age or younger. Infection occurs in spring and fall of the year, and symptoms include circular patches or large gray areas in the turf. These areas may not die completely in the center, but they can develop into connecting patches that can cover an entire green. Sometimes severe turf injury can occur in greens with large dead areas.

There are a couple of facts to be noted about the take-all disease. There are some patterns that may help you to figure out if your greens are suffering from this disease. Generally speaking, the disease will become more severe in the second or third year after establishment of either new greens or a green renovation. The severe decline phenomenon kicks in sometime after the third season of the disease. The greens that are at the highest risk include newly constructed golf greens carved out of areas that have not typically been supported by turf. These areas might include timber areas, wetlands, or even bog areas. There is also a direct link showing that it is more likely to show up where methyl bromide was used during the renovation process. Control for take-all is to use Banner Maxx™ starting around April 10 in the transition zone.

TurfTip

If you are experiencing a disease problem in a ryegrass lawn or some other ryegrass turf areas, do not count out the possibility of gray leaf spot.

Gray leaf spot is the last cool season disease that I want to mention. It is increasingly common in golf course turf. This disease seems to target perennial ryegrass for the most part and is becoming quite a problem in golf course fairways. Symptoms vary from small circular spots on the leaf blades to complete yellowing of the plant from the tip down. The variations in symptoms can result in misdiagnosis. I have seen many cases where this disease has been mistaken for something else and treated incorrectly. If you are experiencing a disease problem in a ryegrass lawn or some other ryegrass turf areas, do not count out the possibility of gray leaf spot. Control of this disease should begin with the use of a fungus treatment program. Unfortunately, the best way to treat this fungus is on a preventive basis instead of after it occurs. I would suggest the use of Daconil™ with an alternating program of Heritage™ and Banner Maxx™. The use of these chemicals should be timed in two-week intervals if possible starting around August 1.

Zoysia grass was once thought of as a disease-free low-maintenance turf, and it still is advertised as such in many magazines and trade publications. Zoysia is now more popular than it used to be, and it is much more widely used throughout the transition zone and even in northern regions. As with the case of any plant species, once a disease problem occurs it is possible for that problem to become much more widespread when numerous host plants are available. This may have something to do with the recent problems that have been discovered in zoysia due to its increased use in golf course fairways and lawns.

One problem that has recently become quite widespread in zoysia grass through parts of the Midwest is a disease called downy mildew. Downy mildew is also referred to as yellow turf partially because of its appearance in infected zoysia plants. Downy mildew is another disease that is being misdiagnosed and is not widely understood. In some cases people are treating this disease as a nitrogen deficiency in hot summer months, which means that the disease problem becomes much worse by the time is actually diagnosed correctly. The best treatment for this disease is the use of Subdue Maxx™. This disease can occur in cool-season grasses, and the control would be the same. It is not uncommon in the cool-season grasses, but it is unusual that it is now becoming common in zoysia.

The most important conclusion you can draw from the last few pest problems I have listed is that they are examples of why continuing education is important within the turfgrass field. As new diseases are discovered, everyone needs to be able to recognize, diagnose, and treat for potential epidemics. Always rely on the resources available to you in the industry to keep you up to date on current pest and disease problems. Also contact chemical company representatives to find out if your control strategies and chemicals are the best ones available for your problem.

INSECT CONTROL

Insect control is very important to a golf course superintendent but unfortunately not nearly as big a priority to lawn care professionals. Insecticide is not always used in a lawn care program until damage is visible. I think that a yearly insecticide application should

TURFTIP

Insecticide is not always used in a lawn care program until damage is visible. I think that a yearly insecticide application should be performed to at least control grub populations in the lawn.

be performed to at least control grub populations in the lawn. Grubs are one of the more common insects to be found in the lawn, but I have only seen a handful of situations where severe damage has occurred (Figure 18.3). With golf courses the problem insects are more numerous, and there are many insects that can be encountered on any given day.

Grubs in a lawn can be controlled with an application of a general-purpose insecticide such as diazinon, but resistance to older chemicals continues to be an increasing problem. There are also an increasing number of insecticides that are being banned each year, which may make the future control of grubs and insects in general a little more difficult.

There are a couple of preferred treatments for grub control. If you are looking for instant control, the chemical Dylox™ works in a 24-hour period, giving you fast and immediate results. Incidentally, it also controls many other lawn insects that can be a nuisance in and around the home. If you are looking for season-long control of grubs, there are a couple of good products. The first one is Merit™ which will give you season-long control (120 days) of grubs when applied at the correct time. The best time for any grub application is usually

FIGURE 18.3 It is hard to believe that the adult stage of a grub worm can be such a mobile insect.

late May to early June for most of the Midwest. If you are looking for a slightly more modern or organic approach, there is a fairly new product that works in a nontypical mode. That product is Mach II ™, which is a hormone that causes the bugs to molt prematurely. What happens is the feeding parts of the grub are covered by premature growth and shedding of the skin so it starves to death. Any of the above products listed will provide exceptional grub control.

Another insect that might be encountered in cool-season turfgrass in the Midwest especially would be chinch bugs (Figure 18.4). Chinch bugs are not nearly so common as grubs, but when they occur in turfgrass the damage can be devastating. One mistake I see in the treatment of chinch bugs is to neglect the fact that they often overwinter in offsite locations and not in the lawn, which makes their control a little more difficult. I have seen many professionals not take into consideration that the bugs can harbor in and around walkways and structures so that treatments of just the turf areas may not be successful. A general-purpose insecticide at a high dose will probably rid most turf of chinch bugs, but multiple treatments are often necessary. Dylox™ or even a chemical such as Seven™ will successfully get rid of these bugs.

Cutworms can be a very destructive insect in terms of the short time frame it takes for them to do significant damage to the turf. Cutworms are a favorite food for many bird species, which can also cause turf injury. The only good thing about the birds feeding on the bugs is that they can be used as an indicator of an insect problem. Cutworms are root feeding insects that can decimate an area in a very short period of time. Cutworms are more commonly known for the destruction they cause to agricultural crops, but they can cause significant turfgrass injury as well.

With a golf course or sports turf there are a few other insects that I have fought here in the Midwest. One insect that is very problematic in golf greens is black turfgrass ataenius. Black ataenius is a shiny black beetle only $3/16$ inch long and $1/8$ inch wide. Black ataenius often feeds in the roots of annual bluegrass, bentgrass, and

TURFTIP

Cutworms are root feeding insects that can decimate an area in a very short period of time. Cutworms are more commonly known for the destruction they cause to agricultural crops, but they can cause significant turfgrass injury as well.

FIGURE 18.4 Chinch bugs are sucking insects that reproduce very quickly.

Kentucky bluegrass. First symptoms of injury occur in June or July and may appear as signs of wilt similar to isolated dry spots. Black turfgrass ataenius is very quick to reproduce, which is why it can cause problems in golf greens. The bugs themselves may look like large grains of sand or hard black specks on a shortly mowed bentgrass green. The other problem with these insects is that they can also draw hordes of birds onto your greens. Often the birds can be as destructive to the turf as the insects. A good dose of Dylox™ applied to the turf will bring up thousands of dead bugs the following day. The chemical Dursban used to be a great control for black ataenius control but has been banned for use in the ornamental turf category.

An interesting insect I want to mention that can occur in turfgrass is the mole cricket. Mole crickets are usually associated with warmer climates, but I am starting to see these bugs in northern Missouri and central Illinois. Mole crickets are burrowing insects that create tunnels that look like tiny mole runs in a golf green. They are especially bad in cases where the runs interfere with the putting surface. Mole crickets are also difficult to control and require fairly high rates of insecticides. The size of the mole cricket is impressive if you ever come across one of these bugs. I have seen them as large as 2 to 3 inches in length but commonly they are in the $1^1/_2$-inch range. Mature adults can develop wings that allow them to transport to new locations. I still have not figured out a pattern to the progression of mole crickets in more Northern regions. Some believe it is due to many years of mild winters.

I mentioned previously the fact that zoysia grass was once advertised and promoted to be the perfect disease- and pest-free turf. Not only has this grass been just recently plagued with disease problems, but there is also a new problem due to an insect. There is now a zoysia grass mite that is being found in zoysia turf, especially in golf course settings. This problem, in combination with the yellow turf disease I mentioned earlier, is finally making people stop and realize that all species can have susceptibilities and that pests can cause damage when least expected. The zoysia mite is treatable, but I think that its progression needs to be monitored to effectively map out future control strategies.

OTHER PEST PROBLEMS IN TURFGRASS

What most pest problems have in common when it comes to turfgrass is that they can be solved by using some type of chemical treatment. There are, however, many pests that cannot be controlled with chemicals. Some of them can be controlled with chemicals but not very successfully. This is when management strategies become your only defense. I want to alert you to a few of the more common pests you might encounter and how I would suggest you handle them.

Probably the most common pest that comes to mind is the mole. The methods for controlling voles, moles, and gophers are all very similar. Each of these pests has one distinctive feature that makes it very difficult to control. All of these animals are mammals, which makes chemical usage unpredictable, dangerous, and unsuccessful. Since humans, pets, and livestock are also mammals, non-target injury could be dangerous or potentially deadly. Most of the products that are available for controlling moles and other underground-dwelling varmints are in the fumigant family. This means that in most states a special fumigant license is required to use the types of chemicals needed to control these pests even though they are considered to be lawn pests.

Moles are very difficult to control even with the use of such fumigants (Figure 18.5). Fumigants are used in a trial-and-error fashion and may require several attempts to even hit any of the target pests. In the meantime there is repeated risk to non-target animals encountering these materials. The problem and danger with most baits, poisons, and fumigants is that they contain potentially toxic active ingredients that could kill a human if exposure was bad

TURFTIP

Fumigants are used in a trial-and-error fashion and may require several attempts to even hit any of the target pests.

enough. All pesticides have potential risks to humans and off-target pests, but the products used to control moles could be lethal in very small quantities to any mammal that encounters them. Just one accidental exposure to these products could be fatal, and that means that use of these products should be done very cautiously. Some of the products used contain active ingredients like strychnine, arsenic, aluminum phosphate, or some other hazardous or poisonous substance.

If a fumigant is used, it must be used in the areas that are showing the most recent signs of mole activity. Moles are one of hardest pests to control that you will encounter in a lawn. They are a frustration for homeowners as well as the people who try to control them. A fresh mole run is the first place you should start treating for moles, because they do not always travel through older runs. Poison pellets or fumigant tablets are deposited on the inside of the runs every 5 to 10 feet in hope that the moles will come into contact with the material or the gas that is released by the products. If you are treating older runs or areas where the moles are not active, then you will not be successful. This is where the trial and error comes into effect. You must continue to treat the fresh runs until you do not see any further activity or damage to the lawn in that area. While you are treating these runs you run the risk of a pet digging into the run and coming into contact with the material that was intended to kill the moles. A human could also encounter this material.

Pocket gophers (Figure 18.6) are one of the more destructive animals to get into the lawn. They make huge mounds of dirt in the yard and cause the most havoc for lawnmowers. These mounds are the result of the gophers cleaning out all of the runs and rooms and excavating the soil out of an opening at the soil level. This large mound causes a lot of difficulties for mowers and creates an unattractive appearance in the lawn. It does make it a little easier to treat for pocket gophers in terms of finding the more active mounds and runs. One good mowing or leveling off of the ground will quickly reveal which runs are being used on a daily basis. The pocket gophers will visit what are considered to be underground room areas regularly. These rooms are sometimes very low in the soil profile, and they can provide an excellent place for the animals to survive during the winter. Gophers are often mistaken for moles by homeowners, but they are two completely different animals.

Traps are another option for the control of moles, but they have very limited success against pocket gophers. Since moles make long runs in the lawn, there is a way to set a trap so the mole must run through it. With the case of pocket gophers it is hard to get the animals to encounter the trap just by the way they behave. There is one mistake that people make when setting traps for moles. The best

TURFTIP

The best way to set a mole trap is to take your foot and smash down 3 to 4 feet of the run to close it off. Then put either a snare or spike trap in the middle of the area you have smashed down. When the mole tries to open the run back up, it will encounter the trap and set it off.

Pest Management for Turfgrass 391

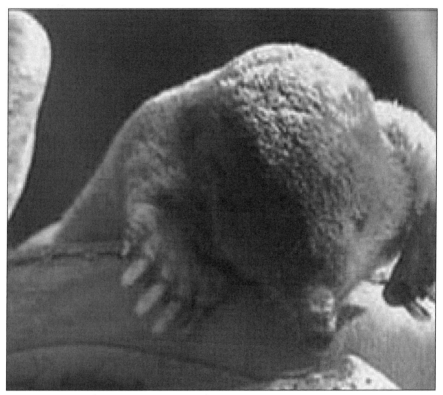

FIGURE 18.5 There is a lot of confusion about mole activity in the yard, but moles leave only shallow runs under the sod.

way to set a mole trap is to take your foot and smash down 3 to 4 feet of the run to close it off. Then put either a snare or spike trap in the middle of the area you have smashed down. When the mole tries to open the run back up, it will encounter the trap and set it off. It is very unlikely that the mole will encounter the trap if the soil is not smashed down first.

Most moles and gophers feed on the grubworms in the lawn, so there is a way to discourage moles and gophers from entering the lawn. Moles and gophers may move off site by eliminating the

FIGURE 18.6 You seldom see pocket gophers out of the ground. Typically you will see the large mound of dirt that they leave while clearing out their run.

food source in a particular area. One good grub treatment each year is a great way to discourage moles from coming into an area and tearing up the lawn. One way to judge mole presence in the lawn is by monitoring the run activity. Moles will leave two types of run patterns when they enter an area. If the mole leaves a long straight run, then it may not be finding much for a food source. On the other hand, if the mole is making a zigzag pattern in the lawn, chances are there is a plentiful source of grubs in that area and the mole may be planning on sticking around for a while. If you have applied a grub treatment to the lawn, moles may enter the lawn briefly and then quickly exit once it is discovered that there is not a good food source available. This may be evident by a long straight run entering the yard and quickly turning around and exiting the same way.

Ground squirrels are animals that could cause damage to the lawn but they are difficult to control by using a fumigant or poison bait. Trapping can be successful. Use a cage-type trap with bait that attracts them into the trap where they are captured. This method can also work on groundhogs, skunks, and raccoons. Many aboveground animals eat bugs, and while they are furrowing around looking for food they can cause a great deal of damage to the turf. Skunks are one of the more notorious animals for damaging turf while looking for grubworms, especially on golf course tees and collars. This is still another reason to make sure you have eliminated grubs in your turfgrass.

There are many pests that can be encountered in the lawn or on a golf course. Some people consider deer to be a pest on a golf course due to the damage that is caused by their hoofed feet. Whatever the measure of a pest is to you, be sure that you are managing your problems in an effective manner that uses chemicals only when necessary. As turfgrass professionals it is our job to limit excessive pesticide usage, so try to incorporate management strategies into a program along with chemical applications.

Chapter 19

Weeds Found in Turf and Landscapes

Most people, when they envision what the term "pest" means to them, will think of nuisance insects, rodents, or perhaps that annoying little neighbor kid. What turfgrass practitioners envision as a pest differs considerably. They may also think of insects, but of those that feed on turf. They usually think of diseases that can cause numerous management headaches, especially during summer. Lastly, turfgrass managers think of weeds when they think of what pests are. Many people will envision illegal plants when they hear the term "weed" (Figure 19.1). In this context, weeds are some of the most problematic pests found in managed turfgrass and landscapes. Surveys conducted at local, state, and national levels have shown that the majority of turfgrass professionals consider weeds to be the most troubling pest problem they deal with. How can this be? Weeds don't appear overnight, as some disease and insect problems do. Weeds don't have the potential to be lethal to turfgrass, as diseases and insects can be. So what is it about weeds that make them so notorious? The answer to that question will differ, depending on whom you talk to. Some common answers, however, might be:

- *Many weeds can't be prevented with pesticides, like diseases or insects.*

396 Turfgrass Installation: Management and Maintenance

FIGURE 19.1 Wild Cannabis sativa plants, growing in a natural setting.

- *Weeds are easy to spot and their presence draws criticism.*
- *More pesticide products are needed to control the spectrum of weeds I must deal with.*
- *Some weeds are highly invasive and the problem gets worse every year.*
- *Pesticides aren't available to control some weeds without harming my turf.*

Do any of these responses sound familiar or have particular pertinence to your situation? If so, you may be one of the many turfgrass professionals out there who view weeds as the worst of their pest problems. Each example response listed above represents a circumstance where weed control differs and is more challenging than is the case for other turfgrass pests.

Fundamentally, weeds are defined as plants out of place. Therefore, what plants are viewed as weeds will depend upon the context of the situation. Some turfgrass managers may view a particular plant as a desired turf while others may view the same plant as a weed. Determining what plants are weeds is at the discretion of the decision makers at a particular facility. Weed control is inherently more challenging than control of other pests because weeds are uninvited plants growing in a plant ecosystem. The biological similarity of weeds to the crop or landscape plants they invade makes managing and controlling them particularly difficult. This is because control strategies are more likely to cause harm to desired turf than control measures that target non-plant pests like diseases or insects.

The purpose of this chapter is to introduce the reader to some of the most common weeds they might experience in a turfgrass or landscape setting. The weeds discussed in this chapter may not be all-inclusive for your area but do represent the majority of the more common weed problems in turf and landscape settings. As with other pests, control of weeds may be a necessary but time consuming and

TURFTIP

Fundamentally, weeds are defined as plants out of place. Therefore, what plants are viewed as weeds will depend upon the context of the situation.

costly portion of your management budget. Because time and money are valuable commodities in any operation, we must focus on fundamentals before getting to the specifics of weed control. The core fundamentals of weed control are proper identification of weed species and their life cycles. Because these two issues are cornerstones of a solid weed management program, they will be discussed prior to any discussion of specific weed control strategies.

WEED LIFE CYCLES: ANNUALS, BIENNIALS, AND PERENNIALS

Understanding the life cycle of a weed is less complicated than for disease or insect pests but is absolutely critical to optimizing management and control of weeds. A life cycle for a pest is best described as the period of time between birth/germination and death from natural causes. The life cycles for plant pathogens that cause disease symptoms or for insects can vary considerably in duration for different species and can often occur multiple times in a given growing season. Weeds tend to be more basic, falling into one of four life cycle categories: summer annual, winter annual, biennial, and perennial. Proper recognition of a weed's life cycle not only helps in selection of a good control method but also assists with and sometimes dictates the timing of applied control measures. We'll now proceed through the basic life cycles and describe each in more detail.

TURFTIP

Weeds tend to fall into one of four life cycle categories: summer annual, winter annual, biennial, and perennial.

Annual weeds most closely resemble the life cycles for other pests. Insects and disease organisms, with few exceptions, have short life cycles and rely upon prolific reproduction for species persistence. Annual weeds, as the name implies, usually do not live more than twelve months, even though some annuals may live during portions of two calendar years. They characteristically rely upon seed production to fuel future generations of weeds. Seed production varies among species but can be highly prolific. Some weed species can produce a million or more seeds per plant! Thankfully, most common annual turf weeds are not this productive, in terms of seed production. Also, common cultural practices like mowing can reduce or prevent seed production. Because seed production is the primary means of reproduction for annual weeds, seeds produced from a previous generation tend to germinate readily when climatic conditions become suitable.

Annuals have two primary growth phases: a vegetative phase and a reproductive phase. The vegetative phase begins at the time of germination and represents the biggest portion of the plant's life cycle. During the vegetative phase, annual weeds can produce large amounts of plant biomass and tend to grow fairly rapidly. As such, they can be aggressive competitors with turf for nutrients, water, and other resources. The reproductive phase involves flowering and seed production and is triggered primarily by temperature and daylength. However, the reproductive phase can also be induced by environmental stresses such as drought. Drought-induced flowering and seed production are part of a survival mechanism that helps the plant fuel

TURFTIP

Annuals have two primary growth phases: a vegetative phase and a reproductive phase.

> **TURFTIP**
>
> Within the larger group of annual weeds that are found in turf and in other cropping systems, there are two primary subgroups: summer annuals and winter annuals.

later generations of weeds under adverse conditions. Seeds produced by annuals can germinate as soon as the following growing season but can persist for longer periods of time if conditions are adverse or if they are buried deeper in the soil by practices such as tillage. Within the larger group of annual weeds that are found in turf and in other cropping systems, there are two primary subgroups: summer annuals and winter annuals.

Summer annuals are so named because summer represents the peak or the central season that comprises their life cycle. They, unlike winter annuals, carry out their life cycle within a single calendar year. Most summer annuals germinate in late winter or spring but are capable of germinating well into the summer. A common example of this ability is late summer breakthrough of weeds like crabgrass (Figure 19.2). Preventive herbicides applied in spring can eliminate many of the weeds that are available for germination but some can persist until the herbicide wears down and sprout later in the summer. Herbicide application strategies must therefore account for these capabilities as much as possible.

When summer annual weeds actually germinate can vary considerably among species but is most always a direct function of soil temperature. Placing a month or date of germination to certain weeds is often unreliable as climatic conditions vary from year to year and certainly among different regions of the country. For example, summer annuals like crabgrass germinate in January or February in

Weeds Found in Turf and Landscapes 401

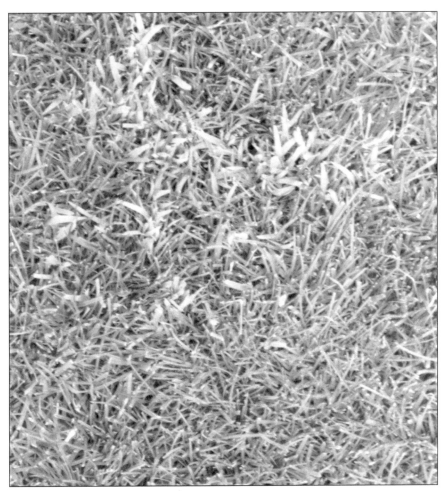

FIGURE 19.2   Crabgrass infestation in managed turfgrass.

many southern regions of the country but will not germinate until May or even June in some northern states. Many cropping systems rely upon model units called growing degree-days, which factor in temperature data from a particular year and help growers plan activities like planting. Such precision has not been pursued with most weed species so always be sure to consult peers or state-based extension literature if you are unsure when certain weeds emerge in your area. Summer annuals often remain in a seedling or immature stage for the first month or two after germination because air temperatures have not yet reached their seasonal peaks. Growth of summer annuals can become very rapid when it does peak and these weeds usually will grow at a rate much faster than for desired turf. As a group, summer annual plants can tolerate longer daylengths and higher temperatures than many other plants so they thrive during the peak of summer. This gives them a distinct competitive advantage if they are not managed or controlled.

A common feature among all summer annuals is induction of the reproductive phase in late summer or fall. Induction of this phase of growth is usually quite noticeable:

- *Broadleaf plants produce a flower.*
- *Grasses produce long stalks that will bear the seeds.*
- *General decline in the vigor of vegetative portions of the plant*
- *Leaves will tend to lose color*
- *Many broadleaf summer annuals will become rigid and stemmy (Figure 19.3). This trend is largely due to the plant's increased allocation of vital resources to its flowering and seed producing parts.*
- *Much of the plant's foliage will senesce or die back, much akin to the leaf dropping process seen with deciduous trees.*

This trend is sometimes preceded by purple coloring in leaves, which is caused by breakdown of the green chlorophyll pigment that gives actively growing leaves their color and greater expression of other naturally occurring plant pigments.

As for germination, specific timings for induction of the reproductive phase in summer annuals will vary among weed species but they will tend to vary less across regions of the country. This is because the summer annual reproductive phase is in part triggered by daylength. This mechanism helps promote flowering and seed production, even

Weeds Found in Turf and Landscapes 403

FIGURE 19.3 A mature prostrate spurge plant, with numerous branching stems.

when conditions may still be fairly hot in many locations. Annual plants have internal sensors that can detect the daylength they are exposed to. Once daylength drops to a certain level, the plant is triggered to initiate the reproductive phase. The daylength trigger helps signal the plant that colder weather is on the way and that the seed production process needs to begin. Different areas of the country experi-

> **TURFTIP**
>
> Practitioners in areas such as the south where daylength fluctuations are less severe may experience a gradual reproductive phase for summer annuals while northern areas that see faster reductions in fall daylength may experience more rapid flowering and seed production with the same weed species.

ence greater daylength fluctuation. Practitioners in areas such as the south where daylength fluctuations are less severe may experience a gradual reproductive phase for summer annuals while northern areas that see faster reductions in fall daylength may experience more rapid flowering and seed production with the same weed species.

Temperature is also intimately related to induction of the reproductive phase in summer annuals. Flowering and seed production tend to begin once night temperatures begin to cool in late summer or early fall and are accelerated by brief freezing events like the first fall frost. This effect of temperature will again occur sooner and will be more rapid in northern climates while it may be more gradual as one moves south. Because of the more consistent daylengths and annual temperatures that exist in many subtropical or tropical climates, some traditionally annual weeds like goosegrass can assume more persistent characteristics and may behave more like perennials in these areas (Figure 19.4). Practitioners in Hawaii or parts of Florida may encounter this additional challenge in their weed management programs but such an occurrence is rare elsewhere.

Winter annuals are usually not active during winter months but, as is the case for their summer annual cousins, winter is the central season that the life cycle always includes. Winter annual plants include many weeds but also some crops like winter cereals. Many

FIGURE 19.4 Goosegrass, an annual grass species that can be perennial in more tropical climates.

winter annuals germinate in the fall, pass through winter as a juvenile, and continue developing to maturity once conditions are again suitable in spring. However, fall germination is not always a requirement as members of a winter annual species can germinate in fall, early spring, or both. Some winter annuals, including winter cereal crops like wheat, must be planted in fall so the plant can experience winter vernalization, a necessary exposure to cold temperatures that is essential for flowering and seed production in the spring. Vernalization is not a requirement for all winter annual plants, especially wild plants like weeds that would suffer as species if they experienced mild winters and could subsequently not reproduce.

As a group, winter annual plants all prefer cool to moderate temperatures and so thrive outside the confines of summer conditions. The duration of a winter annual life cycle is comparable to that for summer annuals but will often occupy portions of two calendar years. Northern regions, which favor relatively short life cycles for summer annual plants, will recognize longer life cycles for winter annuals. Conversely, regions of the country like the south, where summer annual life cycles may be longer, will also tend to experience shorter life cycles for winter annuals. Subtropical or tropical areas have few winter annuals since there is an extremely short or nonexistent cool period in which these plants are able to thrive. In some cases, annual weeds can be either summer or winter annuals, depending upon the particular climate in an area.

Much like summer annuals, winter annuals will undergo both a vegetative growth phase and a reproductive growth phase. Even when winter annuals germinate in fall, they experience relatively

TURFTIP

As a group, winter annual plants all prefer cool to moderate temperatures and so thrive outside the confines of summer conditions.

little growth prior to spring, when growth conditions are much more suitable. The active vegetative growth phase for winter annuals can begin as soon as January in more temperate climates and can last well into the month of May in northern climates. Leaf and root development are the primary emphases of winter annuals during the vegetative phase of growth. Grasses can produce secondary shoots and leaves (called tillers) while broadleaf plants will tend to branch out and become more robust. Both broadleaf and grassy winter annuals in turf will remain fairly compact during the vegetative phase.

Winter annuals undergo a transformation to a reproductive phase that is biologically similar to that for summer annuals but which occurs at a different time of the year. Daylength and temperature are again the most important environmental stimuli that induce the reproductive phase. Depending upon conditions at a particular location, the reproductive phase for most winter annuals can occur as soon as March and as late as June. Typically, the longer days and warmer temperatures that indicate summer is around the corner will stimulate this transformation. However, early spring warm spells or dry spring conditions can induce the reproductive phase earlier than normal. Also, cool and moist conditions may delay the onset of the reproductive phase in winter annuals. The reproductive phase typically lasts four to eight weeks for most weeds but can be much longer for some species. Annual bluegrass is a winter annual grass that is known as a progressive seed producer, meaning it can consistently produce seed over an extended period when the conditions are right (Figure 19.5). In transition zone areas, annual bluegrass might produce seed for a conventional six week period and is limited in this regard by the hot temperatures that are customary in this climate. However, in a maritime climate like the Pacific Northwest, annual bluegrass is capable of producing seed over a three to four month period!

Many of the biological changes that accompany transition into the reproductive phase are similar between summer and winter annuals. Winter annual grasses can produce large quantities of seed but tend not to produce the large seed-bearing stalks that many sum-

FIGURE 19.5 Annual bluegrass in the flowering stage of growth.

mer annual grasses do. Many winter annual grasses are adapted to coexist with managed turf. By virtue of this adaptation, they can often still produce seed while in a compact form that can tolerate closer mowing practices. Annual bluegrass is perhaps the best example of a winter annual grass species that invades and occupies turf. In many climates, annual bluegrass is so prolific that it is managed as the featured turf species. It can tolerate mowing heights as close as those associated with putting greens and can still produce flowers and seeds at these kinds of mowing heights. Because of this adaptation, management or control of annual bluegrass is often better served by trying to chemically suppress seed production. Winter annual broadleaf weeds in turf are often different from summer annual broadleaf weeds in that they are more compact in the vegetative phase but tend to grow much taller as they mature and flower (Figures 19.6 and 19.7). Common turf weeds like henbit and chickweed are both examples of this trend. Vertical growth during the reproductive phase subjects these weeds to practices like mowing but they are often very dense, allowing them to still reproduce effectively. By contrast, summer annual broadleaf weeds in turf (e.g. spotted spurge) tend to remain prostrate, even as they mature into the reproductive phase (Figure 19.8).

A less common life cycle for weeds is a biennial. The term translates roughly to "two years" and, true to the translation, these weeds usually have a two-year life cycle. These weeds will characteristically germinate in the first year and develop into a prostrate, vegetative

FIGURE 19.6 A juvenile or seedling henbit plant.

FIGURE 19.7   A mature henbit plant, which is taller and in the flowering stage.

form called a rosette (Figure 19.9). As for some winter annuals, biennials require a vernalization period (during winter) between the first and second year of growth in order to flower and reproduce. The second year of growth proceeds much like a winter annual, in that warmer spring temperatures and longer days will induce flowering and seed production in the plant. Comparatively few weeds feature this life cycle but they can often be confused with winter annuals since the flowering period is similar.

TURFTIP

The common ground that all perennials share is some ability to withstand all seasons.

Weeds Found in Turf and Landscapes 411

FIGURE 19.8 Prostrate growth is still present in spotted spurge, even at a more mature stage of growth.

The final primary life cycle for weeds is perennial. Many of the most common weeds in turf and landscapes fit into this category, which is unfortunate because many perennials are more difficult to manage and control than annual or biennial weeds. The term

412 Turfgrass Installation: Management and Maintenance

FIGURE 19.9 A common rosette form found in the first year of a biennial life cycle.

perennial implies longevity and many vigorous perennials can indeed persist for many years. By definition, a perennial must be able to persist for more than two years. Some persist for little more than this required period of time while others might last indefinitely under adequate growth conditions. Managed turfgrasses are themselves perennials, which of course is desirable from a maintenance standpoint. The common ground that all perennials share is some ability to withstand all seasons.

Most perennial weeds can and will produce seeds to further the development of the species (Figure 19.10). This serves the function of allowing perennial weeds to spread but doesn't result in the death of the original plant, as would be the case for annual weeds. What distinguishes perennials is the existence of some sort of storage structure that allows them to store food reserves and survive both summers and winters. These structures will vary among biological groups

FIGURE 19.10   Perennial weeds like yellow nutsedge still produce seeds as a means of survival.

of weeds and even among species in a similar group. Storage structures common in perennial weeds are listed with example species in Table 19.1.

Roots are common storage organs for all plants. They are therefore one of the most common means by which perennial weeds persist from year to year. A highly robust and vigorous root system is a common characteristic of most perennial grasses. Orchardgrass, as exampled above, relies exclusively upon its root system while other perennial grasses may have rhizomes, stolons, or both to supplement the storage capabilities of the roots.

Taproots are exclusive to broadleaf weeds and are thus present in both annual and perennial species. However, the taproot in an annual serves simply to optimize the plant's ability to grow during its life cycle while the same structure in perennials serves additionally as the primary food reserve storage area during winter months.

Rhizomes and stolons are generally more common in grasses than in broadleaf weeds but can exist in both types of weeds. Both structures are essentially lateral or horizontal stems. Rhizomes develop underground while stolons develop at the soil surface (Figure 19.11). Since either type of structure serves both to facilitate spread of the weed during the growing season and to store food reserves, they are almost exclusively found in perennial weeds. The short life cycle of annual weeds would make unnecessary the creation and devel-

TABLE 19.1 Common perennial weed storage structures with an example weed species.

Storage Structure	Example Species
Vigorous root system	Orchardgrass
Taproot	Dandelion
Rhizome	Bermudagrass
Stolon	Field bindweed
Bulb/Corm	Wild garlic
Nutlet	Yellow Nutsedge

Weeds Found in Turf and Landscapes 415

FIGURE 19.11 Patches of white clover develop from horizontal stolons.

opment of such structures. Rhizomes and stolons emerge from the original growing point of perennial weeds and spread outward. The extent to which a rhizome or stolon can spread is determined by whether the plant has determinate or indeterminate growth.

416 Turfgrass Installation: Management and Maintenance

Determinate growth is more limited, with stolons or rhizomes able to progress for a distance and then produce a secondary shoot or "daughter plant" from a new growing point called a node. The node can generate both roots and new leaves. Determinate weeds produce only a single node from a particular rhizome or stolon and progress no further. An example of this kind of growth is found in the turfgrass species Kentucky bluegrass, which produces determinate rhizomes. Weeds with indeterminate rhizome or stolon growth can produce these same structures with many nodes, allowing for rhizomes and stolons that can extend for much larger distances from the original plant. Aggressive spreading species like field bindweed or bermudagrass have this indeterminate form of growth (Figure 19.12).

Bulbs, corms, and nutlets are all underground storage structures that are exclusive to monocot weeds other than grasses. They are

FIGURE 19.12 Perennial weeds like bermudagrass, which have indeterminate stolon and rhizomes, can aggressively encroach into established turf.

therefore commonly found in cultivated plants like irises or tulips but also in some weeds, like wild garlic or many of the sedge species. Because these structures are not found in grasses or broadleaf weeds, their relative importance pales in comparison to other perennial storage structures. Bulbs and corms are very similar in appearance and are more common in cultivated plants than in weeds. Weedy species like wild garlic tend to have much smaller bulbs than cultivated plants like tulips or edible garlic, most likely a function of years of breeding efforts with these domesticated plant species. Bulbs in weeds are round or elliptical in shape and store ample food reserves to produce new plant shoots when growing conditions are suitable (Figure 19.13). Nutlets are exclusive to perennial sedge species and are most common in yellow and purple nutsedge. They are produced as extensions from the roots of these species and, like bulbs, give rise to new shoots when conditions permit it.

Storage structures in perennial weeds make controlling these weeds more difficult than for annuals. Because new plant growth can arise from these structures, destroying a perennial weed means one must also destroy the storage organs. While some perennials have only a single means of storing food reserves, others have multiple means. Not coincidentally, those weeds that have more than one place to store food reserves are also those that are more difficult to manage and control. More on this topic will be discussed in the next chapter.

TURFTIP

Storage structures in perennial weeds make controlling these weeds more difficult than for annuals.

418 *Turfgrass Installation: Management and Maintenance*

FIGURE 19.13   Bulbs from a wild garlic plant.

WEED BIOLOGY: BROADLEAF WEEDS, GRASSES, AND SEDGES

The biology of weeds is just as important to understand as the life cycle if weed control is to be optimized. As will be discussed in the next chapter, chemical weed control is founded upon using herbicides that are toxic to weeds but comparatively safe to turf. Much of the basis for this has to do with the biology of the target weeds. The more biological similarity a weed has to turf, the more difficult it will be to control. As we proceed through discussion of the dif-

TURFTIP

Weeds will tend to fall in one of three primary biological categories: broadleaf, grass, or sedge.

ferent biological groups that weeds can fall into, think of how these plants resemble or differ from turfgrass as these similarities/differences have tremendous bearing on what chemicals will effectively control weeds. Weeds will tend to fall in one of three primary biological categories: broadleaf, grass, or sedge.

Broadleaf weeds are a diverse group of plants that represent many taxonomic families. All of these families and species do fall within the dicot order of plant taxonomy. The term dicot is often used to describe broadleaf plants but actually refers specifically to the seeds of these plants. Dicot (actually dicotyledonae) is a Latin term that crudely translates to "two seed leaves". Essentially, seeds of broadleaf plants have two halves that develop into two seed leaves as the plant germinates. This is why young broadleaf seedlings have two leaves. Another practical way to visualize dicot seed structure is to think of a broadleaf seed crop like a peanut, which can be easily split into two halves. Taxonomy aside, broadleaf plants have numerous common identifying features that can easily separate them from other weeds like grasses or sedges. Examples of these features include:

- *Branching stems with leaves situated away from the primary stalk*
- *Netted (think of a maple leaf) vein arrangement in the leaves*
- *Taproot-based root system*
- *Larger and often showy flowers*

- *Leaves that are usually broad or more round in shape*

As might be expected from Mother Nature, there are always exceptions to the above examples. However, these characteristics can provide a fundamental basis for differentiating dicot or broadleaf plants from other plants that may occur as weeds in turf.

To the well-trained eye, grasses and sedges that are considered weeds in turf are fairly easy to differentiate from desired turf. However, because they have more visual similarities to many turfgrass species, they are sometimes more difficult to recognize than broadleaf weeds. As was discussed previously, broadleaf weeds fall into the dicot order of plant taxonomy. By contrast, grasses (including turfgrass species) and sedges fall into the monocot order of plants. By definition, the term monocot means the seed has one primary food storage compartment and therefore will sprout as a single leaf. However, this taxonomic distinction has less practicality for the practitioner so we will focus on some more recognizable characteristics of these plants.

Grasses and sedges, as a group, have thinner, more elongated leaves than broadleaf plants so they are often easily distinguishable in this regard. In addition to the general shape of the leaves, some other common identifying characteristics for grasses and sedges are:

- *Non-branching and compressed stems with leaves attached to the primary stalk*
- *Parallel vein arrangement in the leaves*
- *Fibrous or netted based root system*
- *Small and usually non-showy flowers*
- *A basal growing point (crown) near the surface of the ground*

Unlike broadleaf plants, grasses and sedges offer few exceptions to the guidelines highlighted above. However, there are also some fundamental differences between grasses and sedges, even though they often appear to very similar at first glance. Sedges are usually shinier and have more rigid leaves than grasses. When conditions

TURFTIP

The biological differences between grasses and sedges often result in different herbicide recommendations for what will effectively control these plants as weedy invaders.

are good for both types of plants to grow, sedges will tend to grow faster than grasses. Using a hypothetical example, one week after a lawn with sedge weeds in it has been mowed, the sedges will stand taller than the turfgrass, often by a significant margin. Sedge plants have triangular stems while grasses have round stems. Pulling up a plant from each group and rolling the stem in your fingers will help assist in identifying this biological difference. Another more visual difference is how leaves are arranged on the stem. Grasses will develop leaves that emerge from two sides of the stem while sedges develop leaves that emerge from three sides of the stem. Looking down from directly above a growing grass or sedge plant should help expose this difference to a practitioner (Figure 19.14). As will be discussed in the next chapter, these biological differences between grasses and sedges often result in different herbicide recommendations for what will effectively control these plants as weedy invaders.

BROADLEAF WEEDS COMMONLY FOUND IN TURF AND LANDSCAPES

Broadleaf weed control encompasses the greatest proportion of annual dollars spent on turfgrass and landscape pest control. This may or may not be the case in commercial turfgrass systems but is true when one considers expenditures from the residential con-

FIGURE 19.14 Note how the leaves on a yellow nutsedge plant emerge in three directions. Grasses, by contrast, only develop leaves in two directions.

sumer sector. Table 19.2 lists some of the most common broadleaf weeds found in turfgrasses and landscapes, sorted by their life cycles. While they may seem cumbersome, scientific names are important because common names can vary considerably across regions.

Because this text is not devoted exclusively to weed control, it is not feasible to discuss each of the weeds listed in Table 19.2 in detail. However, I will attempt to cover the spectrum of the weeds in the list as completely and as illustratively as possible. For weeds listed that aren't covered in the sort of detail that meets your needs, I encourage you to consult with your local turfgrass and/or weed control university expert for more guidance.

Annual broadleaf weeds are some of the most common weeds found in turfgrass and landscapes. Figures 19.15 through 19.22 depict some of the common summer and winter annual broadleaf

Weeds Found in Turf and Landscapes 423

TABLE 19.2 Common broadleaf weeds found in turf and landscape systems.

Weed (common name)	Weed (scientific name)	Life cycle
Carpetweed	Mollugo vercillata	Summer annual
Prostrate knotweed	Polygonum aviculare	Summer annual
Common purslane	Portulaca oleracea	Summer annual
Prostrate pigweed	Amaranthus blitoides	Summer annual
Common lespedeza	Lespedeza striata	Summer annual
Black medic	Medicago lupulina	Summer annual
Eclipta	Eclipta prostrata	Summer annual
Florida pusley	Richardia scabra	Summer annual
Spotted spurge	Euphorbia maculata	Summer annual
Purple cudweed	Gnaphalium purpureum	Summer/Winter annual
Catchweed bedstraw	Galium aparine	Summer/Winter annual
Prickly lettuce	Lactuca serriola	Summer/Winter annual
Common groundsel	Senecio vulgaris	Winter/Summer annual
Mayweed chamomile	Anthemis cotula	Winter/Summer annual
Horseweed	Conyza canadensis	Winter/Summer annual
Common mallow	Malva neglecta	Winter/Summer annual
Knawel	Scleranthus annuus	Winter/Summer annual
Hairy bittercress	Cardamine hirsuta	Winter/Summer annual
Common chickweed	Stellaria media	Winter annual
Henbit	Lamium amplexicaule	Winter annual
Purple deadnettle	Lamium purpureum	Winter annual
Small hopclover	Trifolium dubium	Winter annual
Lawn burweed	Soliva pterosperma	Winter annual
Yellow rocket	Barbarea vulgaris	Winter annual/Biennial
Wild carrot	Daucus carota	Biennial
Bull thistle	Cirsium vulgare	Biennial
Wild geranium	Geranium carolinianum	Biennial
Common yarrow	Achillea millefolium	Perennial
Mugwort	Artemisia vulgaris	Perennial
Florida betony	Stachys floridana	Perennial

Continued on next page

TABLE 19.2 *(continued)* Common broadleaf weeds found in turf and landscape systems.

Weed (common name)	Weed (scientific name)	Life cycle
English daisy	Bellis perennis	Perennial
Oxeye daisy	Chrysanthemum leucanthemum	Perennial
Chicory	Cichorium intybus	Perennial
Dichondra	Dichondra carolinensis	Perennial
Dogfennel	Eupatorium capillifolium	Perennial
Dandelion	Taraxacum officinale	Perennial
Mouseear chickweed	Cerastium vulgatum	Perennial
Birdseye pearlwort	Sagina procumbens	Perennial
Field bindweed	Convolvulus arvensis	Perennial
Birdsfoot trefoil	Lotus corniculatus	Perennial
White clover	Trifolium repens	Perennial
Ground ivy	Glechoma hederacea	Perennial
Healall	Prunella vulgaris	Perennial
Common blue violet	Viola papilionacea	Perennial
Yellow woodsorrel	Oxalis stricta	Perennial
Broadleaf plantain	Plantago major	Perennial
Buckhorn plantain	Plantago lanceolata	Perennial
Red sorrel	Rumex acetosella	Perennial
Curly dock	Rumex crispus	Perennial
Moneywort	Lysimachia nummularia	Perennial
Virginia buttonweed	Diodia virginiana	Perennial
Bulbous buttercup	Ranunculus bulbosus	Perennial
Slender speedwell	Veronica filiformis	Perennial
Poison ivy	Toxicodendron radicans	Perennial
Silvery thread moss	Bryum argenteum	Perennial

weeds found in turfgrass and landscape systems. One of their common characteristics, although they and those not pictured from Table 19.2 represent numerous plant families, is that they are fairly low growing in the vegetative phase. This feature allows these annuals to thrive, even under regular mowing practices.

Weeds Found in Turf and Landscapes 425

FIGURE 19.15 Carpetweed, a common summer annual broadleaf weed in turf and landscapes.

Common summer annual broadleaf weeds represent many different plant families. However, there are groups of these weeds that are taxonomically similar to one another and, where relevant, these weeds will be discussed together. Carpetweed is a summer annual with long, narrow leaves that emerge in a circular arrangement from the stem (Figure 19.15). Carpetweed, like many other summer annuals, has a prostrate growth form when in the juvenile phase of

426 Turfgrass Installation: Management and Maintenance

FIGURE 19.16 Prostrate knotweed is a low-growing species that is one of the earliest summer annual weeds to germinate.

FIGURE 19.17 Black medic is similar to clover but has an extended central leaflet that sets it apart.

FIGURE 19.18 Spotted or prostrate spurge is usually identifiable by the dark watermark found on the leaves.

growth. Similar species are catchweed bedstraw, which also has the circular leaf arrangement and is best known for its tendency to stick to clothing, eclipta, which is a common weed in southern regions, and horseweed (sometimes called marestail), which is a more upright growing species that thrives in lower maintenance or native areas (Figure 19.20).

Some of the best-known summer annual broadleaf weeds have earned their reputations by thriving in areas that are not as suitable for growth of turf and other plants. Among the most note-

428 Turfgrass Installation: Management and Maintenance

FIGURE 19.19 Prickly lettuce looks a lot like thistles but is less spiny to the touch.

worthy of these is prostrate knotweed, a prostrate summer annual that can form dense patches, especially in highly compacted areas (Figure 19.16). This plant germinates very early in the growing season but is not easily identifiable until it is mature. The leaves of knotweed are small throughout its life cycle and it is perhaps best

Weeds Found in Turf and Landscapes **429**

FIGURE 19.20 Horseweed is a common invader of low maintenance turfgrass areas or landscapes.

defined via the ochrea, a papery appendage that appears at the point where leaf-bearing stems develop. The ochrea is a feature only common to plants in the Polygonum genus, which also includes field crop weeds like smartweed and ladysthumb. This unique characteristic can thus be used to help differentiate knotweed from other low-growing weeds. A species often confused with prostrate knotweed is prostrate or spotted spurge. Spurge will also have a low and spreading growth habit and can similarly form dense patches in compacted areas (Figure 19.18). It is best told apart from knotweed by its leaves, which are both more round in shape and have a distinct dark watermark. Both of these species can be solid indicators of compacted soils that may need to be aerated before turf will again successfully grow. Other annual species that can develop in compromised areas include prostrate pigweed and puncturevine, the latter of which can be a hazard to applicators and equipment due to its sharp spines.

Some common broadleaf annual weeds are close relatives of the well-known white clover, which is itself a perennial. Such annual species include summer annuals like common lespedeza and black medic, and the winter annual hop clover. These species all look like clovers in that they have leaves with three smaller leaflets but can be distinguished from white clover upon closer inspection. Common lespedeza is a species that is similar in appearance to clover but that has noticeable veins that extend towards the perimeter of the leaflets. It also has purple flowers, instead of white as seen on white clover, and is more common in southern regions. Both black medic and hop clover have yellow flowers but black medic can be distinguished by the fact that its middle leaflet is born on a short stalk (Figure 19.17).

Black medic is biologically more similar to alfalfa than it is to clovers. Hop clovers are close relatives of white clover and have a similar prostrate growth habit, so use of flower color can be an excellent means of telling them apart from their perennial cousin.

Common winter annual broadleaf weeds in turf include chickweed and henbit. Common chickweed is a pale green winter annual with smooth, teardrop-shaped leaves (Figure 19.21). Its vertical growth during the later part of its life cycle can make it a visual distraction but mowing can help minimize the visual impact of this weed.

TURFTIP

Common winter annual broadleaf weeds in turf include chickweed and henbit.

Weeds Found in Turf and Landscapes 431

FIGURE 19.21 Common chickweed is a winter annual with small, teardrop-shaped leaves.

A similar species is mouseear chickweed, a perennial that resembles common chickweed but is a much hairier plant. Henbit is a unique weed in that it has numerous identifiable characteristics. It has showy purple flowers in the mature form and, in sufficiently large populations, can thus have some ornamental attributes (Figure 19.22).

FIGURE 19.22 Henbit is a winter annual with square stems, showy purple flowers, and a minty fragrance.

Henbit is a member of the mint family and thus has square stems, rendering it unique among weed species. Its leaves are also very closely attached to the stem, which differs from the leaf arrangement on most broadleaf weeds. The heavily lobed leaves of henbit are similar to those of ground ivy, a creeping perennial that is a common turf invader. Other less common winter annual broadleaf weeds include common groundsel, mayweed chamomile, knawel, hairy bittercress, and common mallow, the last of which is readily identified by its branching, maple leaf-style leaf vein arrangement.

Figures 19.23 through 19.25 show some common biennial broadleaf weeds that are commonly found in turfgrass or landscapes. These weeds are again often confused for winter annuals because both will tend to flower in the spring. They also may be confused for perennial weeds, due to their persistence in the vegetative form during the first year of their life cycle. Since this life cycle includes so few weeds, proper recognition and identification of these species is critical to properly plan a weed control program.

Weeds Found in Turf and Landscapes   433

FIGURE 19.23   Yellow rocket, a broadleaf weed that is a winter annual in some climates but more commonly has a biennial life cycle.

Common biennial broadleaf weeds can be placed into two categories to assist with their identification. The first is the true weedy category, within which can be placed yellow rocket and bull thistle. Yellow rocket is a member of the mustard family but has no value as a source of edible mustard. The plant forms a dense, low-growing rosette in the first year of its life cycle and then undergoes a vertical growth spurt when it enters its reproductive phase (Figure 19.23). The yellow flowers and rapid growth spurt contribute to the naming of this weed species. The rosette form can tolerate mowing during the earlier portions of the life cycle. Bull thistle is one of several thistle species that have the biennial life cycle. It and other biennial thistles also form a dense rosette during the first year of their life cycle

FIGURE 19.24 Bull thistle, a biennial broadleaf weed, displays the classic rosette vegetative form during the first year of its life cycle.

but are easily distinguishable from other weeds by the leaf spines common to most thistles (Figure 19.24). Bull thistle, like yellow rocket and other biennials, will bolt or grow rapid vertical in its second year and produce showy flowers before releasing seeds. The weedy nature of thistles is both due to their low, dense growth form at the juvenile stage and the spiny leaves that can be painful to the touch. A species resembling many thistles is prickly lettuce but this annual is less spiny than the thistles it is often confused with (Figure 19.19).

Weeds Found in Turf and Landscapes 435

FIGURE 19.25 Wild geranium, a biennial broadleaf weed that has similar physical characteristics to its cousins in the flower and nursery industry.

The second category for biennials is an ornamental category, within which wild carrot and wild geranium can be placed. The reason for this categorization of these weeds is their use as or similarity to common ornamental plants. Wild carrot is often called Queen Anne's lace and is a common decorative addition to many floral arrangements. As a weed species, wild carrot can be easily identified by its umbrella-shaped flower head that gives the plant ornamental value. This shape of the flowering structure is also an identification aid for the perennial species, common yarrow. The vegetative form is also very similar to the leaves of cultivated carrot or other broadleaf species like dogfennel. Wild geranium has no ornamental value but is very similar in appearance to the geranium varieties that are sold as flowering plants (Figure 19.25). Wild geranium leaves are smaller than those of their commercial cousins but the shapes of the leaves can help identify this plant.

Some of the most widely recognized and also most troublesome broadleaf weeds in turf and landscapes are perennials. Figures 19.26

FIGURE 19.26 Dandelion, one of the most commonly recognized perennial broadleaf weeds in the green industry.

Weeds Found in Turf and Landscapes 437

FIGURE 19.27   Field bindweed is one of the most troublesome broadleaf weeds to manage, due to its invasiveness.

FIGURE 19.28 White clover, another of the most easily recognized perennial broadleaf weeds, can form large patches in managed turf areas.

Weeds Found in Turf and Landscapes **439**

FIGURE 19.29 Yellow woodsorrel (often called oxalis) can often be mistaken for clover because of its similar leaf structure.

FIGURE 19.30 Broadleaf plantain is unique among broadleaf weeds in that it has parallel leaf veins, normally a characteristic of grasses.

FIGURE 19.31 Curly dock is one of the larger broadleaf weeds found in turf and landscapes and is often found in areas with lower maintenance.

Weeds Found in Turf and Landscapes 441

FIGURE 19.32 Speedwells are a family of low-growing broadleaf weeds that can easily escape the routine practice of mowing.

through 19.32 depict some common perennial broadleaf weeds found in turf and landscapes. These are the most difficult broadleaf weeds to control effectively, due to their perennial storage structures and sometimes invasive characteristics. The next chapter will discuss in detail how control strategies must differ to specifically and effectively target these types of weeds.

Without a doubt, the most commonly recognized and the most widespread perennial broadleaf weed in turf is dandelion. Dandelion has a low growth habit that gives it excellent tolerance to mowing. It exists primarily in the vegetative form but can flower prolifically in either fall or spring. Flowers are bright yellow and can regenerate rapidly if they are mowed off. Eventually, flowers develop into a seed producing puffball that relies on wind or young children for seed dispersal (Figure 19.26). Dandelion persists with the help of a deep taproot that allows it to survive stressful periods. Wind-dispersed seed can result in dense dandelion infestations if existing weeds are not controlled.

Another very common perennial broadleaf weed species is white clover. It has a very low growth habit and can thus also escape the effects of regular mowing. White clover has the classic trifoliate leaf structure common to many legume plants and often has a whitish watermark on the leaflets (Figure 19.28). It spreads with the help of stolons, and can form dense patches in turf when left uncontrolled. White clover produces small, white, ball-shaped flowers that additionally assist in its development. Other common perennial weeds in turf that are similar to clover include birdsfoot trefoil and yellow woodsorrel. The latter species, sometimes referred to simply by its genus name oxalis, also has the trifoliate leaf structure but leaflets are more heart-shaped and the plant produces yellow flowers (Figure 19.29). While the dense development of white clover can gradually choke out desired turfgrasses, its stolons do not render it highly invasive. A stoloniferous species that can be highly invasive is field bindweed. Its rapid spreading ability and ability to thrive in dry areas can make it extremely difficult to manage. Field bindweed has been a problem in field crops for years but has more recently become a growing issue in many turfgrass and landscape areas. Its arrow-shaped leaves make it relatively easy to identify (Figure 19.27).

Some perennial broadleaf weeds are more sensitive to mowing and can be more common in less maintained areas or in landscapes. Examples include plantains, curly dock, red sorrel, mugwort, Florida betony, and chicory. Plantains are unique among broadleaf weeds in that they have a parallel leaf vein arrangement that is more common in monocot weeds. Two common plantain species exist in turf and landscapes. Broadleaf (sometimes called blackseed) plantain has shorter, thicker leaves (Figure 19.30) while buckhorn plantain has longer, narrower leaves. Both species can tolerate some mowing but tend to thrive in areas receiving less maintenance. Curly dock is a species with large leaves that have curled edges (Figure 19.31). It and its close cousin red sorrel have little tolerance to mowing but can become very robust in landscapes or native areas adjacent to turf stands. Mugwort, Florida betony, and chicory are less common in most turf areas but are very common in both landscapes and the container nursery industry.

A low growth habit is a common feature of many perennial broadleaf weeds and the speedwell family is an excellent example of this trait. A common weedy species of speedwell in turfgrass is slender speedwell, which has small, lobed leaves and can form thick patches if left uncontrolled (Figure 19.32). Other small-leaved prostrate perennial broadleaf weeds include daisies, the southern species Virginia buttonweed, buttercup, dichondra, and blue violet. Dichondra and violet can both be identified by their leaves, which are roundish and cup towards the stems. Dichondra is often used as a ground cover in southern regions but its ability to form dense patches can make it a troublesome weed. Perennial violet is one of the most troublesome broadleaf weeds in landscapes, due to its persistence even when control measures are employed. An additional low-growing problematic weed is moss, which falls into its own unique taxonomic class. Because mosses are more primitive and don't have the characteristics of many higher plants, control options must also be unique. Mosses are common in moist, shaded areas and have become a growing problem on golf courses, especially on putting greens.

GRASSY WEEDS AND SEDGES COMMONLY FOUND IN TURF AND LANDSCAPES

Grasses and grass-like plants are often more challenging weed problems for turfgrass practitioners because they have more biological similarity to the turfgrasses we manage. These similarities can also make proper identification of these weeds difficult. Most broadleaf weeds have numerous identifiable characteristics and can often be identified based solely on leaf shape. The grassy weeds often do not offer such clues and it may require closer inspection or consultation with experts or peers to properly identify these types of weed species. There are generally fewer grassy weeds that turfgrass and landscape managers must contend with than broadleaf weeds but control options are also more limited. It is for this reason that this aspect of weed management is usually regarded as one of the most difficult

TABLE 19.3 Common grass weeds found in turf and landscape systems.

Weed (common name)	Weed (scientific name)	Life cycle
Field sandbur	Cenchrus longispinus	Summer annual
Large crabgrass	Digitaria sanguinalis	Summer annual
Smooth crabgrass	Digitaria ischaemum	Summer annual
Goosegrass	Eleusine indica	Summer annual
Yellow foxtail	Setaria glauca	Summer annual
Crowfootgrass	Dactyloctenium aegyptium	Summer annual
Broadleaf signalgrass	Brachiaria platyphylla	Summer annual
Annual ryegrass	Lolium multiflorum	Winter annual
Annual bluegrass	Poa annua	Winter annual
Tufted hardgrass	Sclerochloa dura	Winter annual
Little barley	Hordeum pusillum	Winter annual
Bermudagrass	Cynodon dactylon	Perennial
Orchardgrass	Dactylis glomerata	Perennial
Tall fescue	Festuca arundinacea	Perennial
Nimblewill	Muhlenbergia schreberi	Perennial
Dallisgrass	Paspalum dilatatum	Perennial
Bahiagrass	Paspalum notatum	Perennial
Roughstalk bluegrass	Poa trivialis	Perennial
Quackgrass	Elytrigia repens	Perennial
Torpedograss	Panicum repens	Perennial
Johnsongrass	Sorghum halepense	Perennial

TABLE 19.4 Common sedge or sedge-like weeds found in turf and landscape systems.

Weed (common name)	Weed (scientific name)	Life cycle
Annual sedge	Cyperus compressus	Summer annual
Texas sedge	Cyperus polystachyos	Summer annual
Spreading dayflower	Commelina diffusa	Summer annual
Yellow nutsedge	Cyperus esculentus	Perennial
Globe sedge	Cyperus globulosus	Perennial
Purple nutsedge	Cyperus rotundus	Perennial
Green kyllinga	Kyllinga brevifolia	Perennial
Slender rush	Juncus tenuis	Perennial
Wild garlic	Allium vineale	Perennial
Wild onion	Allium canadense	Perennial
Star-of-Bethlehem	Ornithogalum umbellatum	Perennial

for practitioners. Tables 19.3 and 19.4 list some of the most common weedy grass species and grass-like weeds, respectively, that turfgrass and landscape practitioners must contend with.

We will now cover some details concerning these weeds, placing them in common groups where it is applicable. Again, it will not be feasible to cover them all in detail but, as you encounter unknown grassy or sedge-like weeds, focus first on whether the plant is a grass or a different kind of monocot plant. From there, proceed to looking for structures that would point to the plant as being annual or perennial. These decisions will be critical to determining a proper course of action. If the identification of the plant remains a mystery, use your local extension resources to firm up the identity of the plant before pursuing control strategies.

Figures 19.33 through 19.39 show some common summer and winter annual grasses found in turf and landscapes. Because these grasses are annual, they can be much easier to control effectively than perennial grasses. However, as a group, they produce large quantities of seed and can be perennial problems for turfgrass managers.

FIGURE 19.33 Field sandbur can pose particular difficulties because of the spiny burrs that can be a nuisance to those who encounter them.

FIGURE 19.34 Crabgrass is one of the most widespread summer annual grassy weeds and represents a large proportion of annual herbicide expenditures.

Weeds Found in Turf and Landscapes **447**

Some of the annual grasses listed in Table 19.3 and pictured may be more familiar to you than others. Let's briefly cover some of the environments and/or areas where they are most common. Crabgrass is one of the most widespread and recognizable grassy weeds in turf (Figure 19.34). Its broader and shorter leaf shape makes it easily distinguishable from turf. The two most common species of this weed are large and smooth crabgrass. They may appear at first glance to be the same but large crabgrass has hairy leaves, upon closer inspection, while smooth crabgrass does not. The good thing is that control strategies for the two species do not differ. Another similar species is southern crabgrass, which exists only in more tropical areas of the country. Crabgrass has excellent tolerance to mowing and can be an invader of a wide variety of turfgrass settings, from lawns to golf course putting greens.

Field sandbur is most notorious for its spiny burrs that can plague people and animals that encounter it (Figure 19.33). It does very well in hot, dry climates such as the Great Plains states. It is highly preva-

FIGURE 19.35 Goosegrass is easily distinguishable by its open-faced growth habit and "wheel spoke" center.

lent in waste areas or roadsides but is also common in managed turf. Goosegrass is a summer annual grass more common to golf course turf than in many other areas relevant for turfgrass and landscape managers (Figure 19.35). It tends to develop later in summer than crabgrass and prefers hotter climates, thus making it more prevalent in the transition zone and in southern areas. It's open and prostrate growth habit when mature makes it highly undesirable as a weed. Goosegrass thrives on compacted soils and can often indicate that compaction is a problem for a particular location. Foxtails are a group of summer annual grasses that are common invaders of field crops and are becoming more prevalent weeds in turf and landscapes (Figure 19.36). The soft seedhead gives foxtails their common name and is a highly recognizable trait, although some species can handle reasonable mowing heights and still persist. Yellow foxtail is one of the more common foxtail species in turf but others include green or giant foxtail. Like for sandbur, they are most common in low maintenance areas but are a growing concern in managed turf, especially when new housing developments and golf courses expand into previously rural areas. Other summer annual grass species, like crowfootgrass and broadleaf signalgrass, are more prevalent in southern regions of the country.

Among the common winter annual grassy weeds in turf, annual bluegrass is the most recognized and most widespread species (Figure 19.37). It struggles in very hot climates so is usually limited to the transition zone or northern states. Its prolific growth in cool climates makes it often the turfgrass species of choice since removal of it from other species would be too great and too perennial a challenge. Two unique biotypes exist for annual bluegrass. One is a true annual that is a prolific seed producer while the other is a weak perennial that produces less seed and is most commonly found as a competitor of creeping bentgrass on golf course putting greens. The widespread management of annual bluegrass as a turf in the northern US has resulted in breeding efforts to select varieties that can be used more deliberately. Annual ryegrass is another winter annual grass that has a history of turf use, especially as a species for overseeding warm-season grasses in the southern US. Higher quality species like perennial ryegrass and rough bluegrass have replaced

Weeds Found in Turf and Landscapes 449

FIGURE 19.36 Foxtails are traditionally thought of as field crop weeds, but are a growing problem in turf due to the progression of urban landscapes into rural areas.

annual ryegrass for this purpose in the modern era. Often called Italian ryegrass, annual ryegrass is less of a problem in turf than in field crops but can often be found in turf areas such as sod farms, where crops used to be grown.

450 Turfgrass Installation: Management and Maintenance

FIGURE 19.37 Annual bluegrass is managed as a turfgrass in many areas but is a troublesome weed for most turfgrass managers.

Other lesser-known winter annual grass weeds include little barley and tufted hardgrass. They are not as widespread as some of the other annual grasses mentioned earlier but they can still present management problems. Little barley is a wild cousin of the barley varieties grown commercially for grain (Figure 19.38). When uncontrolled, little barley can form dense, low-growing clumps and will eventually produce seedheads that resemble those from cultivated barley. This species is less pronounced in urban, developed areas and tends to be more common in turf and landscapes that occupy once rural areas. Tufted hardgrass is a winter annual most prevalent in the central regions of the US (Figure 19.39). It is often confused for annual bluegrass in the juvenile stage of growth. As the plant matures, it develops an open-faced and prostrate growth habit that is similar

Weeds Found in Turf and Landscapes 451

FIGURE 19.38 Little barley is a wild cousin of the cultivated winter cereal crop.

to goosegrass. While tufted hardgrass is thus sometimes prone to be mistaken for goosegrass, its compact seedhead and spring flowering period help differentiate the two species. Another potential cause

FIGURE 19.39 Tufted hardgrass is a growing problem on athletic fields with higher levels of soil compaction.

for confusion between tufted hardgrass and goosegrass is that they both thrive on compacted soils. Hardgrass was discovered as a problem in trafficked areas of athletic field turf by being mistaken for early-season goosegrass by practitioners who were unfamiliar with the species.

Perennial grassy weeds present the biggest challenges to turfgrass practitioners. This challenge is rooted in the fact that practitioners must control an unwanted perennial grass species within turfgrasses that are themselves perennial. This close biological similarity usually translates into few viable options for effective control. Some of the most significant turfgrass renovations or species changes at a facility are based upon this very problem. Figures 19.40 through 19.44 illustrate some of the more prevalent perennial grasses that pose major challenges for turfgrass managers. These same weeds can also be very difficult in landscape settings but more herbicide options exist in these areas with greater numbers of broadleaf plantings.

Some of the perennial grasses listed in Table 19.3 and/or pictured may be more problematic in your area than others. Let's briefly cover some of the environments and/or areas where they are most common. Bermudagrass represents one of the most ongoing love/hate relationships that exist in turfgrass management:

- *Bermudagrass is the most desired and most commonly used turfgrass species.*
- *Tolerance to low mowing*
- *Superior heat tolerance*
- *Drought tolerance*
- *Excellent recuperative ability*
- *Ideal option for southern athletic fields, golf courses, and lawn-type turf.*
- *An annoying weed problem where other turfgrasses are being grown (Figure 19.40).*
- *An aggressive network of both rhizomes and stolons make unwanted bermudagrass both robust and invasive.*

Nimblewill is a perennial grass that can sometimes be mistaken for bermudagrass, due to similar leaf structure. However, nimblewill is less invasive and is more prevalent in shaded areas that the sun-loving bermudagrass might not occupy.

Orchardgrass is a robust perennial species that is not invasive but forms thick clumps that are unsightly in most turfgrass settings (Figure 19.41). The species is a common forage crop and is usually introduced to turfgrass via seed contamination. Many regional seed production facilities, particularly those that produce the Kentucky 31 (or K-31) variety of tall fescue, have large popu-

FIGURE 19.40 Bermudagrass can be a managed turfgrass or a highly invasive weed species, shown here encroaching into zoysia grass.

FIGURE 19.41 Orchardgrass is a common forage grass but can be a persistent weed species in managed turf.

> **TURFTIP**
>
> Tall fescue is recognized as an excellent turfgrass species for lawn turf, particularly in the transition zone. However, it can also form unsightly clumps when it appears as a weed in other turfgrass species.

lations of orchardgrass. Both tall fescue and orchardgrass can be and are used as forages so the existence of both species at these seed production facilities is common. Similar seed size between orchardgrass and tall fescue makes it difficult to remove orchardgrass from commercially available seed and results in the weed's introduction to new turf areas. Few chemical options are available for control of orchardgrass so it can be difficult to get rid of once it has become established. Tall fescue is recognized as an excellent turfgrass species for lawn turf, particularly in the transition zone. However, it can also form unsightly clumps when it appears as a weed in other turfgrass species.

Perennial weedy grasses that are more prevalent in southern regions of the US include dallisgrass, bahiagrass, and torpedo-

> **TURFTIP**
>
> Bahiagrass is sometimes used as a turfgrass species but is perhaps most prevalent as a roadside species that is an effective ground stabilizer.

FIGURE 19.42 Dallisgrass is a very robust perennial weed species in the Southern USA.

grass. Dallisgrass is a robust perennial that has very thick rhizomes (Figure 19.42). It is very commonly found on roadsides in the southern region but can also be a contaminant in close-mowed turf. Under mowed conditions, dallisgrass can form thick clumps and can very difficult to control. Bahiagrass is sometimes used as a turfgrass species but is perhaps most prevalent as a roadside species that is an effective ground stabilizer. It has a very robust rhizome system, which can make it difficult to control when it appears as an undesirable weed species. The characteristic V-shaped seedhead is a classic indicator of this plant's existence during its late summer flowering period.

Torpedograss is a perennial weed species that only exists in subtropical to tropical regions that experience little to no cold weather. Its name is derived from its pointed rhizomes that facilitate its invasive tendencies. This species can be an aggressive invader of any managed turfgrass systems, including aggressive species like bermudagrass. Torpedograss can tolerate a variety of mowing heights and thus can be a problem even on golf course putting greens.

Perennial weedy grasses that present greater challenges for northern practitioners include quackgrass and rough bluegrass. Quackgrass is a species that can spread via rhizomes and can thus be fairly invasive (Figure 19.43). It has some visual similarity to common turfgrass species like tall fescue or perennial ryegrass but can be distinguished by its clasping auricles, which are white, hook-like appendages that circle around the stem at the point where leaves develop. Quackgrass

TURFTIP

Torpedograss can tolerate a variety of mowing heights and thus can be a problem even on golf course putting greens.

458 Turfgrass Installation: Management and Maintenance

FIGURE 19.43 Quackgrass is a spreading perennial that can be a stiff competitor for cool-season turfgrasses.

naturally prefers environments where it can grow taller but it can tolerate lower mowing heights, even those found on golf courses. Rough bluegrass, often referred to by its scientific name of Poa trivialis, can be difficult to differentiate from annual bluegrass. It is not

> **TURF TIP**
>
> Many overseeded golf course putting greens in states like Arizona or Florida use rough bluegrass, since it can adequately tolerate the close mowing that is necessary.

as aggressive, in terms of seed production, as its annual cousin but also is used in many areas as a desired turfgrass.

Many overseeded golf course putting greens in states like Arizona or Florida use rough bluegrass, since it can adequately tolerate the close mowing that is necessary. In similar fashion to annual bluegrass, poor tolerance to heat limits the use of rough bluegrass as a desired turf. Its prevalence as a weed species is similar to that for orchardgrass in that it is a common contaminant of commercial turfgrass seed, particularly seed produced in western Oregon. Rough bluegrass spreads via stolons and can form large patches within other turfgrass species. Its appearance is not always objectionable but its texture differs considerably from species like tall fescue and its poor heat tolerance can result in declining or dead patches in desired turf during summer. Another perennial species that can grow throughout much of the country is johnsongrass. It does not tolerate close mowing but can be a robust contaminant of ornamental areas or in unmowed native grass areas found at many facilities (Figure 19.44).

Sedges and sedge-like species all fall into the same monocot family of plants that also includes grasses. However, this group of plants can present unique challenges for turfgrass managers because many grow at a rate much faster than most turfgrasses. While some of these species tolerate close mowing better than others, their faster growth rate allows them to remain competitive in many managed environments. Figures 19.45 through 19.49 depict some of the common

460 Turfgrass Installation: Management and Maintenance

FIGURE 19.44 Johnsongrass is a tall-growing species that can aggressively invade low maintenance turf areas or native landscape areas.

monocot weed species, other than grasses, which are common in turf and landscape systems.

The most common types of non-grass monocot weeds are sedges. Sedges are biologically more similar to rapid-growing species like rushes and horsetails than they are to grasses. One of the rush species, slender rush, is actually fairly common as a weed in turf and landscapes.

FIGURE 19.45 Yellow nutsedge is one of the most widespread sedge species and can impact both field crops and the green industry.

Numerous sedge species exist as weeds, especially in turfgrass. The most common of these is yellow nutsedge, which thrives in many parts of the country (Figure 19.45). Common also in many field crops, yellow nutsedge is a perennial that can aggressively compete for available resources. The species can reproduce via seed but more commonly from underground tubers or nutlets, which serve as the sources for new plants each year. Generation of new plants from nutlets can increase the density of yellow nutsedge populations but the plants don't otherwise spread laterally. Other annual or perennial sedge species, including annual sedge, Texas sedge, and globe sedge, reproduce primarily through seed production.

Some sedge species are more invasive than the ones mentioned above. Purple nutsedge is a perennial species that looks similar to yellow nutsedge (Figure 19.46). It also relies upon tuber or nutlet production but the rhizomes on which the nutlets are formed can develop into long chains, making this species much more aggressive than yellow nutsedge. Purple nutsedge is more common in the southern US but many agricultural and turfgrass practitioners consider this weed to be one of the most difficult they have to contend with. Another group of sedges that is more common in turf than in production agriculture is the kyllingas. Several kyllinga species have been identified but the most common of these is green kyllinga (Figure 19.47). Its darker green foliage and propensity to form dense patches can make it more difficult to distinguish from desirable turf species

TURFTIP

Purple nutsedge is more common in the southern US but many agricultural and turfgrass practitioners consider this weed to be one of the most difficult they have to contend with.

FIGURE 19.46 Purple nutsedge is a highly invasive sedge species and can be very difficult to eradicate.

and, although it is most commonly found in southern areas, it may be more widespread than scientists and practitioners realize. Green kyllinga develops from seed but also rhizomes that allow the formation of dense patches. The species prefers moist conditions such as low spots where moisture may tend to accumulate.

FIGURE 19.47 Green kyllinga can be easily mistaken for a grass, due to its low growth habit and narrow leaf width.

Other non-grassy monocot weeds fall outside the sedge family. Examples include the summer annual spreading dayflower and the perennials wild garlic, wild onion, and Star-of Bethlehem. Spreading dayflower can form into low-growing large patches in moist areas

TURFTIP

Star-of-Bethlehem is commonly sold as a flowering ornamental bulb species but is prone to escape managed ornamental beds and can develop into patches in turf.

Weeds Found in Turf and Landscapes 465

FIGURE 19.48 Wild garlic is a smaller cousin of cultivated garlic but the smell tells you they are related.

and may resemble broadleaf species like Virginia buttonweed that have a similar growth habit. However, close inspection of spreading dayflower will reveal the parallel leaf vein formation common to all monocots. Wild garlic and wild onion are very similar to one another and both are wild cousins of the cultivated herbs that we may be more familiar with. Wild garlic can grow much taller than

FIGURE 19.49   Star-of-Bethlehem is often sold as an ornamental plant but can be troublesome in undesired locations.

turf it infests and can be further identified by its hollow stems and characteristic garlicky odor (Figure 19.48). Wild onion is similar in appearance but has flat, rather than hollow, stems. Both plants produce small bulbs underground that can give rise to future generations of plants. Star-of-Bethlehem is commonly sold as a flowering ornamental bulb species but is prone to escape managed ornamental beds and can develop into patches in turf (Figure 19.49). While the flowers may be showy, Star-of-Bethlehem can continue to proliferate in turf if uncontrolled and is poisonous if it is consumed.

WEED ECOLOGY AND CULTURAL WAYS OF CONTROLLING WEEDS

There's an old adage that states the best defense against weeds is a good healthy stand of turf. As practitioners, we must remember that weed management is founded upon solid turf management. Insects and diseases directly feed upon or damage turf, and thus seem to prefer turf that is healthy or well managed. Weeds exact a more indirect effect upon either turfgrass systems or landscape plantings. Their role as pests is founded upon their ability to compete with desirable plants for common resources like moisture, sunlight, and nutrients. When turf is disadvantaged or poorly maintained, weeds are usually the first pest problems that result. Is that a coincidence? As was stated earlier, weeds are simply defined as plants out of place. When we provide them with space to occupy, they are very adept

TURFTIP

Insects and diseases directly feed upon or damage turf, and thus seem to prefer turf that is healthy or well managed.

TABLE 19.5 Local origins for common turfgrass species.

Species	Warm or Cool-Season	Origin
Perennial ryegrass	Cool-season	Northern Europe
Creeping bentgrass	Cool-season	Northern Europe
Kentucky bluegrass	Cool-season	Northern Europe
Tall fescue	Cool-season	Northern Europe
Fine fescues	Cool-season	Northern Europe
Bermudagrass	Warm-season	Africa
Zoysiagrass	Warm-season	Eastern Asia
St. Augustinegrass	Warm-season	Africa/Caribbean
Centipedegrass	Warm-season	Southeast Asia
Buffalograss	Warm-season	Central Great Plains

at using that opportunity to their advantage. Over the next few paragraphs, we will discuss/review some of the basics of weed ecology and turfgrass management, and how breakdowns in management can result in weed infestations.

Turfgrasses and landscapes represent a departure from what one could call a natural setting. Species are selected for these areas based upon reasonable adaptation to the local climate. Does this mean they are well suited for the area in which they are planted? This is

TURFTIP

When turf is disadvantaged or poorly maintained, weeds are usually the first pest problems that result not necessarily the case. Table 19.5 lists some of the major turfgrass species grown in the US and where they originated.

As you can see, buffalograss is the only common turfgrass species that can be considered indigenous or native to the United States. What does this mean to the practitioner who manages this or the assortment of imports they must choose from? From an ecological standpoint, most turfgrasses are out of place where they are planted. The same could be said for many of our ornamental landscape plantings. Stated another way, if nature was the sole driving force behind the plant composition of an area, turfgrasses and landscape plantings would likely not exist. We, as practitioners, force the issue and insist upon species native to other areas to be the prevalent plants at a managed location. This so-called defiance of nature and the laws of ecology should not be seen as a bad thing. However, this concept is presented so that we may better understand why turfgrasses and landscapes are often difficult to manage. Climatic stress and pest problems represent nature holding us accountable for growing species beyond where they originated. As such, we now have entire scientific fields that are devoted to optimizing turfgrass and landscape management, due to these ever-present challenges.

The role of weeds in our managed turfgrasses and landscapes has a lot to do with this concept of ecology. We know weeds are opportunistic. This attribute boils down to the fact that most of our common weeds are better adapted to the area in which they are found than our desired plantings. Weeds rely upon reproduction and

TABLE 19.6 Species composition of an undisturbed area in the Southeast US over a 100 year period.

Time Frame	Primary Species
Years 0-1	Commercial corn (field then abandoned)
Years 1-10	Wide array of annual and perennial weeds
Years 10-50	Pine forest, annual species phase out
Years 50-100	Hardwood forest, pines become less prominent

> **TURFTIP**
>
> From an ecological standpoint, most turfgrasses are out of place where they are planted.

proliferation of their species to ensure survival. This gets to the core of what the laws of nature dictate. Table 19.6 explores a hypothetical setting in the Southeast US.

What we see in this example is the gradual tendency for an area that is maintained for agriculture to revert back to what was likely there before humans arrived. This transition will of course vary considerably in duration and the ultimate species composition across areas but the idea is clear. The occurrence of annual and perennial weeds in our managed turfgrass and landscape systems is a sign that nature is trying to shift back to a more natural ecosystem. Practices like turfgrass maintenance and weed control practices help us avoid these natural tendencies. As it stands, nature is a powerful foe and we must focus on the management tools that allow us to maintain and grow turf or landscape plantings where we want them.

How do turfgrass management practices affect the occurrence and proliferation of weeds? Our four primary cultural practices are mowing, fertilization, irrigation, and aeration. Each can affect the development of weeds and are collectively the cornerstone of a solid weed management program. Mowing has a direct impact on the types of weeds found in turf. Not coincidentally, species that are more prostrate in growth succeed more in turf than in taller crops like corn, where they would be shaded out and not able to compete for available resources. Some taller weeds can succeed in turf and landscapes but they typically are found in less maintained areas like native grass stands. Shorter mowing heights, such as those on golf courses, can further limit the types of weeds found in turf but can also predispose adaptable weeds to be more competitive. Examples are summer annual grasses like crabgrass or goosegrass that, once established, are generally more competitive for available resources than the turfgrasses they inhabit. Shorter mowed turf has a shallower root system but tends to be denser. Weeds thus have a more difficult time getting started but, once they do, they can be more vigorous.

The challenge facing turfgrass managers is to optimize mowing heights, such that turf is as healthy and competitive against weeds as possible. Proper turfgrass species selection is a key component of this strategy. For example, many varieties of Kentucky bluegrass are adapted to be mowed at fairway heights of cut and can provide a dense, competitive stand. By contrast, tall fescue is better adapted to taller mowing heights and, when it is mowed too short, it weakens and is more susceptible to weed invasion. Placing a species or variety under conditions they are not well suited to will diminish turfgrass competitiveness and create additional weed management challenges.

Cultural practices like irrigation and fertilization are critical to the development and maintenance of quality turfgrass stands. They also can present challenges to weed management in that these practices provide resources that can be used by both turf and weeds. It

---

TURFTIP

The challenge facing turfgrass managers is to optimize mowing heights, such that turf is as healthy and competitive against weeds as possible.

is thus critical to focus these inputs on times when the turf can best utilize them. For example, early spring fertilization of warm-season grasses or summer fertilization of cool-season grasses is usually not recommended. These mistimed applications can create an unwanted stimulation of growth when temperatures do not support active growth. While such circumstances can be potentially detrimental to turf, they can also increase the incidence of weeds. Summer fertilization of cool-season lawn turf like tall fescue may not benefit the stand and will undoubtedly provide a source of nutrition for opportunistic summer weeds like crabgrass. Similarly, watering too often or in too much volume for what turf needs can be just what moisture-craving weeds like yellow nutsedge need to develop. Focus these key cultural practices on periods of active turf growth to achieve optimum benefit to the turf stand and avoid unintentionally supporting weeds.

The last key cultural practice in turf is aeration. Aeration is a critical soil conditioning practice that is unique to turf because it is a perennial ground cover. Aside from new establishment projects, we don't have the option of annually tilling soil to loosen it and prepare it for planting. Aeration serves to relieve soil compaction, can improve water infiltration through soil, and is an excellent tool for managing thatch development. All of these benefits are critical to maintaining soils that turf can effectively grow in. Lack of regular aeration can result in excessive thatch levels in turf, which can weaken the stand and open doors for more competitive weeds. Compacted

TURFTIP

Watering too often or in too much volume for what turf needs can be just what moisture-craving weeds like yellow nutsedge need to develop.

> **TURFTIP**
>
> Lack of regular aeration can result in excessive thatch levels in turf, which can weaken the stand and open doors for more competitive weeds.

soils, common to athletic fields and some golf course or lawn areas, discourage the development of most turfgrasses and can result in proliferation of weeds that thrive in compacted areas. Example species are goosegrass, tufted hardgrass, prostrate knotweed, and spotted or prostrate spurge. Controlling the weeds becomes a necessity in these cases but loosening the soil is a better long-term solution to problems related to compacted soils.

Recognition of the relationship between certain weeds and an adverse environment for turfgrass can be critical to developing a solid weed management program. Weeds sometimes occur in healthy, vigorous turf stands but are especially prevalent in compromised areas where turf growth is poor. Weeds can thus serve as a tool to help identify problems that exist and that usually can be corrected. Consider the example species from the previous paragraph:

- *A persistent and aggressive outbreak of goosegrass may be a coincidence but will usually point to a soil compaction problem that can be remedied with aeration.*
- *Prolific nutsedge outbreaks often point to overwatering or an irrigation leak as the culprits, which are again fixable problems.*
- *Annual bluegrass can grow in many areas but is often more concentrated in areas that are heavily watered and fertilized.*
- *Weeds that succeed in areas where turf performs poorly are often called indicator weeds.*

> **TURFTIP**
>
> Weeds can serve as a tool to help identify problems that exist and that usually can be corrected.

Knowing what weeds fit into this category and what problems they point to can help you develop management strategies that both promote better turf growth and help reduce populations of the weed species.

The next chapter discusses in greater detail the theory behind and principles of chemical weed control. While herbicides are an essential component of most weed management programs, it can become easy to rely upon chemical control as the principal means of eliminating weeds. When herbicides become a crutch, rather than a component of a total turf management program, the basics of turf management inevitably become compromised and weed problems can actually worsen. Adherence to the basics of weed control: proper weed species and life cycle identification, site remediation, and proper turf management techniques, will optimize the performance of herbicides that do need to be used and result in a better overall approach to weed control at your facility.

Chapter 20

Basic Chemical and Pesticide Principles

Understanding the behavior of pesticides, why some are potential environmental contaminants, why some are toxic, and why so many people fear them, goes well beyond the scope of the practitioner. The old saying goes, "Familiarity breeds contempt." On the flip side, things we don't fully understand tend to cause fear and skepticism. Many practitioners and even more of the general public do not understand the science that is behind pesticides and other turfgrass chemicals. Because of this lack of understanding, pesticides are popular targets for public concern and environmentalist scorn. As a member of the scientific community, I can empathize with the bad rap pesticides get because they are simply not understood. As a turfgrass specialist who must present scientific information in an understandable format, I also recognize that the information out there is often complicated, boring to read, or otherwise cumbersome. Members of the green industry need to know why this knowledge gap exists and how best to overcome it.

Pesticide manufacturers employ Ph.D.-educated biologists, pesticide scientists, toxicologists, and chemists. Their roles within these companies extend to depths of detail that far exceed what the practitioner needs to know in the workplace or what the end consumer

> **TURFTIP**
>
> Practicing members of the green industry represent a link between the end consumer and the companies that develop, manufacture, and market pesticides.

would ever want to know. Practicing members of the green industry represent a link between the end consumer and the companies that develop, manufacture, and market pesticides. Therefore, the exhaustive amount of detailed information that is available for a particular pesticide needs to be condensed down into a user-friendly format that has the essential core elements to it. That is what herbicide labels and Material Safety Data Sheets (MSDS) represent. The information contained in a four-page pesticide label and an MSDS of similar length represents years of research and millions of dollars spent to bring a pesticide out of the laboratory and onto the shelf. So, the next time you find yourself scratching your head over the contents of one of these forms, remember that what you see is the highly condensed version. Scary perhaps, but this information is vital to applying chemicals safely and properly.

The perception the public has of pesticides often does not take into account the science behind these materials. Commercial practitioners and consumers alike take into account the basic instructions for use and perhaps some basic safety information. What is lost in the shuffle is that these instructions and safety guidelines establish a reputable basis for a pesticide to be marketed. Were there not this information, a pesticide could never be manufactured and sold. Directions for use represent the results from dozens of scientific studies investigating how a particular pesticide should be best used. Safety guidelines may imply danger but, if a pesticide is marketed, it has endured rigorous testing mandated by the EPA and thus has

Basic Chemicals and Pesticide Principles 477

TURFTIP

While any pesticide can be dangerous if used improperly, many household products can be much more dangerous than pesticides.

EPA's "stamp of approval." Many environmentalists refer to pesticides as "poisons" and would have the public believe that pesticide manufacturers have released these materials on society like the plague. While any pesticide can be dangerous if used improperly, many household products can be much more dangerous than pesticides (Figure 20.1). The reason why pesticides are subjected to a

FIGURE 20.1   Many common household products are more dangerous than most pesticides.

greater amount of criticism is not because they are scrutinized less by the EPA. It is simply because the public knows less about them.

This chapter is designed to help bridge the gap between the complexities of pesticide chemistry and the applications of these materials in the consumer market. A detailed understanding of chemistry, soil science, toxicology, or plant physiology is not required. We can leave that to chemists and researchers. However, by being able to grasp some basic principles and to directly see how they apply to the process of selecting and applying pesticides, the practitioner will be better able to comprehend their own actions and to explain them to their clients. The goal herein is to provide ample detail but to present it in such a way that it has meaningful relevance to the workplace. Let's get started!

PESTICIDE FORMULATIONS

The physical form that we see for pesticides is often a far cry from how they would naturally occur. All pesticides are fundamentally composed of an active ingredient, which is the chemical compound directly responsible for acting against a particular pest. The means by which a pesticide or chemical active ingredient specifically affects its target is called a mode of action. Mode of action for specific pesticides and other chemicals will be discussed further in later chapters. The

TURFTIP

All pesticides are fundamentally composed of an active ingredient, which is the chemical compound directly responsible for acting against a particular pest.

TURFTIP

Inert ingredients to help pesticides get into plant tissues may include water, organic solvents, additives, or even fertilizer.

way a pesticide active ingredient is packaged so that it has shelf life and so that it can be efficiently applied is via a pesticide formulation. The formulation includes the active ingredient and all other inert ingredients in a pesticide product. Inert ingredients to help pesticides get into plant tissues may include water, organic solvents, additives, or even fertilizer. All pesticides are formulated. None occur naturally in a form in which they could be efficiently applied to plants. This makes sense since most active ingredients in turfgrass chemicals and pesticides are synthetic, originating in the laboratories of the manufacturers. Even natural or organic products must be formulated so that they can be effectively packaged and distributed to consumers for use in the workplace. While many pesticide and chemical products originate in the laboratory, the role of the chemists is often preceded by a market analysis. The market analysis is highly critical to determine whether or not a developed product will be sufficiently marketable to justify the costs of discovery, development, and manufacturing.

Most pesticide manufacturers actually have two different types of chemists. The first type is a chemist who searches for good active ingredients. These chemists take a chemical compound and modify it until it has the properties they are looking for. Active ingredient chemists may spend years searching for a compound that is usable for pesticide purposes. Often, a compound they are interested in will have properties they are looking for but may not be stable. They must then change the chemical properties so that the compound doesn't break down. Otherwise, the compound wouldn't be worth

> **TURFTIP**
>
> Active ingredient chemists may spend years searching for a compound that is usable for pesticide purposes.

producing and marketing. Another challenge faced by active ingredient chemists is the reproducibility of the process involved in creating a compound. Pesticide manufacturers must be able to produce a compound efficiently in order for it to be cost effective. Much like for the automobile industry, a pesticide needs to be set up for assembly line production to be profitable. Every year, thousands of promising compounds developed by these chemists are dismissed because the chemical procedures required to produce the compound are too cumbersome or too expensive.

The discovery of an active ingredient is certainly an important part of developing a pesticide, but it is really only the beginning. Once an active ingredient compound is identified, it then must be formulated. Formulation chemists must take the active ingredient and package it in a form that is usable to the practitioner. This process of developing a suitable formulation is not easy and is another reason why some active ingredients never become products available

> **TURFTIP**
>
> Formulation chemists must take the active ingredient and package it in a form that is usable to the practitioner.

TURFTIP

Unstable compounds may not have adequate shelf life if they are packaged in a liquid formulation. These materials are often packaged as dry formulations such as powders or granules so that they retain stability until the practitioner uses them.

for purchase and use. The characteristics of an active ingredient compound heavily influence how it is formulated. For example, many active ingredient compounds are not soluble in water. For a pesticide that needs to be applied as a liquid, an insoluble active ingredient can pose a major problem to the formulation chemist. The chemist must determine how to package the active ingredient or the product will not be marketable. Another factor that influences pesticide formulations is the stability of the active ingredient. Unstable compounds may not have adequate shelf life if they are packaged in a liquid formulation. These materials are often packaged as dry formulations, such as powders or granules, so that they retain stability until the practitioner uses them. Marketing can also influence the choice of a pesticide formulation. If a manufacturer wants to be able to sell a combination product (e.g., a herbicide on a fertilizer carrier), the active ingredient needs to be properly formulated to accommodate this purpose. Just what different formulations are out there?

All matter in the universe can occur as a solid, liquid, or gas. Formulations can generally be broken down into these same three categories. Gaseous formulations are the least common for turf pesticides. They would include aerosols, foggers, and fumigants. Aerosols and foggers are fairly common pesticide formulations for in-home use (Figure 20.2) or use by exterminators but aren't common for turf

FIGURE 20.2 Insect repellent, a common household aerosol pesticide.

applications since they don't allow for adequate coverage of the affected areas. Fumigants are sometimes used in turf situations. An excellent example is methyl bromide, which is often used for golf course renovation. Methyl bromide effectively sterilizes the soil, killing all existing vegetation and even seeds. However, the fumigation process in turfgrass systems is cumbersome and is usually restricted to high profile areas like putting greens. Plastic covers must be secured over the target area to keep the methyl bromide gas trapped in once it is inserted. Fumigation can also be dangerous. Methyl bromide is highly toxic to applicators so its use must be handled with extreme caution. Without a doubt, the demands of applying pesticides to larger areas of turf make necessary more user-friendly formulations like solids or liquids.

Solids and liquids are the most common formulations for pesticides and chemicals used in agriculture, including turfgrass systems. Some common formulations for turfgrass pesticides and chemicals are listed in Table 20.1. Solid or dry formulations are manufactured, packaged, and sold in the dry form but may be applied as a dry material or in a liquid spray solution. Dry formulations that are applied dry, or sold in a ready-to-use form, include dusts, baits, and granules. All three of these formulations have low concentrations of active ingredient since they are ready to apply as packaged.

Dusts are not commonly used for turfgrass applications but have utility in smaller areas like gardens or ornamental landscapes (Figure 20.3). They are usually applied to cover the surfaces of plants that a particular pest is prone to target. Many fungicides and insecticides come in dust formulations but dusts are rare for herbicides or other chemicals. Dusts used to be more common for broad scale applications of materials like DDT (Figure 20.4) but this method is no longer considered safe or practical.

TURFTIP

Dusts are advantageous in that they are easy to use and treated plants are easy to identify. However, dusts are not useful for application to larger areas and the fine particle size can represent a hazard to the applicator if the dust is inhaled.

TABLE 20.1 Common dry and liquid formulations used for turfgrass chemicals and pesticides, or for facilities pest management.

Formulation	Abbreviation	Dry/Liquid	Method of Application	Description
Dust	D	Dry	Ready to use	Very fine particles.
Bait	B	Dry	Ready to use	Larger pieces, designed for pests to eat.
Granule	G	Dry	Ready to use	Pellet size particles.
Water dispersible granule	WG/WDG	Dry	Liquid spray	Similar in appearance to granules, break down and suspend in spray solution.
Water-soluble granule	WSG	Dry	Liquid spray	Similar in appearance to granules, dissolve readily in spray solution.
Wettable powder	WP	Dry	Liquid spray	Fine powdery material, suspends but doesn't dissolve in spray solution.
Soluble powder	SP	Dry	Liquid spray	Similar to wettable powder but dissolves in spray solution.
Dry flowable	DF	Dry	Liquid spray	Similar in concept to wettable powder but measured by volume, not by weight.
Flowable	F/FL	Liquid	Liquid spray	Thick, concentrated liquid that is diluted in spray solution.
Emulsifiable concentrate	EC	Liquid	Liquid spray	Organic liquid with ingredients to allow for "oil and water" to mix effectively.
Microencapsulated concentrate	M/MC/MEC	Liquid	Liquid spray	Active ingredient particles are encased in a polymer coating so they can suspend in spray solution.
Solution concentrate	SC	Liquid	Liquid spray	Active ingredient is packaged in a solvent that naturally mixes well with water.
Liquid	L/SL	Liquid	Liquid spray	Concentrated liquid solution that is easily diluted in water.
Ready to use spray	RTU	Liquid	Liquid spray	Diluted liquid solution for immediate use on target pests.

Basic Chemicals and Pesticide Principles **485**

FIGURE 20.3 A dust insecticide coats the surface of a treated ornamental plant.

FIGURE 20.4 Dust application of DDT in mid-20th century

Dusts have specific characteristics:

- They are easy to use.
- Treated plants are easy to identify.
- They are not useful for application to larger areas.
- The fine particle size can represent a hazard to the applicator if the dust is inhaled.

Baits are also not common for application to large areas but can be useful for household or maintenance building management of small animal pests such as rodents. Baits have specific pros and cons:

- They are easy to use.
- Evenness of application is not a concern since ingredients in the bait attract the pest to it.
- It can be uncertain if the target pest has been eliminated.
- If the pest dies in an inaccessible location, odors can become a problem.
- Baits can also be hazardous to children or pets if they are placed in areas where they can be found and eaten.

For turfgrass facilities like golf course maintenance shops, traps or use of domestic predators like cats have become preferred ways of managing the types of pests that baits would ordinarily target.

Among ready to use dry formulations, granules are the most common for turfgrass applications. Granules have an inherent advantage over other formulations (Figure 20.5):

- They are similar in appearance to many fertilizers.
- They can be applied without the need for separate or specialized equipment like liquid sprayers.

Basic Chemicals and Pesticide Principles 487

FIGURE 20.5   Particles of a common granular pesticide (left) are similar in size to those for synthetic fertilizers (right).

- *Most turfgrass operations already have spreaders for fertilizer applications so granular pesticides represent a user-friendly format for application (Figure 20.6).*
- *Granular pesticides contain lower concentrations of active ingredient since they are not diluted before they are applied, like many other formulations.*
- *The inactive or inert part of a granule is usually clay or an organic material that binds and stabilizes the active ingredient until the granule is applied. This type of formulation is ideal for pesticides that might be less stable in a liquid formulation.*

Upon application, exposure of the granule to water and other climatic elements gradually releases the active ingredient so that it can reach its target pests. While granules have many advantages, there are also some drawbacks. Care must be taken that granules be applied evenly since they naturally do not provide the type of coverage that can be expected from a liquid application. Because active ingredient concentrations in granules are low, larger amounts of the formulated product are needed for an application. From a manufacturer's standpoint, the costs of producing granules are higher

FIGURE 20.6 Application of granular pesticides requires similar equipment as for fertilizer applications.

than for other formulations, so the user-friendly format granules offer often comes with higher product costs to the consumer.

Dry formulations that are applied as liquids in a spray solution include water-dispersible granules, wettable powders, soluble powders, and dry flowables. Unlike ready to use dry formulations like granules, these dry materials are packaged and sold in a more concentrated form with higher proportions of active ingredient. For example, granular pesticides used in turf usually have 5 percent or less active ingredient and rarely more than 10 percent. By contrast, dry formulations like wettable powders typically have 50-90 percent active ingredient with the expectation that the material will be diluted in the spray tank.

Basic Chemicals and Pesticide Principles 489

There are cost benefits to using the more concentrated materials as they usually cost less to manufacture, but these materials require more effort to mix and apply properly. Most solid formulations designed for liquid applications require a weighing scale to accurately define how much needs to be placed in the spray tank.

Water-dispersible granules (WDG) are packaged in a form that is similar in appearance to ready-to-use granules (Figure 20.7). They are sometimes also labeled as wettable granules (WG) or water-soluble granules (WSG) but the idea is similar. These products are all designed for application in a liquid spray solution. Whether this type of product is a WDG, WG, or WSG depends upon how soluble the material is in water. WDG and WG formulations do not readily dissolve in water, but the particles break down and are suspended in the spray solution. Because they don't actually dissolve, the spray solution needs to constantly mixed or agitated to keep the particles suspended evenly. Otherwise, they will tend to settle to the bottom of the spray tank. WSG formulations look much like WDG or WG materials but dissolve readily in water and don't require as much agitation once the spray solution has been initially mixed. These formulations are advantageous because the particle sizes are larger and are thus fairly

FIGURE 20.7   Particles from two different wettable granule (WG) products.

TURFTIP

Water dispersible granules (WDG) are packaged in a form that is similar to appearance to ready to use granules. They are sometimes also labeled as wettable granules (WG) or water-soluble granules (WSG) but the idea is similar.

easy to handle and measure without spillage. The large particle size also offers safety because particles are not prone to drift during handling and mixing and don't pose risk of inhalation to the person preparing for an application. Because these formulations are designed for liquid applications, they must be stored in a dry location and kept dry until they are placed into a spray solution.

Powders are another common class of dry formulations intended for use with liquid spray applications. As the name implies, powders feature much smaller particle sizes than do wettable granules (Figure 20.8). There are two primary types of powders, wettable powders (WP) and soluble powders (SP). The difference between them, as was

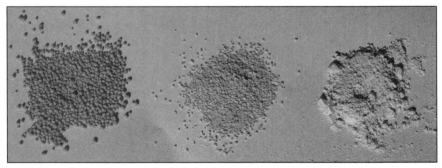

FIGURE 20.8 Powder particles (right) are much finer than those from wettable granules (middle/left).

the case with the liquid-applied granules, has to do with how the product behaves in water. WP formulations do not dissolve readily in spray solution so they require agitation to maintain an even concentration. Think of wettable powders as kicking up some silt while wading in a stream. Initially, the water would become foggy with silt particles dispersed in it. However, in time, the water becomes clear again as the silt particles settle back to the bottom. Wettable powders behave much like this. If the spray solution is not agitated, the particles will settle out and the spray solution will contain less pesticide. The leftover powder residue can also contaminate later applications with the same spray equipment if it is not adequately cleaned. Soluble powders are similar in appearance to wettable powders but dissolve readily in water and do not require as much agitation.

Because powders are much finer particles, they do pose risk of inhalation during the mixing process. Many manufacturers have helped address this risk by packaging powders or other dry formulations in dissolvable packets that can be inserted into the spray tank without having to directly handle the product (Figure 20.9). These packets typically contain defined amounts of the pesticide, based upon recommended application amounts, so they can help reduce the time required in preparing for an application. A dry flowable (DF) formulation is similar to the powder formulations in appearance but is unique because the amount required for mixing is measured by volume rather than by weight. This simplifies the mixing process by making it more like that for liquid formulations without the problem with messiness that liquids can sometimes present.

FIGURE 20.9 The herbicide Plateau® is one that is sometimes packaged in dissolvable pouches to minimize product handling.

Liquid formulations are also very common for turfgrass pesticides and chemicals:

- *Liquids are always applied in a liquid spray but their composition can vary considerably.*
- *How liquid formulations vary depends greatly on how readily the active ingredient dissolves in water.*
- *Highly soluble materials are easy to formulate, as they can simply be packaged and sold as a concentrated water-based solution or as a diluted, ready to use product.*

Many common residential or "over the counter" pesticides are formulated in the ready to use or RTU fashion because these products require no mixing and are easy for the untrained applicator to use (Figure 20.10). Water-based concentrates, sometimes abbreviated L or SL, may or may not be colored but characteristically do not change their appearance in spray solution, other than perhaps appearing more diluted and thus paler in color (Figure 20.11). Water-based formulations don't require any agitation to avoid product settling but a thorough initial mixing ensures an even concentration of the spray solution.

The other liquid formulations used for turfgrass pesticides and chemicals are not water-based, usually because the active ingredient does not dissolve in water. These formulations must somehow package the insoluble active ingredient so that it can be applied in a water-based spray solution. Sound tricky? Formulation chemists have cleverly come up with means to get this difficult job done. Examples include:

- *Flowables*
- *Emulsifiable concentrates*
- *Microencapsulated concentrates*
- *Solution concentrates*

Flowables, abbreviated F or FL, are best described as liquid versions of their cousins, the dry flowables or wettable powders.

Basic Chemicals and Pesticide Principles   493

FIGURE 20.10   Ready to use (RTU) formulations are very user-friendly.

Essentially, the active ingredient is suspended in spray solution with the help of the other ingredients in the formulation. The packaged formulation appears as a thick liquid (Figure 20.12) and, like dry flowables or wettable powders, will settle out if the spray solution is not consistently agitated. Flowables often contain higher amounts of active ingredient than other liquid formulations because of their thick, concentrated form. While this increases the use efficiency for the consumer, flowables are difficult to manufacture and therefore not as common as other formulations.

494 Turfgrass Installation: Management and Maintenance

FIGURE 20.11 A true water-based formulation, shown diluted in water (left) and undiluted (right).

Basic Chemicals and Pesticide Principles **495**

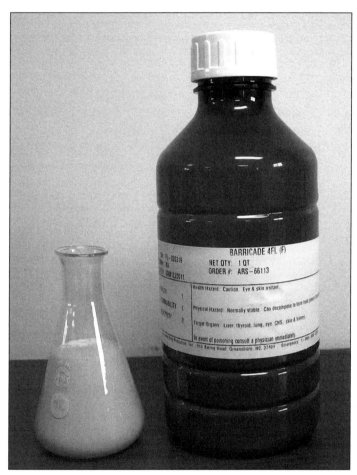

FIGURE 20.12 Barricade 4FL®, a liquid flowable herbicide prod-

Emulsifiable concentrates (ECs) are very common liquid formulations for pesticide active ingredients with low water solubility. In many cases, active ingredients may not dissolve in water but will readily dissolve in an organic liquid. Petroleum compounds like gasoline are organic liquids, as are laboratory solvents like toluene or xylene. An EC contains active ingredient that is dissolved in one of these solvents, rather than water. However, another hurdle must be crossed because organic liquids themselves don't mix well with water. Think of an oil slick floating on water as a visual reference. To solve this problem, chemists add what is called an emulsifying agent to the formulation. The emulsifying agent allows the active ingredient and organic liquid component to be evenly dispersed in water, forming a mixture of two different types of liquid known as an emulsion. A common household example of an emulsion is chicken soup broth, which is fundamentally water but has visible oil droplets suspended within it. ECs are usually colored liquids in the packaged form but their reaction when put into water is striking. Once an EC is combined with water, white clouds immediately form in the water (Figure 20.13) and, as the new emulsion is mixed, the spray solution becomes evenly white (Figure 20.14). Once an even emulsion between water and the EC is formed, it remains stable and is not prone to separate or settle out.

FIGURE 20.13 An emulsifiable concentrate (EC) forms thick clouds when first added to water.

Basic Chemicals and Pesticide Principles **497**

FIGURE 20.14 Once mixed thoroughly, an EC formulation results in a white spray solution.

A microencapsulated concentrate (MC, M, or MEC) is similar to an EC in that there is an emulsifying agent in the formulation to help the tiny pesticide-containing particles disperse evenly in the spray solution. Because there is a similar ingredient as with the EC, MC formulations tend to form white liquids when mixed with water. The difference is that the active ingredient in an MC is usually even less dissolvable in water than that for an EC. This situation can be remedied but would require greater quantities of more toxic organic liquids to get the job done. To combat this problem, MC formulation chemists encase microscopic portions of the stubborn active ingredient in a tiny plastic shell that releases the pesticide once the spray application is made. Concerns regarding the environmental effects of organic liquids used in EC formulations have made the MC formulation a common alternative or replacement for EC chemistry. However, as you might suspect, the microencapsulation process can be expensive for the manufacturer and may lead to higher consumer costs.

A final category of liquid formulations is the solution concentrate (SC). SC formulations are less complicated than ECs or MCs but also are not feasible for as many active ingredients, limiting their availability. The logic is similar to an old law of mathematics that states that if A=B and B=C, then A must equal C. Here's how a SC formulation works. The active ingredient does not dissolve in water, just like for flowables, ECs, or MCs. However, the active ingredient does dissolve in a liquid that, in turn, can be dispersed evenly or dissolved in water. SC formulations are often already milky liquids in the packaged form (Figure 20.15) so they don't undergo the radical visual change from jug to tank that we see for EC or MC formulations. Finding the right liquid that can serve this "mediator" role is challenging and is why SC formulations are not more common. However, in some cases, this ideal scenario is feasible and can result in both a more cost efficient manufacturing process and the use of less hazardous ingredients.

Now that we've discussed the types of formulations that one might encounter in the workplace, it is necessary to discuss how the formulation is identified to the consumer and what it means to the applicator or to someone preparing a spray solution for application. The formulation is often identified or indicated as an attachment to the name of the product (Figure 20.16). This attachment will include a number and then the abbreviation for the formulation. What exactly does it mean to the practitioner? The meaning of the number will vary, depending upon whether a material is dry or a liquid. For dry materials like granules or wettable powders, the number indicates

Basic Chemicals and Pesticide Principles   499

FIGURE 20.15 Packaged EC formulations are often colored liquids (left) while solution concentrates (SC) are often milky liquids

the percent of active ingredient that is contained in the formulated product. For example, a 2 G formulation is a granule that contains 2 percent active ingredient while a 50 WP is a wettable powder that contains 50 percent active ingredient. This same reasoning also holds true for SP, WG, WSG, and WDG formulations.

Liquid formulations have a different numbering scheme. Some liquids don't have an identified number or formulation abbreviation in their title (Figure 20.17). Usually, these are materials that are simply diluted in water and for which the manufacturer recommends a particular volume of the product that is to be applied to a given size area or for a known quantity of spray solution. Other liquids specify in their title a number and a formulation abbreviation (Figure 20.18). In these cases, the number indicates how many pounds of active ingredient are contained in one gallon of the formulated prod-

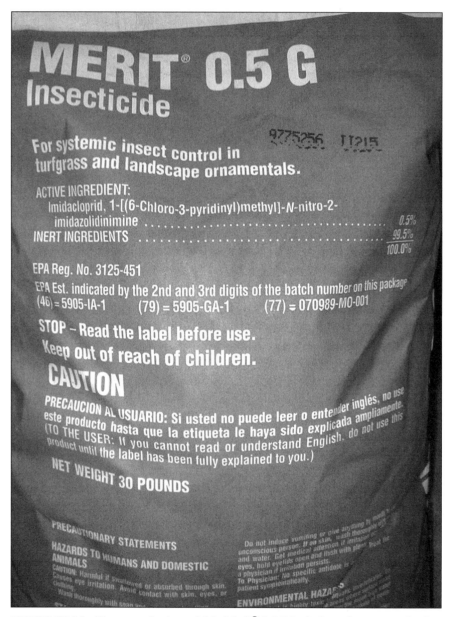

FIGURE 20.16 The granular insecticide Merit® 0.5G includes reference to the formulation in the title.

Basic Chemicals and Pesticide Principles 501

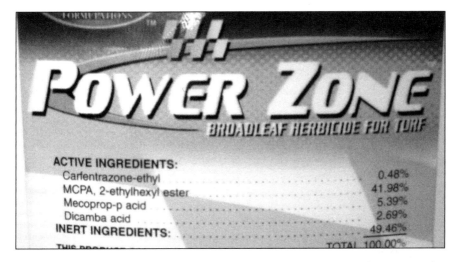

FIGURE 20.17 Many water-based pesticides don't include a reference to the formulation.

FIGURE 20.18 Liquid products created with organic solvents usually include the formulation in their title.

TURFTIP

Knowing what the numbering scheme for formulations means can help you properly calculate the correct amount of a formulated material to use for a given application.

uct. For example, a 3.3 EC is an emulsifiable concentrate containing 3.3 pounds of active ingredient per gallon while a 2 SC, as pictured in Figure 20.18, is a solution concentrate with 2 pounds of active ingredient per gallon of product. Most product labels will indicate recommended amounts of product for a given size area, but they usually also indicate recommended amounts of active ingredient for that same area. Knowing the numbering scheme for formulations means can help you properly calculate the correct amount of a formulated material to use for a given application. More on proper calculations for pesticide and chemical applications will be covered in a later chapter.

Many pesticides or other chemicals are formulated in more ways than one to offer flexibility to the consumer. Is there an advantage to using one formulation instead of another? The answer to that question is usually based upon the specific needs at your facility:

- *Granular formulations can usually be applied to larger areas in less time, thus influencing labor costs associated with pesticide applications.*

- *Granular pesticides can also be advantageous in residential lawn care where close proximity of sensitive landscape plantings to turf may increase the risk of drift injury with liquid applications.*

- *Granular formulations are also the most common for combination products containing pesticides and fertilizer.*

Despite these advantages, granular materials are not always the best choice for a pesticide application.

Many pesticides are not available in a granular form so they must be applied as a spray solution. Liquid applications such as this also have their advantages. Sensitive turf areas such as putting greens are not well suited for granular pesticide applications as granules can affect playability of the surface, are highly visible, and can damage turf by introducing a concentrated amount of pesticide to a small area. Liquid applications are essentially discreet in that there is little evidence of the application once it has been made. Liquid applications also result in more even coverage to the target area, decreasing the chances of unintended injury. The mode of action, or the means by which a pesticide's active ingredient targets a pest, can also be better suited to a liquid application. Herbicides that are absorbed by leaves of target weed species are usually more effective when applied as liquids since there is more even pesticide coverage on the leaves. Many fungicides function best when they are able to coat turfgrass leaves and thus be evenly present when the disease-causing organism attempts to feed on the turf plant. These types of materials are also best applied as liquids.

TURFTIP

Many fungicides function best when they are able to coat turfgrass leaves and thus be evenly present when the disease-causing organism attempts to feed on the turf plant.

Lastly, the toxicity of many pesticides may point to using a liquid application. Innocent bystanders like birds may consume granular insecticides and be unintentionally harmed or killed. Liquid applications may be justified in such a case to minimize the impact of the pesticide on organisms it does not target. While all of these issues pertaining to either solid or liquid applications need to be considered, cost issues and equipment limitations also need to be a part of the decision process for selecting formulations.

CHEMICAL AND PESTICIDE SAFETY INFORMATION—USING AN MSDS FORM

Material Safety and Data Sheets, better known as MSDS, are required to be included with pesticides at the time of sale. Most commercial turf and landscape facilities or organizations keep an active database of MSDS, both to satisfy regulations from groups like OSHA or EPA and to serve as a handy reference in case the information is needed in the workplace. Just about any chemical or material used will feature an MSDS. All chemical reagents used in laboratories have MSDS and there is even one for water. Sound crazy? Well, the truth of today's workplace is that we must have precautionary strategies for most any type of situation to minimize the occurrence of problems. We can no longer presume that common sense is adequate to prevent mishaps and other problems. The MSDS is a highly detailed docu-

TURFTIP

The EPA classifies pesticides with one of three signal words: caution, warning, and danger.

ment with information most of us will rarely have to use or reference. Nonetheless, we must have a basic understanding of what information the MSDS does contain so, when the time does come, you can use it appropriately.

Just what does an MSDS contain? Unlike pesticide labels, which can feature information often requiring six to eight pages of text, the MSDS is a detailed but condensed (usually two pages) document that allows for similar presentation of data, be the topic material water, a chemistry kit reagent, or a pesticide. MSDS have 16 general sections, each of which will be discussed briefly in this section. The intent is to decipher what may seem to be complicated terminology, so that you can better understand what the information really means to you as the practitioner.

All MSDS have a brief heading which identifies the trade name, the formulation, and the type of material (e.g. a herbicide). Also included in the header is the name of the manufacturer, along with an address and phone number so they can be contacted in case of a problem or emergency. Immediately following the heading is Section 1, which provides basic information about the material. Let's use a turf fungicide as an example (Figure 20.19). Section 1 of a herbicide MSDS would provide the trade name, the given name and concentration of the active ingredient, the chemical name of the active ingredient, and the name of the fungicide family to which the active ingredient belongs. The EPA also classifies pesticides with one of three signal words: caution, warning, and danger. The level relevant to the fungicide in Figure 20.19 is caution. Most herbicides also fit into the caution category, since they tend to be less toxic than other pesticides. Herbicides with more toxic properties and many other fungicides are given the signal word warning while the most toxic pesticides, including many insecticides, are given the signal word danger to maximize caution used in their handling and use. Section 2 lists all of the components that go into the pesticide product. This, of course, includes the active ingredient but also the other constituents that make up the pesticide formulation. Remember that all pesticides are formulated so a product may only contain a small proportion of

> **HERITAGE FUNGICIDE**
> Syngenta Crop Protection, Inc.
> Post Office Box 18300
> Greensboro, NC 27419
> In Case of Emergency, Call 1-800-888-8372
>
> **1. PRODUCT IDENTIFICATION**
> Product Name: HERITAGE FUNGICIDE
> Product No.: A12704A
> EPA Signal Word: Caution
> Active Ingredient(%): Azoxystrobin Technical (50.0%)
> CAS No.: 131860-33-8
> Chemical Name: Methyl (E)-2-{2-[6-(2-cyanophenoxy)pyrimidin-4-yloxy]phenyl}-3-methoxyacrylate
> Chemical Class: A beta-methyoxyacrylate fungicide
> EPA Registration Number(s): 100-1093 (formerly 10182-408)
> Section(s) Revised: 3, 5, 7, 16

FIGURE 20.19 Introductory portion of the MSDS for Heritage fungicide.

the active ingredient. Identification of the other ingredients is important since they may also have potentially toxic properties that should be listed in an MSDS.

Sections 3 through 8 on an MSDS cover the hazards associated with exposure to and use of a pesticide and how to deal with potential problems. Included in these sections are proper first aid strategies, fire-fighting recommendations, methods for cleaning up spills, proper pesticide storage, and proper protective equipment for handling and use. This information can be critical when a problem arises. For example, Sections 3 and 4, which cover the symptoms associated with exposure to the pesticide and first aid measures, often contain useful information that can be passed along to a doctor in the event of a medical problem. As a practitioner, this information can expedite the medical care a person may require and, in an extreme case, could help save a life.

Fire fighting recommendations, listed in Section 5 of the MSDS, not only indicate whether or not the pesticide is flammable but also indicate what sort of fire extinguisher is proper to put out a fire and what personnel evacuation procedures should be followed. Spill control recommendations provide optimum ways to handle and minimize the spread of product spills. Access to either type of information can be essential in the event of an emergency to minimize both property damage and personal injury. Handling and storage recommendations are useful to manufacturers, transporters, and end users alike. Making sure that anyone who may encounter a pesticide knows how to handle or store it properly and what hygiene practices are necessary to avoid exposure maximizes worker safety. The recommendations for personal protective equipment (PPE) on an MSDS are often for the benefit of the manufacturer or for personnel involved in transport of the pesticide. Additional PPE recommendations for application of a pesticide are usually included in the product label, which is discussed in the next section of this text.

Sections 9 and 10 on an MSDS cover physical and chemical properties that are pertinent to the practitioner:

- *A description of the appearance*
- *Melting point*
- *Boiling point*
- *Physical density*
- *Solubility in water*
- *Volatility or vapor pressure*

The scale used ranges from 0-4 where:

- *0 represents the least amount of hazard potential*
- *1=slight hazard potential*
- *2=moderate hazard potential*
- *3=high hazard potential*

- *4=severe hazard potential*

Aside from the appearance description, this information might not seem practical but it does offer some insight as to why the product is recommended to be stored in a certain way. Whether they are chemically stable or not, it is recommended that most products be stored in a dark and dry location where the effects of water, sunlight, and heat or cold are minimized. These storage conditions promote maximum shelf life of a product, many of which are not used in their entirety during a single application period. Section 10 of an MSDS briefly comments on the chemical stability of a product and indicates whether or not it will react with other chemicals. For instance, oxidizing agents like nitrogen fertilizers or flammable materials may pose hazards if brought into contact with certain pesticides. Section 10 points out that these types of materials should not be stored with or near the pesticide described on the MSDS.

Sections 11 and 12 of the MSDS cover the toxic potential for the subject pesticide. Section 11 addresses the potential for toxicity to humans who may be handling or using the pesticide. As you might expect, humans are not the test organisms used to determine this information. Usually, scientists will use small mammals like rats in tests to determine toxicity. Examples of information in this section are shown in Figures 20.20 through 20.22. Acute or immediate toxicity is the concentration at which the pesticide can cause injury or death when eaten or swallowed, absorbed through the skin, or inhaled. The units used to express acute toxicity are either an LD50 or an LC50, which indicate the lethal dose or lethal concentration of a pesticide, respectively, at which 50 percent of a tested population of organisms are killed. Note that the oral LD50 for the example herbicide and fungicide (Figures 20.20, 20.21) are greater than 5,000 mg/kg while the oral LD50 for the insecticide (Figure 20.22) is only 173 mg/kg. This tells us that the insecticide is much more toxic if consumed because a much lower quantity is required to kill 50 percent of a test population. Acute toxicity also addresses whether or not the pesticide is prone to cause eye or skin irritation. You can again see from the three example figures that the extent to which pesticides can irritate the eyes or skin can vary. Mutagenic potential is

11. TOXICOLOGICAL INFORMATION

Acute Toxicity/Irritation Studies
Ingestion: practically non-toxic
Oral LD50 (Rat): >5,000 mg/kg body weight
Dermal: slightly toxic
Dermal LD50 (Rat): >2,000 mg/kg body weight
Inhalation: Not Available
Inhalation LC50 (Rat) : Not Available
Eye Contact: mildly irritating (Rabbit)
Skin Contact: mildly irritating (Rabbit)
Skin Sensitization: sensitizing (Guinea Pig)
Mutagenic Potential
Prodiamine: None Observed
Reproductive Hazard Potential
Prodiamine: Fetal toxicity at high dose levels (rats); developmental and maternal toxicity observed at 1g/kg/day.
Chronic/Subchronic Toxicity Studies
Prodiamine: Liver (alteration and enlargement) and thyroid effects (hormone imbalances) at high dose levels (rats); decreased body weight gains.

FIGURE 20.20   Toxicological MSDS information for the herbicide product Barricade®.

the ability of the pesticide to cause mutations. Reproductive hazard potential assesses the concentration at which a pesticide can harm either a developing fetus or the mother. Chronic toxicity assesses how a pesticide may build up in tissues and what, if any, organs are affected by the buildup. Carcinogenic potential identifies the potential for tumor formation and in what parts of the body. While most of the toxicity information in Section 11 refers to the pesticide active ingredient, there is also information there that identifies potential toxicity of the other ingredients that make up a pesticide product.

Section 12 follows up the human toxicity information with both ecological toxicity and environmental impact data. Ecological toxicity covers the potential impact for the pesticide to harm wildlife. For example, concentrations are often listed which are lethal to test fish species. Again, the term used is the LC50, which identifies the

> **11. TOXICOLOGICAL INFORMATION**
>
> Acute Toxicity/Irritation Studies (Finished Product)
> Ingestion: Practically Non-Toxic
> Oral (LD50 Rat): > 5,000 mg/kg body weight
> Dermal: Slightly Toxic
> Dermal (LD50 Rat): > 2,000 mg/kg body weight
> Inhalation: Moderately Toxic
> Inhalation (LC50 Rat): > 4.67 mg/l air - 4 hours
> Eye Contact: Moderately Irritating (Rabbit)
> Skin Contact: Slightly Irritating (Rabbit)
> Skin Sensitization: Not a Sensitizer (Guinea Pig)
> Reproductive/Developmental Effects
> Azoxystrobin Technical: Shows weak chromosomal damage in mammalian cells at cytotoxic levels. Negative in whole animal assays for chromosomal and DNA damage at high dosages (> or = 2,000 mg/kg).
> In rabbits, no effect was observed up to the highest dose level (500 mg/kg/day). In rats, developmental effects were seen only at maternally toxic doses (100 mg/kg/day).
> Chronic/Subchronic Toxicity Studies
> Azoxystrobin Technical: In a rat 90-day feeding study, liver toxicity was observed at 2,000 ppm. This was manifest as gross distension of the bile duct, increased numbers of lining cells and inflammation of the duct.

FIGURE 20.21 Toxicological MSDS information for the fungicide product Heritage®.

concentration (in this case, in a body of water) at which 50 percent of the test population of a fish species are killed. The lower the LC50, the more toxic the pesticide is to the test organism. LC50 values represent acute toxicity potential while chronic toxicity looks more at how a pesticide might accumulate over an extended period of time and cause long-term health problems in the test organism.

This topic will be covered in greater detail in a later portion of this chapter. The environmental fate of a pesticide relevant to an MSDS would include information such as pesticide responsiveness to sunlight and how long it can remain in a stable form in soil. Sunlight

11. TOXICOLOGICAL INFORMATION

Acute Toxicity/Irritation Studies (Finished Product)
Ingestion: Moderately Toxic
Oral (LD50 Rat): = 173 mg/kg body weight
Dermal: Slightly Toxic
Dermal (LD50 Rat): > 2,000 mg/kg body weight
Inhalation: Slightly Toxic
Inhalation (LC50 Rat): > 0.764 mg/l air - 4 hours
Eye Contact: Severely Irritating (Rabbit)
Skin Contact: Moderately Irritating (Rabbit)
Skin Sensitization: A weak skin sensitizer.
Reproductive/Developmental Effects
Cypermethrin Technical: There were no cypermethrin-induced effects in fertility in two separate two-litter three (filial) generation studies in the rat.
Chronic/Subchronic Toxicity Studies
Cypermethrin Technical: NOEL (2-yr) for dogs 5 mg/kg, rats 7.5 mg/kg.
Nervous system effects typical of pyrethroids (motor incoordination, gait abnormalities) in a range of repeated dose studies (dog and rat). Possible nerve fiber degeneration in 14-day study in rats.

FIGURE 20.22 Toxicological MSDS information for the insecticide product Demon.

can break down many pesticides, especially if they are mixed into a liquid solution. The duration a pesticide can last in solution before it breaks down gives us an idea of how stable it is. The process by which sunlight can break down a pesticide in solution is called photolysis. The unit used for this process is the half-life, or the time required for 50 percent of the pesticide to break down. Half-life is also important when evaluating how long pesticides remain active in soil. Soil half-life gives us an idea of how long the pesticide can actively control pests or potentially impact the environment. For example, a herbicide applied on April 1 with a soil half-life of two months would be at maximum concentration at the time of application, a 50 percent concentration by June 1, a 25 percent concentration by August 1, and down to a 12.5 percent concentration by October 1. Temperature and rainfall extremes can affect the soil half-life but you get the idea.

> **TURFTIP**
>
> Much of this information might be deemed common sense, but remember that common sense can not be presumed and we must recognize that documentation of this information is a necessity for these products to be available to us.

Sections 13 through 16 of the MSDS address the regulatory issues, which govern the handling of a particular pesticide. Section 13 outlines any specific instructions for proper disposal of a pesticide product, especially those that supplement local and regional laws for pesticide disposal. Section 14 outlines any specific rules associated with proper transportation of the pesticide. Much of this information might be deemed common sense but remember that common sense can not be presumed and we must recognize that documentation of this information is a necessity for these products to be available to us. Section 15 lists any additional regulatory information that may pertain to a pesticide while Section 16 is the safety net known as "Other Information," which includes anything not covered earlier in the MSDS. One useful piece of information often found in this last section is the National Fire Protection Association (NFPA) hazard rating for the pesticide (Figure 20.23). These ratings use a numeric scale for three hazard categories:

- *Health hazard*
- *Flammability hazard*
- *Chemical reactivity hazard*

The scale used ranges from 0-4 where:

- *0 represents the least amount of hazard potential*

Basic Chemicals and Pesticide Principles 513

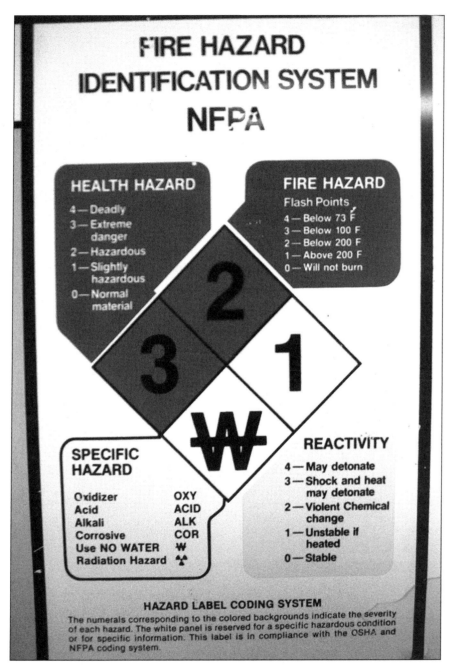

FIGURE 20.23 National Fire Protection Association (NFPA) hazard ratings posted show information for potentially hazardous chemicals.

- *1=slight hazard potential*
- *2=moderate hazard potential*
- *3=high hazard potential*
- *4=severe hazard potential*

I mentioned earlier that many herbicides are given a caution rating by the EPA, the lowest hazard category possible. The EPA looks primarily at toxicity or environmental contamination potential when rating pesticides. The hazard scale used by NFPA is similar but is more pertinent for proper storage and handling of a pesticide. A pesticide labeled caution by EPA might commonly be assigned a 1 (slight hazard potential) in all three hazard categories by NFPA while stored fuels might be assigned a 3 or 4 (high to severe hazard potential) in the same categories. Precautions such as these are commonly featured on the doors of pesticide storage buildings or rooms to indicate the hazard potential of the contents (Figure 20.24).

Remember that, while the MSDS might not seem as important or meaningful as the pesticide label to the average practitioner, it can be critical in the event of an emergency. Knowing where the MSDS for a particular pesticide is located and how to interpret its contents can help you make proper decisions in an emergency situation and, in some cases, might help save someone's life. This is

TURFTIP

Many organizations that employ people who use and therefore are exposed to chemicals include safety training as an early part of the employment process.

Basic Chemicals and Pesticide Principles 515

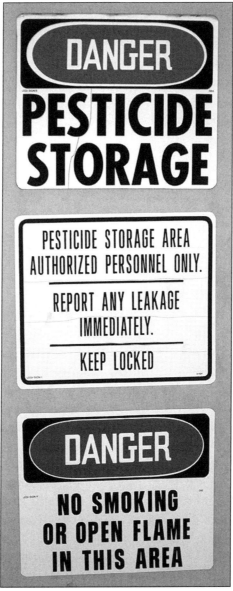

FIGURE 20.24 Precautionary statements posted outside a pesticide storage facility.

important for both managerial personnel and for employees. Many organizations that employ people who use and therefore are exposed to chemicals include safety training as an early part of the employment process. Many of the readers of this text may require such training or have received it themselves. However cumbersome or boring the training may seem, a typical and critical part of it is knowing where to find safety documents like MSDS.

UNDERSTANDING THE CHEMICAL/PESTICIDE LABEL

The label for a chemical or pesticide is perhaps the most important reference document that exists for the benefit of the practitioner. MSDS are critical for safety issues but don't offer much in terms of how to actually use a particular product. Think of a pesticide label as both the warranty and the owner's manual for the product it references. Most questions that you may have regarding use of a pesticide will be answered somewhere within the contents of the label. It is the ultimate responsibility of the practitioner to apply a material properly but most potential problems can be avoided by following label specifications. As with the MSDS, the label contains a lot of detailed information. However, as a practitioner, you will find the label more meaningful and informative because it addresses the actual use of the pesticide. The label is typically a bigger document than an MSDS, sometimes requiring 10 or more pages. However, by systematically describing the contents of the label, it is my aim to help you make more sense out of it and allow you to effectively use the information contained within it for your specific needs in the workplace.

Just what is contained in a product label? Perhaps a better question to ask would be, "What is not contained in a product label?" Labels do vary somewhat but their diversity is attributable to the wide variety of turf chemicals and pesticides that are out there. There is some common ground among product labels and this discussion will focus on what is similar among the labels you might encounter in the workplace. First and foremost, the trademark name of the product is featured at the beginning of the label. Some product names will include reference to the formulation. Other products, particularly liquid formulations, may not indicate the formulation in the title but will usually feature a "fine-print" description of the formulation in the early portion of the label. Brief summaries of the uses for the product and, for pesticides, the sorts of pests they control are also listed at the beginning. The name of the active ingredient in the product and its concentration in the formulated product are indicated on all product labels.

Following the introductory segment, labels will feature condensed versions of the information that is also included in the MSDS. Examples of this are:

- *First aid recommendations*
- *Recommended personal protective equipment (PPE)*
- *Potential hazards to humans or animals*
- *Storage/disposal recommendations*
- *Potential environmental hazards*

Because this safety information is not as detailed as it is in the MSDS, the label should not be viewed as a complete substitute for the MSDS in making safety-related decisions. However, the presence of this information in the product label does serve as a convenient reference for the practitioner and can minimize, in many cases, the number of documents a practitioner needs to use when preparing for and making an application.

TURFTIP

Because safety information is not as detailed as it is in the MSDS, the label should not be viewed as a complete substitute to the MSDS for making safety-related decisions.

CONDITIONS OF SALE AND LIMITATION OF WARRANTY AND LIABILITY

NOTICE: Read the entire Directions for Use and Conditions of Sale and Limitation of Warranty and Liability before buying or using this product. If the terms are not acceptable, return the product at once, unopened, and the purchase price will be refunded.

The Directions for Use of this product should be followed carefully. It is impossible to eliminate all risks inherently associated with the use of this product. Crop injury, ineffectiveness or other unintended consequences may result because of such factors as manner of use or application, weather or crop conditions, presence of other materials or other influencing factors in the use of the product, which are beyond the control of SYNGENTA CROP PROTECTION, Inc. or Seller. All such risks shall be assumed by Buyer and User, and Buyer and User agree to hold SYNGENTA and Seller harmless for any claims relating to such factors.

SYNGENTA warrants that this product conforms to the chemical description on the label and is reasonably fit for the purposes stated in the Directions for Use, subject to the inherent risks referred to above, when used in accordance with directions under normal use conditions. This warranty does not extend to the use of the product contrary to label instructions, or under abnormal conditions or under conditions not reasonably foreseeable to or beyond the control of Seller or SYNGENTA, and Buyer and User assume the risk of any such use. SYNGENTA MAKES NO WARRANTIES OF MERCHANTABILITY OR OF FITNESS FOR A PARTICULAR PURPOSE NOR ANY OTHER EXPRESS OR IMPLIED WARRANTY EXCEPT AS STATED ABOVE.

In no event shall SYNGENTA or Seller be liable for any incidental, consequential or special damages resulting from the use or handling of this product. THE EXCLUSIVE REMEDY OF THE USER OR BUYER, AND THE EXCLUSIVE LIABILITY OF SYNGENTA AND SELLER FOR ANY AND ALL CLAIMS, LOSSES, INJURIES OR DAMAGES (INCLUDING CLAIMS BASED ON BREACH OF WARRANTY, CONTRACT, NEGLIGENCE, TORT, STRICT LIABILITY OR OTHERWISE) RESULTING FROM THE USE OR HANDLING OF THIS PRODUCT, SHALL BE THE RETURN OF THE PURCHASE PRICE OF THE PRODUCT OR, AT THE ELECTION OF SYNGENTA OR SELLER, THE REPLACEMENT OF THE PRODUCT.

SYNGENTA and Seller offer this product, and Buyer and User accept it, subject to the foregoing conditions of sale and limitations of warranty and of liability, which may not be modified except by written agreement signed by a duly authorized representative of SYNGENTA.

FIGURE 20.25 Sample legal disclaimer found on a pesticide product label.

Product labels all include some sort of legal disclaimer offered by the manufacturer as a condition for sale and use of the product (Figure 20.25). While most of us, author of this text included, are not fond of fine print that is written in "legalese", this legal disclaimer does establish what rights you do and don't have as a practitioner. The disclaimer, regardless of what product label it is a part of, will place most of the responsibility for proper product use on the practitioner. This may seem unfair but most of the responsibility should be placed on the practitioner. Only the practitioner can choose when to apply a product, how to prepare for and make an application, and determine what results should be expected. Chemical manufacturers are responsible for providing the formulated product as indicated on the product label. Beyond the point of sale, the manufacturer's legal obligations are usually limited. These terms are stipulated in the product disclaimer and should be understood before a product is purchased.

What rights do practitioners really have? If a practitioner has reason to believe a product is tainted, contaminated, or damaged at some point prior to when the product is purchased, the disclaimer clearly states that the remedy for such a situation shall be either product replacement or a refund. Liability by the manufacturer usually ends there, although most manufacturers and/or distributors will generously adhere to these guidelines and proactively inform consumers that a batch of product may be defective. Otherwise, customers may lose faith in a product or any products offered by a particular manufacturer and choose to take their business elsewhere. Further legal obligations may exist, at the manufacturer level, if a practitioner actually uses defective or contaminated products. When these applications result in a negative outcome (e.g. turfgrass injury or death), the manufacturer may be liable. However, the practitioner must be able to prove that the product contamination was solely responsible for the undesirable injury. Ruling out other reasons for such injury, such as applicator error or environmental conditions, is essential for a practitioner to have a legal claim against the manufacturer. Proof of manufacturer negligence, usually knowledge of a product defect before the product is sold, is also a helpful and sometimes necessary component of a legal claim against a manufacturer. Proper documentation of each and every application made is thus critical for a practitioner to maintain their consumer rights.

The above-described components of a product label are given more discussion because they are common components of all labels. The remainder of most labels, and usually the biggest proportion, details how a product is to be used. This part of the label represents the most important information to the majority of practitioners. Because products vary so much, in terms of what pests they target and for what situations they are most useful, it is impossible to generalize this part of a product label in a few paragraphs. However, common examples of information contained in the instructional portion of a product label include:

- *Proper mixing instructions for a chemical or pesticide, either alone or tank-mixed with other materials*
- *How to determine compatibility of a chemical or pesticide with other products*
- *Whether or not a chemical or pesticide requires supplemental ingredients, such as adjuvants, for proper application*
- *Whether or not a liquid spray solution containing a particular pesticide or chemical requires special treatment like agitation*
- *What volume of water per unit area a pesticide or chemical should be applied in (liquid applications only)*
- *What amount (rate) of a pesticide or chemical should be applied per unit area and how this amount varies for different turfgrasses or for different levels of management*
- *What turfgrass species a pesticide or chemical can be safely applied to*
- *What alternative plant species (e.g. ornamentals) a pesticide or chemical can be applied to or in proximity to*
- *Areas where a pesticide or chemical should not be applied (e.g. near bodies of water, sensitive areas like putting greens)*
- *Recommended timing(s) for application of a pesticide or chemical*

- Specific pests that a pesticide will target and to what extent
- Specific recommendations for how to treat or maintain an area following an application of a pesticide or chemical
- Environmental conditions in which a pesticide or chemical should, or should not, be applied
- Duration of pesticide or chemical activity that should be expected and how that varies with specific amounts used or with different pests
- How much total pesticide or chemical can be safely applied during a growing season or calendar year
- How many total pesticide or chemical applications can be safely applied during a growing season or calendar year
- Geographical variations in recommendations for pesticide or chemical use (e.g. climatic differences, regulatory issues)

The above list covers much of what might be expected in the instructional portion of a product label but may not be all-inclusive. Therefore, it cannot be emphasized enough that a thorough understanding of the instructional portion of a product label is essential to your successful use of that product. As much as following label instructions can help an application, ignorance of or ignoring label instructions can also mean poor results. Being sure you understand the label yourself and also familiarizing your staff with label contents will help ensure that pesticide and chemical use are positive components of a total approach to turfgrass management.

ENVIRONMENTAL IMPACT OF CHEMICALS AND PESTICIDES

The environmental impact of pesticide applications is always a sensitive issue and is the basis for many of the criticisms that are levied

TABLE 20.2 Possible fates of turfgrass chemicals and pesticides.

Fate	Description
Plant metabolism	Pesticide is absorbed into plant tissue and broken down by plant enzymes.
Pest consumption	Target pests consume pesticide and metabolize it.
Soil absorption	Pesticide becomes bound to soil particles.
Organic matter absorption	Pesticide becomes bound to organic matter in thatch and soil.
Microbial degradation	Soil microorganisms like bacteria consume pesticide and metabolize it.
Photolysis	Pesticide is broken down by the energy contained in sunlight.
Chemical degradation	Pesticide is broken down via chemical reactions. with water and other compounds contained in the soil.
Volatility	Pesticide is lost to the atmosphere via evaporation from plant and soil surfaces or via drift.
Surface runoff	Pesticide moves with flowing water across turf, soil, or impermeable surfaces before it enters tar get plants or soil.
Leaching	Pesticide moves vertically (percolates) through soil, reaching ground water or subsurface drainage sources.

against the use of these materials. As practitioners, understanding how pesticides may influence other organisms or the natural environment is a key component of responsible pesticide use. To understand the possible consequences of pesticide use, we must first understand what fates may lie in store for a pesticide after it is applied. Stated another way, where can a pesticide go once it leaves the sprayer or spreader? Table 20.2 lists some possible fates of a pesticide one it has been applied. There are a number of such possible fates but it is also important to remember that rarely does one possible fate account for 100 percent of an applied pesticide. Combinations of these fates are more common so practitioners must try to maximize the desirable fates and minimize those that result in loss to the environment.

The most desirable fate of a turfgrass chemical or pesticide is usually uptake by the target plant or pest. How this happens can vary significantly, depending upon what type of chemical is used and what the target pest or plant is. Some pesticides are applied to target turfgrass plants, some are applied to target weeds, and some are applied with intent that they reach the soil. Most herbicides are applied as a broadcast application, meaning they will contact both turfgrass and weeds. Absorption or uptake into both kinds of plants must therefore be expected. Turfgrasses tolerant to herbicides that target broadleaf weeds are able to harmlessly metabolize the herbicide while the sensitive weeds are injured or killed when they metabolize the same material. In either case, uptake by plants diminishes the chance for the herbicide to experience alternative fates. However, some persistent herbicides can still experience other fates if they are removed from the target site with turfgrass clippings during a mowing event. Proper disposal of clippings must include consideration of what pesticides may be contained in them.

Some herbicides and many other pesticides are applied with pest consumption as the intended fate. For example, preemergence herbicides are applied to form a thin layer just beneath the soil surface so that emerging weeds grow into the layer and then absorb the herbicide. Many fungicides are applied to coat the surfaces of turfgrasses susceptible to a particular disease. When the pathogen attacks the turf, it also consumes fungicide and is thus controlled. Other pathogens and many insects feed on turfgrass roots so pesticides that target these pests must be situated in the thatch or soil so the pests will consume them. Whether or not these pesticides are actually absorbed into turfgrass tissues depends upon the specific characteristics of the pesticide. Those that are not taken up by turfgrass plants may be more subject to alternative fates, in addition to pest consumption.

Alternative Fates of a Pesticide

Something must happen to the proportion of the pesticide that does not get taken up or used by plants or pests in the target area. What are some of these possible alternative fates of a pesticide? Pesticides can be subject to volatility or loss into the air. Volatility can happen in one of two ways: evaporation from the target surface or drift prior to reaching the target surface. Drift is the more common means by which pesticides are lost to the air and is an avoidable problem. High spray pressures and windy spray conditions are two common reasons why drift occurs. Evaporation from the plant or soil surface is less common because most pesticides applied as liquids are formulated to rapidly adhere to absorb into the plant or the soil. Some active ingredients are unstable and are thus chemically prone to dissociate into the air. However, this shortcoming is usually addressed by formulating these active ingredients in more stable forms like dry granules.

Both the characteristics of the target area and the chemical properties of the pesticide can influence the appearance of a pesticide in a non-target area. Site characteristics most relevant to this discussion include slope, proximity to sensitive areas, soil texture, and the type of vegetation that is present. Sloped areas are more prone to pesticide loss, usually surface runoff, than flatter areas because water from heavy rain or irrigation will often flow downhill faster than it can percolate into the soil. Turfgrass areas near sensitive areas like ponds or wetlands present more risk for pesticide contamination simply because of the close proximity. (Figure 20.26)

Conservative or avoided use of pesticides and alternative plantings like buffer strips are common solutions to this potential problem. Water may infiltrate into coarse soils like sands faster than others soils but coarse soil particles also do not bind pesticides as well. Therefore, turfgrass growing on sandy soils may be less prone to surface runoff but more prone to pesticide loss via leaching or percolation through the soil. The opposite is true for heavier soils like clays. Slow water infiltration rates may increase the likelihood of surface runoff

FIGURE 20.26 A turfgrass area near a pond often includes a protective buffer strip with taller vegetation.

with heavy rainfall or irrigation events. Vegetation can also affect the potential for pesticide loss. Turfgrass is a consistent vegetative surface that can intercept and take up a high percentage of applied pesticides. However, landscaped areas with less uniform ground cover may allow for more alternative fates for an applied pesticide.

Pesticide chemical properties that most greatly influence potential for environmental contamination are persistence and mobility. Pesticides that have long half-lives and thus degrade more slowly have more opportunities to move away from where they were applied. Most pesticides naturally dissipate or degrade by several mechanisms. Plant uptake and use is one means by which a pesticide can be degraded. The influence of climate is also important. Solar radiation and water can both help break down a pesticide over time. Chemical reactions in the soil can also either break down a pesticide into smaller chemical components. Lastly, the influence of soil microbes can significantly affect a pesticide's longevity. All soils have populations of microscopic organisms like bacteria and fungi, which feed on organic matter in the soil. Food for these organisms can include plant residue like thatch or pesticide compounds that are in the soil. Persistent pesticides are those that can best resist or withstand these different possible forms of break down.

Pesticide Mobility

Pesticide mobility presents greater risk for environmental contamination than persistence, with the products that are available in today's market. Mobile pesticides are only considered safe (and therefore marketable) by the EPA if it can be shown they have low persistence. Persistent pesticides, by similar reasoning, can only be registered if they have low mobility. This decreases the likelihood that persistent pesticides will move away from the target area. However, let's consider the two primary ways pesticides can move away from a target site. The first is surface runoff, where a pesticide or chemical moves with flowing water across the surface of the ground, eventually arriving in streams, ponds, or other areas where drainage collects (Figure 20.27). Surface runoff can result in contamination with any kind of pesticide if the runoff occurs at the wrong time. The classic example of a situation resulting in surface runoff is a thunderstorm, which deposits heavy amounts or rainfall in a short amount of time. Pesticides that have recently been applied and/or have not moved into the soil are subject to surface runoff, regardless of their persistence. Sloped areas or areas adjacent to bodies of water are the most prone to surface runoff contamination.

The other way pesticides may move away from the target area is through leaching. Unlike surface runoff, leaching contamination requires percolation through the soil with water. Eventually, the contaminant reaches the water table or drainage pipes that can

Basic Chemicals and Pesticide Principles   527

FIGURE 20.27   Surface runoff can be accelerated on slopes and impermeable surfaces like pavement.

lead into collection areas. Leaching is less common than surface runoff but the likelihood of it happening increases with the use of mobile pesticides. By their nature, mobile pesticides more readily dissolve in water and do not bind well to soil particles, allowing them to percolate freely through the soil. Shallow water tables, large volumes of water moving through soil such as with heavy rainfall or irrigation, and coarse soil textures (e.g. sands) are all conditions that can favor leaching contamination. Golf course putting greens are common turfgrass environments that are prone to leaching because many are sand-based; they are regularly irrigated, and subsurface drainage is usually shallow. Use of mobile pesticides in this sort of environment would certainly risk pesticide contamination.

Although environmental contamination is a risk for any pesticide application, proper application techniques and sensitivity to the environment can minimize potential problems. Turfgrass systems are inherently less prone to environmental contamination because there is consistent ground cover, unlike for most field crops. Turf is an excellent interceptor and filter for materials applied to it. As practitioners, we must use our heads and keep our applications away from sensitive areas like bodies of water or impermeable areas like concrete. Beyond these common sense issues, adherence to pesticide application guidelines will take you a long way towards proper environmental stewardship.

Economic and Aesthetic Pest Thresholds

The need to control pests, be it through natural or chemical control measures, is often a matter of consumer opinion. One homeowner, for example, might have minimal expectations for their lawn and therefore tolerate a certain amount of weeds or damage from insects or diseases. Other homeowners might have extremely high expectations for their lawns and therefore have a so-called "no tolerance" approach to any pests their turf may be exposed to. Consumer sentiment can thus have a significant role in how the pesticide practitioner approaches a given turf situation and how the pests are best managed. Money or budgetary issues can also play a key role in the control of pests. Whether it is a home lawn, an athletic field, or a golf course, the amount of funds available for the purchase of pesticides or to pay to have an application made often dictates the level of pest control that is achievable.

Many would perceive that golf courses have the greatest ability to budget for pest control and pesticide use. However, it is more likely that homeowners have the highest amount of flexibility when it comes to pest control. There are no patrons who might complain about pest problems in a home lawn. There also aren't supervisors who may dictate the amount of funds available for pest control pursuits. Whether the limitations to pest control are due to money, aesthetics, or both, the concept of an acceptable pest threshold becomes very important and warrants further discussion.

What is a pest threshold? It might best be defined as "the maximum tolerable level of infestation or injury, by a particular pest, before the infestation or injury becomes objectionable". Pest thresholds differ among the different kinds of pests and the plant systems within which they occur. The agricultural threshold for pests is most often associated with the amount of injury to plants that they cause. However, pest thresholds can also be based upon their potential impact on humans. For example, homeowners or grounds managers might be concerned about bee or wasp damage to tree fruit but are also concerned about the likelihood of bee stings if the insects are sufficient in number. Some weeds, like thistles or sandburs, can be potentially injurious to humans, livestock, or pets, which necessitates a certain threshold for these types of weeds. Despite the potential human impact of some pests, the majority of pest thresholds relevant to agriculture and to turf are still related to either economics or aesthetics.

The concept of acceptable pest thresholds is not a new one but is something that has evolved over the course of history. Early agriculturists did not have access to the pest control methods that we experience today. Farms were small and harvested crops were more intended to ensure survival rather than to achieve economic gain. The destructive impacts of pests could, at times, be severe, and put people at risk of survival. As far back as biblical times, we read of reported locust swarms that could decimate crops and the food source for many people. Under these circumstances, pest thresholds were very low because the stakes were high.

The modern era of agriculture has changed our views of acceptable pest thresholds. Reliable pesticides are available to help control highly destructive pests. Larger farming operations have created surpluses of food crops, such that significant pest problems aren't individually prone to cause starvation or other societal tragedies. However, while this is the case in more developed parts of the world, there are still many places where drought or major pest invasions can have drastic effects. In the third world, pest thresholds are still very low because of the risk of pests to human survival. Back home, modern pest thresholds are more based upon economics.

High yielding cropping systems diminish the per acre value of harvested crops and create a narrower margin between a profitable crop and one which breaks even or results in an economic loss for the grower. Pest thresholds in these cases are established with more precision, and are based upon specific estimates of yield loss due to a particular pest. Let's take a closer look at how thresholds are established for different pests and how these thresholds are modified when we look at turf, rather than conventional agriculture.

Insect damage can influence crop yields in a very visible manner, based upon crop injury. Farmers and turf managers alike must determine how much damage can be tolerated before yields or aesthetics are compromised. Insect populations can be monitored to an extent but many damaging insect pests are migratory, such as locusts or mole crickets, and damage can occur rapidly if control measures are not employed. Growers or turfgrass managers must therefore establish, in advance, tolerance levels for insect pests. These levels are based upon the type of crop, reproductive and behavioral characteristics of the insect, and costs of controlling the pest. Many insect pests have higher threshold levels for large-scale grain cropping systems, due to the high costs of treating the pest. Fruit crops often have very low thresholds for insect pests because any damage caused by the pest is sufficient to reduce the value of the crop. Wormy apples aren't going to be attractive at the marketplace to the end consumer so costs of treating orchards for insects are easily justified.

Turf insect thresholds are similar to that for fruit crops because of the high aesthetic premium placed on turf. However, there is variability in insect thresholds among different types of turf. Homeowners or commercial landscape managers may be able to tolerate some grub damage without needing to apply insecticides. However, thresholds in commercial sod production for insects like sod webworm may be very low, due to the impact of these pests on quality of and ability to sell the sod. Golf course putting greens have very low thresholds for insects, due to their high visibility on the golf course and the immediate visual impact of pest damage (Figure 20.28). Problems with putting greens are the top reason why superintendents lose their jobs so great care must be taken to avoid insect damage on greens.

Basic Chemicals and Pesticide Principles 531

FIGURE 20.28 Golf course putting greens are highly visible and have low thresholds for pest problems.

Diseases are also pest threats that can significantly damage target plants. Be they caused by fungi, bacteria, or viruses, diseases can influence yields of agricultural crops or the aesthetics of ornamental plants such as turfgrass. Whether diseases require treatment or not is at the discretion of the grower, be they a farmer, turf manager, or homeowner. Once again, the concept of a pest threshold becomes very important. Some diseases can be tolerated in grain crops due to the high costs of treatment and acceptable damage levels caused by the disease. Fruit crops have very low disease thresholds for the same reasons low thresholds exist for insects. Moldy or scabbed fruit is not appealing to the consumer and therefore cannot be at all tolerated.

Turf disease thresholds will again vary by circumstance. Minor brown patch outbreaks in a home lawn may not warrant corrective treatment.

When diseases threaten a highly visible golf course, the thresholds drop considerably. Low disease thresholds are why fungicides are one of the biggest costs for highly maintained golf courses. Disease thresholds highly influence the strategy selected for disease control. The primary two options are either preventive or curative. Preventive disease control means treating for a disease at or prior to when conditions are favorable for disease incidence, such that symptoms do not appear. Golf courses or high profile athletic fields with low disease thresholds would usually opt for a preventive control program so symptoms don't detract from the aesthetics and playability of the turf. Curative disease control means treating once disease symptoms have appeared, so as to avoid further damage. Many homeowners view disease from this curative standpoint. If disease appears, they then would decide whether or not to control it. Facilities with low disease thresholds unfortunately do not have this sort of flexibility.

Cost considerations also can influence disease thresholds. It is no secret that preventive disease control is much more costly than is curative control. Preventive control measures are also more reliable and avoid problems associated with visible disease symptoms. High fungicide costs can create dilemmas for turfgrass managers and force critical decisions to be made regarding approaches to disease control. Where budget constraints are a problem, high profile areas like putting greens usually maintain the lowest disease thresholds while less visible areas might be allotted a higher acceptable threshold for some diseases or be switched from a preventive to a curative disease control program. Disease thresholds can, in some cases, stimulate broad changes to a turf facility. Highly damaging diseases like gray leaf spot warrant significant preventive fungicide costs for golf courses that grow perennial ryegrass. Instead of maintaining low disease thresholds and incurring these annual expenses, many golf courses have switched to alternative species like zoysiagrass or Kentucky bluegrass that are not sensitive to this disease.

Weeds may be a more subtle threat among the possible pest groups. The threshold for weeds is usually based more upon the numbers of weeds present in a given area than the direct damage they cause to desired plants. The impact of weeds may not visually compromise the health of a desired plant species. However, over the course of a growing season, competition for space and soil resources can cause costly yield reductions. That is why weeds are the number one pest category in American agriculture, from the standpoint of pesticide use. Because weeds exact a more gradual toll on crop production, the intellectual and financial effort to determine the economic threshold for these pests has been considerable. Factors such as weed emergence, crop planting date, numbers of seeds a single weed can produce, and size of the weed can all influence threshold values and have been the subjects of numerous research studies. The true threshold established for weeds may differ considerably among different cropping systems.

A farmer sees weeds as plants that significantly decrease yield. This yield loss can pertain both to a current growing season and to those in the future. Some weeds are capable of producing hundreds of thousands of seeds per plant, ensuring years of difficulty. Velvetleaf is an example of such a weed and, not surprisingly, its acceptable threshold is very low because of its seed-producing capabilities. Some weeds are recognized as "noxious". State regulatory agencies identify such weeds and establish zero tolerance thresholds for them so that they can be kept in check. Weeds that are most capable of reducing yields, or otherwise interfering with crop harvest, typically have the lowest acceptable thresholds. These thresholds vary significantly among crop species, as management inputs and the herbicide control measures available differ from crop to crop.

A turf manager is more apt to see weeds as diminishing the beauty of a landscape. Because of practices like mowing, turf weeds are not as numerous as for other crops. However, those weeds which do thrive in turf often can tolerate being mowed so alternative strategies must be executed to control them. Much as for insects and diseases, thresholds for weeds in turf will depend on the specific turf and management intensity. Weeds would have a very low threshold on putting greens, both because they stand out visually and because they impact the playability of the surface. In general, the greater visibility a turf area has, the lower the acceptable threshold for weeds. A dandelion infestation on a professional baseball field would not be tolerated while a homeowner might be able to live with a few weeds here and there.

Acceptable weed thresholds in turf can be influenced by the cost of herbicides, the expected impact of herbicide treatment, and the consequences of allowing the weeds to remain. A good example of a threshold dilemma that weeds can create is volunteer bermudagrass. While bermudagrass is a desired turf species in many areas, it can be an aggressive invader of other species, particularly in the transition zone. Because of its aggression, many turfgrass managers would identify a very low threshold for bermudagrass. However, few reasonable selective control options exist for it, resulting in an increase in the acceptable threshold. The spreading nature of bermudagrass can permanently alter the species composition of a turfgrass area. At some point, a turf manager must choose to accept the infestation and perhaps choose bermudagrass as the desired turf, or aggressively pursue controlling it with no guarantee of success. The perennial nature of turfgrass systems is often subject to such difficult situations when it comes to weeds. Because weeds are such a difficult problem for so many turfgrass situations, they will be the focus of the next two chapters.

This chapter has been designed to introduce you to some of the many terms, concepts, and supporting documents that are associated with pesticide and chemical use in turfgrass management. Some of the key points that you should now be familiar with are:

- *The science that goes into discovery and development of a pesticide product*
- *Why pesticides are formulated*
- *The types of formulations that exist for the products you use on the job*

- *Advantages and disadvantages of different formulations*
- *Why a pesticide MSDS is important*
- *What information is contained in an MSDS and how it benefits the practitioner*
- *Why a pesticide label is critical to successful application of a product*
- *What information is contained in a product label and how to use it*
- *What possible fates exist for a pesticide, once it has been applied to turf*
- *What conditions promote undesirable fates for a particular pesticide application*
- *The importance of establishing pest thresholds*
- *What criteria are necessary for determining pest thresholds*
- *How pest thresholds differ for different pests and why*
- *How pest thresholds differ for different turf situations and why*

I encourage you to review parts of this chapter if some of the highlighted key points are still unclear. Subsequent chapters will expound upon the different topics discussed in Chapter 2 so a good understanding of the basics will make later chapters more meaningful to you as a practitioner.

Chapter 21

Disease Problems and Fungicides Used in Turfgrass

Turfgrass diseases are some of the most feared pests in the industry because of their potential for dramatic visual impact and rapid development, and their penchant for attacking turf when it is most vulnerable. This combination of attributes has made disease control a premium concern in fine turf and an entire discipline of study. The importance of diseases in turfgrass management has increased with the level of maintenance expectation that turfgrass is subjected to. However, concurrent improvements in technology and management practices have contributed to the ever-changing face of disease control. University researchers and professors, chemical manufacturers, and turf industry professionals have answered the call by discovering new and improved methods to provide improved disease management practices and better turf conditions.

How best to approach disease management at your facility is the subject of this chapter. Modern turf management experts have adopted the concept known as Integrated Pest Management or IPM. Theoretically, IPM is a sound practice. However, familiarity with the location you are managing is important, as is establishment of how much disease damage is tolerable for your situation. More will be discussed on strategies for developing disease management programs

> **TURFTIP**
>
> Modern turf management experts have adopted the concept known as Integrated Pest Management or IPM.

later in the chapter. However, as this topic is introduced, remember that management of diseases involves many of the fundamental concepts that have already been or will be discussed in this text.

Knowledge of the pests, understanding the turf you are managing, recognition of diverse control options, and proper communication of your control programs will help ensure a safe, responsible, and effective program. With these things in mind, let's get started!

DISEASE PROBLEMS IN TURF

Before discussing control options and strategies for any pest, it is important to understand the characteristics of the pests themselves. For diseases, this key precursory step means proper diagnosis of disease problems, recognition of the key disease threats in your area, and comprehension of where these problems are most likely to occur. Disease identification can be difficult or simple, but understanding the types and the patterns of disease development will make prognosis clearer. Disease pathogens are naturally occurring microorganism entities in the soil. When turf is grown in areas where a pathogen exists, the potential for disease outbreaks is created. In addition to a susceptible turfgrass host and the presence of a pathogen, there is a third criterion that must be in place for disease to occur: a proper environment for disease development. These three criteria are collectively

TURFTIP

Environment affects disease development and explains why diseases seem to appear at the same time each year.

referred to as the disease triangle because, much as the absence of one side of a triangle disrupts its structure, the absence of any of these three criteria prohibits disease development. Usually, environment is the wild card in this "triangular" relationship. Environment affects disease development more than the other two factors and explains why diseases seem to appear at the same time each year. It is likely that both the pathogen and the turfgrass are there the entire time, regardless of the conditions around them.

There are three main types of disease-causing organisms:

- *Fungi*
- *Viruses*
- *Bacteria*

Parasitic nematodes are often included as a fourth member of this group, since their activity may cause similar symptoms. Pathogens attack plants in different ways and require various techniques to identify, control, and prevent them. Diseases can be diagnosed in two ways: signs and symptoms. Disease signs are physical features of a pathogen that can be found on infected turfgrass plants:

- *Mushrooms*
- *Spore bodies*
- *Fungal hyphae*

- *These are often visible to either the naked eye or with a hand lens.*

Disease symptoms are more easily identified. They are the tangible effects of a pathogen on infected turfgrass:

- *Lesions on turfgrass leaves*
- *Rotted crowns*
- *Death*

Being able to recognize disease symptoms is crucial to proper diagnosis and timely treatment of a disease. The following paragraphs discuss some of the most common troubling disease problems a turf manager can be faced with at any given time.

Anthracnose is a fungal disease that readily attacks annual bluegrass but can spread from infected annual bluegrass to desirable species like creeping bentgrass, perennial ryegrass, Kentucky bluegrass and bermudagrass. Typically this disease occurs under warm and humid summer conditions and is especially prevalent with night temperatures above 68°F and 80 percent or more humidity. It can appear and survive with much cooler temperatures. If the disease is allowed to exist and thrive untreated on the leaf tissue, it will persist and infect the crown of the plant. Black fruiting bodies are visible to the naked

TURFTIP

Early applications of a systemic fungicide labeled for anthracnose control is most effective. Members of the strobularin family of fungicides have shown excellent control results, especially when accompanied by a balanced nutritional program.

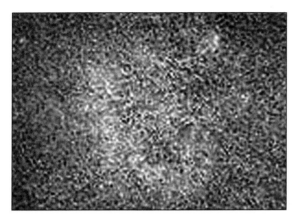

FIGURE 21.1 Symptoms of anthracnose infection are most evident in the crowns of turfgrass plants.

eye when examining leaf tissue, as well as an orange appearance of shoots in infected plants (Figure 21.1). At this stage, it is the point of no return and permanent damage can occur. Early applications of a systemic fungicide labeled for anthracnose control will be most effective. Members of the strobularin family of fungicides have shown excellent control results, especially when accompanied by a balanced nutritional program.

Brown patch, or *Rhizoctonia solani*, has a smoke ring appearance and has been detected in all types of turfgrasses (Figure 21.2). Closer inspection of infected leaves will reveal clear lesions caused by the pathogen (Figure 21.3). Moist morning conditions can also present opportunities to identify this disease by the presence of cottony mycelia (Figure 21.4). It is most severe at low mowing heights, specifically those maintained below $1/2$ inch. It appears in hot, humid weather and on all turfgrass types, especially in conditions with limited or no air movement and high use of nitrogen-based fertilizers. Constant, repetitive mowing will increase the spread of brown patch. In locations that are ripe for brown patch development, the infection can be stubborn and can persist for extended periods until weather conditions are favorable to promote uninfected turfgrass growth. A cousin of brown patch, sometimes referred to as cool-season brown patch or yellow

FIGURE 21.2 Brown patch causes a classic smoke ring appearance in infected turf.

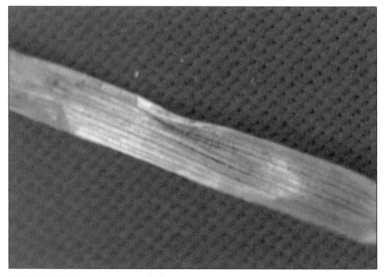

FIGURE 21.3 Brown patch lesion on an infected turfgrass leaf.

Disease Problems and Fungicides Used in Turfgrass 543

FIGURE 21.4 Brown patch mycelia are most visible during cool morning hours.

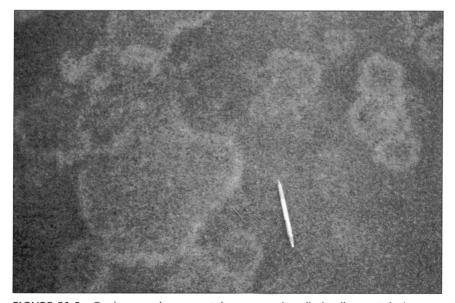

FIGURE 21.5 Cool-season brown patch, commonly called yellow patch, is more prevalent under cooler conditions and can be visually disruptive, even if it doesn't cause permanent damage.

TURFTIP

Contact fungicides tend to work the best on brown patch, although follow up applications of a systemic fungicide may be necessary if disease conditions persist.

patch, also appears in many turf stands during cool damp periods (Figure 21.5). Although it is less prone to cause permanent damage to turf, the presence of the yellow patch pathogen can weaken desirable turf to the point of invasion by a secondary pest. Contact fungicides tend to work the best on brown patch, although follow up applications of a systemic fungicide may be necessary if disease conditions persist.

While several diseases are known by a colorful name, two stand out in this regard. Red thread is a disease known to attack peren-

FIGURE 21.6 Red thread disease gets its name from the thread-like structures seen here among turfgrass leaves.

nial ryegrass, fescues, Kentucky bluegrass, and bentgrasses. It appears under nutrient deficiencies and when turf is usually growing slowly, typically early and late season. It is a disease that a turf manager who seasonally overseeds should be wary of. Red thread infection can be recognized by the characteristic leaf tip die-back and presence of reddish sclerotia, thread-like structures that permit survival of the pathogen in a turfgrass canopy (Figure 21.6). Recognition of these features can help differentiate red thread from other diseases. Because red thread is usually present where turf is nutrient deficient, a turf manager may consider adjusting fertilizer practices to remedy this disease. Another "red" disease is rust, which appears in many turfgrasses, although perennial ryegrass and bluegrasses are most commonly affected. Rust can be easily recognized by the rusty-colored spore structures that are easily visible and also rub off on your shoes

FIGURE 21.7 Rust spore structures, seen here on turfgrass leaves, give the disease its name.

> **TURFTIP**
>
> Because red thread is usually present where turf is nutrient deficient, a turf manager may consider adjusting fertilizer practices to remedy this disease.

when walking on infected turf (Figure 21.7). A severe outbreak of this disease could also be an indication of nutrition problems.

Pythium blight is a highly destructive disease that can wipe out a turf stand in very little time. Often called damping off in new turf stands, pythium will cause germinating seedlings to wither and die. The moist, almost wet environment that is necessary for seedling establishment is ideal for the destructive damping off action of pythium. The pythium disease family is broad based and attacks in a variety of ways, making it one of the most difficult diseases to pinpoint and understand. Proper identification can make it easier to manage because it is a disease that will spread rapidly from mowing or other mechanical equipment to nearby uninfected turf. Strains of pythium diseases can be present in many different settings, seasons, and turfgrasses. All turfgrasses are susceptible to

> **TURFTIP**
>
> During a seeding project, granular pythium-specific fungicides should be applied to the seed bed. Protecting seedlings is vital on highly visible and intensely managed turf.

pythium outbreaks, but cool-season grasses are most affected. The pathogen easily overwinters and infects plants during cool, wet periods, although it seldom displays signs and symptoms at this time. The most recognizable and severe damage to foliage appears during hot and humid weather conditions when a turfgrass stand could be obliterated in less than one day. Infected turf can be covered with a thick, white mycelium web or can appear greasy and lucid under severe infection (Figure 21.8).

Pythium cannot be taken lightly by turf managers, due to its rapid and destructive behavior. During a seeding project, granular pythium-specific fungicides should be applied to the seed bed. Protecting seedlings is vital on highly visible and intensely managed turf. On areas where pythium has historically appeared, a preventative application should be made to the rootzone to protect plants from further infection.

FIGURE 21.8 A close-up picture of pythium infestation in tall fescue turf.

Snow molds are diseases whose names can be a bit misleading because snow is not an absolute prerequisite for infection. Pink snow mold or *Microdochium nivale* is most destructive to annual bluegrass but it infects many turfgrass species under a wide range of environmental conditions. However, it is not active when air temperatures exceed 70°F. It is active in turf areas with:

- *High levels of nitrogen*
- *High humidity*
- *Rain*
- *Snow*
- *Shade*
- *Little or no air movement*

Snow falling on unfrozen ground is a common precursor for this copper colored, leaf-infecting pathogen. It tends to develop best under snow cover, producing a pink halo of damaged turf that is covered by a sparse, cottony mycelium (Figures 21.9 and 21.10).

Gray snow mold or *Typhula incarnata* does have a true snow requirement, needing 40 days or more of snow cover for infection. Gray snow mold can take advantage of weakened turf under snow cover, and can be more damaging than its pink cousin. In advanced stages, matted water soaked leaves, in a distinctive circular pattern,

TURFTIP

Post infection control of gray snow mold is difficult. Preventative measures are most effective when a late fall fungicide application for snow mold protection is made.

FIGURE 21.9 Mid stage of pink snow mold development with dense cottony mycelia.

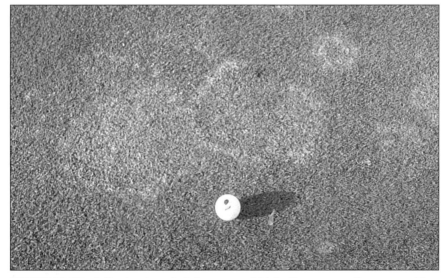

FIGURE 21.10 Fully developed pink snow mold infection on a golf course putting

550 Turfgrass Installation: Management and Maintenance

FIGURE 21.11   Advanced stages of a gray snow mold infection.

are clear indications of gray snow mold (Figure 21.11). Post infection control of gray snow mold is difficult. Preventative measures are most effective when a late fall fungicide application for snow mold protection is made. A curative approach and turf replacement should commence when growing conditions are favorable and active. At that time, a systemic fungicide should be applied when the turf is able to absorb the fungicide. Curative applications of fungicides are ill-advised before the turf breaks winter dormancy.

Summer patch is a root disease that is prevalent in hot humid seasons. It appears most commonly on heavily irrigated and fertilized turf maintained at low mowing heights. Summer patch is most commonly found to infect Kentucky bluegrass, annual bluegrass, and creeping bentgrass. It is compounded by concentrated and heavy traffic from machinery and mowing equipment. Infection from summer patch can be devastating, since afflicted species are often unable to grow successfully at the time damage from the pathogen occurs (Figure 21.12). On

Disease Problems and Fungicides Used in Turfgrass 551

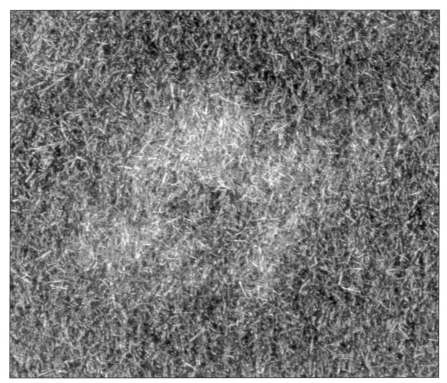

FIGURE 21.12 Summer patch can have injurious effects on susceptible species like Kentucky bluegrass.

intensely managed turf, a preventative application of a systemic fungicide should be made before the onset of severe summer conditions.

Take all patch or *Gaeumannomyces graminis* is a disease that is predisposed by cool wet years and most definitely by the pH of the soil. Adjustments to lower a high soil pH can make a significant difference in the development of this disease. This pathogen wreaks havoc with young bentgrass, as infected plants die in irregular circular patches or rings. Southern turf is less prone but not immune to take all patch. It has been detected on intensely managed bermudagrass and should be monitored on overseeded greens that are sown to cool-season species. Sand-based greens with high soil pH values

> **TURFTIP**
>
> Sand-based greens with high soil pH values are most susceptible to take all patch and a fungicide application of fenarimol drenched into the rootzone is an effective means of control.

are most susceptible and a fungicide application of fenarimol drenched into the rootzone is an effective means of control. Take all patch can be difficult to correctly diagnose. It is a common diagnosis for turf samples with unknown problems, prompting it to be referred to sometimes as "catch all" patch.

Dollar spot is aptly named not for just its appearance of silver dollar size clusters of infected turf but for the great deal of money spent on fungicides, fertilizers, and cultural practices to combat this disease. *Sclerotinia homoeocarpa* is the most common pathogen that causes dollar spot, although other dollar spot strains are common among all turfgrasses in various settings. Cool, damp nights that produce heavy dews, coupled with warm days, provide ripe conditions for dollar spot, especially where turf is nutrient deficient. On closely mowed turf where clippings are collected, infected leaves coalesce and the disease infection forms sunken pockets that produce uneven and unsightly ruts in the thatch. On turf that is maintained above two inches, the acceptable threshold for dollar spot can be set much higher but the disease can still be easily identified in this setting. Close examination of the leaves will show tan lesions with brown margins (Figure 21.13). Because of nutrient deficiencies and stressful conditions, late season infestations are common and can be very destructive until the first heavy frost (Figure 21.14). To successfully combat dollar spot infection, adequate fertilizer must be applied to promote turf growth. When infection does occur, early intervention with fungicides will clear the disease up fairly easily. Many fungicides are

FIGURE 21.13 Dollar spot lesion on a Kentucky bluegrass leaf.

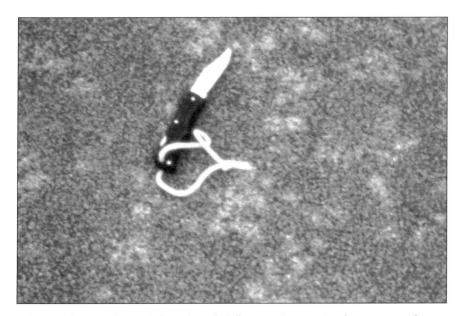

FIGURE 21.14 A heavy infestation of dollar spot in creeping bentgrass turf.

> **TURFTIP**
>
> To successfully combat dollar spot infection, adequate fertilizer must be applied to promote turf growth.

effective against this pathogen, but alternating systemic fungicides with various modes of action work the best. Contact fungicides such as chlorothalonil have also proven to be effective. Be cautious when applying strobularin fungicides to target other diseases because, in areas where dollar spot has been detected, they may actually increase dollar spot incidence.

So aptly named "blast," gray leaf spot exploded onto the scene to ravage perennial ryegrass in the mid-Atlantic, the Northeast, and the Midwest of the United States in the 1990s (Figure 21.15). Previous to recent outbreaks in perennial ryegrass, the pathogen had only been reported as common in St. Augustinegrass. Gray leaf spot weakens the plant from the top down. Leaves begin to blight and become twisted as the plant weakens. The pathogen then moves to the crown and on into the root system. Detection and diagnosis of gray leaf spot may not be definite at the outset, due to some similarities to the bacterial pathogen *Helminthosporium*. During warm and humid periods, symptoms are capable of destroying grass plants quickly. Gray leaf spot thrives in poorly drained areas, with long periods of leaf wetness and with daytime temperatures around 85°F. It can spread via equipment traffic and drainage patterns, much like pythium diseases. It is at this stage that careful diagnosis is necessary because the two diseases may be confused. Above 90°F, gray leaf spot is less infectious, but the pathogen may still be present.

Spring dead spot is a fungal disease that is exclusive in turf to bermudagrass. Spring dead spot establishes itself in the fall, colonizes during the winter, and manifests itself in the spring as large,

Disease Problems and Fungicides Used in Turfgrass 555

FIGURE 21.15 Gray leaf spot or "blast" can cause massive damage in perennial ryegrass turf.

brown patches of turf that fail to green up (Figure 21.16). Re-growth of dormant turf is slow in the infected and damaged areas and can appear year after year in the same locations. Spring dead spot is a more prevalent disease in mature turfs that are intensely managed. Frequently, patches coalesce and the damage is nonuniform, thus appearing similar to some types of winter injury. The disease is less severe or absent on turfgrasses maintained at low fer-

TURFTIP

Late fall applications of a systemic fungicide that are drenched into the rootzone may be an effective means of control for spring dead spot.

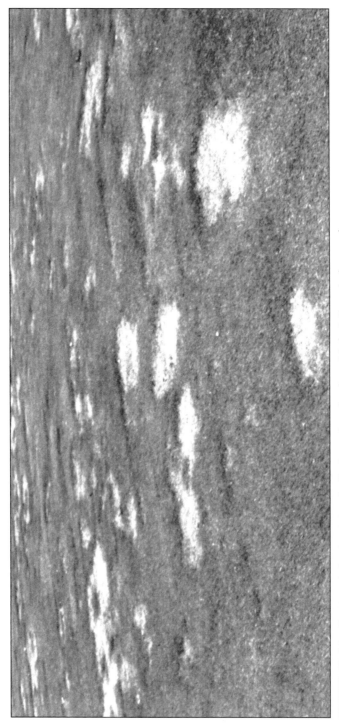

FIGURE 21.16 Scattered dead patches indicate spring dead spot infestations in bermudagrass.

tility levels. Heavy late fall applications of nitrogen fertilizers should be avoided to lessen the severity of spring dead spot. Late fall applications of a systemic fungicide that are drenched into the rootzone may be an effective means of control for spring dead spot. Large patch is a similar disease to spring dead spot but occurs primarily in zoysiagrass turf. Like spring dead spot, large patch is characterized by infected areas that fail to green up in spring (Figure 21.17). However, the naturally slower growth rate of zoysiagrass can make recovery from this disease a challenging problem.

St. Augustine decline (SAD) is unusual among turfgrass diseases in that it is caused by a virus, rather than by the fungi that are responsible for all the diseases discussed up to this point. SAD is caused by a strain of *Panicum* mosaic virus. SAD results in St. Augustinegrass turf that is weak and shows small yellow spots on the leaves (Figure 21.18). Entire lawns can succumb to SAD because of the turf's weakened and vulnerable state, allowing other pests to attack the weakened plants. SAD is a slow moving viral disease that can resemble mite damage or numerous other diseases such as downy mildew. It can

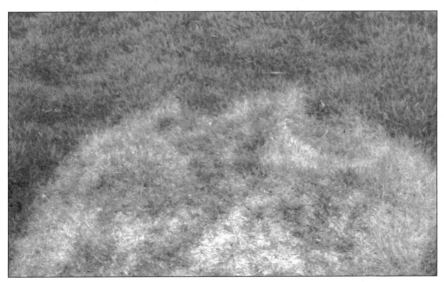

FIGURE 21.17 Large patch is observable in zoysiagrass during the green up phase in spring.

558 Turfgrass Installation: Management and Maintenance

FIGURE 21.18   Small yellow spots characterize St. Augustinegrass decline (SAD) infection.

be spread when grass is mowed wet. The disease is one to watch out for because of the difficulty in effectively controlling it.

Helminthosporium leaf spot or melting out is a bacterial disease, caused by several bacterial pathogens, that infects a variety of turfgrass species. The name *Helminthosporium* reflects an older taxonomic group for these pathogens that has since been split into more specific groups. Many avenues of infection are present with this pathogen. Leaf, crown, and root infections are all possible with *Helminthosporium* pathogens. All turfgrasses are susceptible to forms of this disease. In the spring when cool, damp conditions persist, the leaf spot variety is common and, when the pathogen is allowed to be active, severe infection can occur. If unchecked into summer, the infection migrates to the crown. The severity of the damage increases and will be fatal if not treated before it reaches the root zone.

TURFTIP

Clipping removal, adequate irrigation, and balanced nutrition are measures to combat whichever form of the *Helminthosporium* family of diseases is present. A systemic fungicide provides an effective means of control, especially if allowed to migrate to the rootzone for absorption, thus providing excellent protection against crown and root damage.

Some keys to recognizing the symptoms of *Helminthosporium* are the blighting of the leaf tips that appear red to brown in color (Figure 21.19). During major infections, the crowns of the plant display a spotty, black, greasy algae-like appearance. During warm wet weather, the disease multiplies rapidly, particularly if grass clippings are left to decompose. At this point, if another pathogen is present, the two will work together and devour a turf stand. When *Helminthosporium* reaches the root infection stage, it is nearly fatal

FIGURE 21.19 Symptoms of *Helminthosporium* leaf spot on turfgrass leaves.

and very difficult to control. Turf managers in all regions should not let *Helminthosporium* linger. If conditions are favorable for development, take measures to control and implement appropriate cultural practices. Clipping removal, adequate irrigation, and balanced nutrition are measures to combat whichever form of the *Helminthosporium* family of diseases is present. A systemic fungicide provides an effective means of control, especially if allowed to migrate to the rootzone for absorption, thus providing excellent protection against crown and root damage.

One very different fungal disease that is a recurring topic of conversation with southern turf managers is fairy ring. Fairy rings uniquely result in three distinct types of signs or symptoms. At different points in the life cycle, fairy ring can result in mushrooms, dark or deep green rings, and rings of dying turf (Figure 21.20). Fairy ring spores find a home in a desirable location and then begin to infiltrate and infect turfgrass plants. Fairy rings develop because of concentrated sections of organic matter within the soil environment. It could be soil amendments such as peat moss or composted materials, or even concentrations of decomposing leaves or buried tree stumps. High levels of organic matter can result in such a high concentration of fairy ring fungi that they become toxic to turfgrass plants. Decomposing

FIGURE 21.20 Fairy ring is often identified by the presence of mushrooms.

organic matter is a perfect home for fairy rings and the thatch layer of turf grass is the perfect launching pad for specimen fairy rings.

A good recipe for control of fairy ring is a regular maintenance regime of core cultivation and application of wetting agents to discourage water repellency in the soil. In addition, a fungicide application of azoxystrobularin that is drenched into the root zone may be an effective means of control. The impervious layer of fungal activity is very difficult to control and may appear year after year in the same location, especially when turf appears nutrient deficient. Applications of balanced fertilizers and micronutrients to mask the symptoms of fairy rings can also be an effective control measure.

Nematodes can cause disease-like symptoms but are unlike other pathogens. Nematodes resemble worms but are only visible under a microscope. Many species of nematodes exist but only a few are harmful to turf and other cultivated plants. Nematode infestations are much more widespread and damaging in southern turf, causing damage primarily to bermudagrass, zoysiagrass, and St. Augustinegrass. Nematodes, although present in cooler regions, are less severe in northern climates and do not produce the amount of turf damage as in southern turf. Direct nematode damage via feeding is usually to the roots of turfgrass plants. Nematodes can also bore into and occupy root conducting tissue like xylem, thus restrict-

TURFTIP

A good recipe for control of fairy ring is a regular maintenance regime of core cultivation and application of wetting agents to discourage water repellency in the soil. In addition, a fungicide application of azoxystrobularin that is drenched into the root zone may be an effective means of control.

> **TURFTIP**
>
> Pinpoint diagnosis is very difficult when dealing with parasitic nematodes, so it is wise for a turf manager to enlist the assistance of a plant pathologist to accurately identify the type of nematode present and suitable treatment options.

ing plant water use and resulting in symptoms of drought or heat stress. Pinpoint diagnosis is very difficult when dealing with parasitic nematodes so it is wise for a turf manager to enlist the assistance of a plant pathologist to accurately identify the type of nematode present and suitable treatment options. When turf loss is evident, an application of a nematicide would be advised because fungicides have minimal effect on nematode populations.

FUNGICIDE FAMILIES AND MODES OF ACTION

By definition, a fungicide is a chemical that is used to kill fungi in a short period of time. This distinction helps explain why control is often difficult for bacterial, viral, or nematode infections because these organisms require unique control measures. Fortunately, most turfgrass diseases are caused by fungi so fungicides are reliable tools for most problems you may encounter. Fungicides are classified in groups according to their biochemical and topical modes of action. There are two general classifications. Contact fungicides are those fungicides that are effective only on fungi present on the external tissues of the plant. This contact mode of action can help protect the plant from potential infection, especially if fungicides remain on the plant surface for longer periods of time. Examples of contact fungicides are chlorothalonil, etradiazole, mancozeb, quintozene, and thiram. Contact fungicides kill spores and mycelia by direct engagement with the spore-producing pathogen. Contact fungicides possess the trait of providing a protective coating to the plant creating a barrier of protection from further infection. The following factors will rapidly decrease the fungicide efficacy of contact materials by physically removing them from plant surfaces:

- *Plant growth*
- *Precipitation*
- *Volatilization*
- *Mowing*

Penetrant fungicides are another group of fungicides that are described under the general heading of systemic. These compounds are absorbed into the plant tissue and kill the target pathogen. Because they are most effective when absorbed into plant tissues, systemic fungicides should not be applied to dormant turf or severely injured turf that is unable to effectively absorb the fungicide. Systemic fungicides can function in two ways in that they can kill pathogens on the tissue surface as well as inside the plant. Systemic fungicides tend to offer a longer period of protection against infection, because they are able to reach interior portions of the plant. Table 21.1 lists some common turfgrass fungicides, their chemical families, and their contact/systemic mode of action.

Within the penetrant or systemic group are three sub groups. First, there is the localized penetrant group, which enters the tissue and stays there to control pathogens near the site of fungicide absorption. Examples of localized penetrants are:

- *Vinclozolin*
- *Iprodione*
- *Propamocarb*

Second, there is the acropetal penetrant group. These compounds only move in the water conducting tissue (xylem) of the plant. Examples of acropetal penetrants are:

TABLE 21.1 Common Fungicides Used in Turf Management.

Fungicides	Trade names	Fungicide Families	Contact/Systemic
Azoxistrobularin	Heritage	Strobularin	Systemic
Chlorothalonil	Daconil, Echo, Concorde	Substituted Aromatic	Contact
Cyproconizole	Sentinel	Triazole	Systemic
Etradiazole	Koban	Thiadiazole	Contact
Fenarimol	Rubigan	Pyrimidine/Sterol Inhibiter	Systemic
Fludioxonil	Medallion	Phenylpyrrole	Systemic
Flutalonil	Pro Star	Benzinilide	Systemic
Forsetyl-Al	Signature Alliette	Phosphonate	Systemic
Iprodione	26GT, Fungicide X	Dicarboximide	Systemic
Mancozeb	Fore	Dithiocarbamate	Contact
Mefenoxam/Metalaxyl	Subdue Maxx	Phenylamide	Systemic
Myclobutanil	Eagle	Triazole	Systemic
PCNB	Turfcide, Terraclor	Substituted Aromatic	Contact
Polyoxin D	Endorse	Chitin synthase inhibiter	Systemic
Propamocarb	Banol	Carbamate	Systemic
Propaconazole	Banner Maxx	Triazole / Sterol Inhibiter	Systemic
Pyraclostrobin	Insignia	Strobularin	Systemic
Thiophanate-Methyl	Cleary's 3336	Benzimidizole	Systemic
Triadimefon	Bayleton	Triazole/Sterol Inhibiter	Systemic
Trifloxystrobin	Compass	Strobularin	Systemic
Vinclozolin	Curalan	Dicarboximide	Systemic

- *Fenarimol*
- *Flutalinol*
- *Triademefon*
- *Thiophanates*
- *Propiconazole*

The third group of penetrant fungicides are systemic penetrants. These fungicides are getting the most attention in research for new product development. Systemic penetrants can move in both the xylem and the phloem, meaning they can move in any direction within the plant. Example fungicides are:

- *Ethylphosphonate*
- *Asoxystrobularin*
- *Polyoxin D zinc salt*

These materials are exciting because of both their newness and because most common pathogens have not been previously exposed to this mode of action.

Deciding between a contact and a systemic fungicide usually is not an either/or case. In that light, perhaps the topic of discussion should be contact and systemic rather than contact versus systemic. The two types of fungicides are partners in turf management and should be used in conjunction with one another. Rotation between and use of these fungicides together are effective means for combating disease development. It is always a wise practice, for turf managers planning a disease control strategy, to alternate between fungicide families, rotate between modes of action, and tank mix contact and systemic fungicides for enhanced control. Rotation of fungicides follows the same logic that was described in the previous chapter on herbicides.

Disease pathogens, like other pests, can become resistant to fungicide applications. University research and field experience have

> **TURFTIP**
>
> The two types of fungicides are partners in turf management and should be used in conjunction with one another. Rotation between and use of these fungicides together are effective means for combating disease development.

clearly shown this to be true. Through use of fungicide rotation and tank-mixes of different products, resistance can be avoided and control can actually be enhanced. The theory is that contact and systemic fungicides, applied together, can be collectively more effective than if each were applied separately. It is believed that the contact fungicide makes the target disease-causing organism more vulnerable to the systemic fungicide. This approach is particularly effective for turfgrass disease control programs under a high degree of maintenance, such as on golf courses, where disease control is at a premium.

Fungicide formulations can be numerous. They come packaged as:

- *Dispersible granules*
- *Emulsifiable concentrates*
- *Wettable powders*
- *Water dispersible bags*
- *Flowables*
- *Granular products on inert carriers*

Deciding among these formulations will depend upon your situation. Many of the products you have to choose from are designed for liquid sprays so that even coverage on turfgrass surfaces can be achieved. This ensures that the contact activity of most fungicides will take effect. For systemic fungicides, placement of the fungicide can affect its performance. This decision involves understanding the characteristics of the disease you are targeting. For example, pythium-specific fungicides such as carbamates, acylalanines, and ethyl phosphonates are systemic fungicides that are typically absorbed by the plant either through the leaf crown or the roots. However, to maximize control, it is best that they be irrigated into the soil to enter through the root system. In a case such as this, tank mixing a contact and a systemic fungicide may not be recommended since the necessary irrigation that moves the systemic material where it needs to go would impact the effectiveness of the contact fungicide.

Follow label instructions and think of such issues before any fungicide application is made. Although most fungicide products are manufactured for sprayable functions, many fungicide formulations are also available to turf managers in granular or particle-based products. These products are excellent tools to have around because they are easy to apply for spot treatments, treatments during winter thaws, or during emergency situations. For example, when overseeding small but critical turf areas, a granular pythium fungicide product is a very useful tool.

Lastly, recognize that certain turf fungicides can have detrimental side effects on desirable grasses. Although fungicides aren't designed to kill plants like many herbicides, some fungicides have growth retarding properties. This effect may be more pronounced

---

TURFTIP

Tank mixing a contact and a systemic fungicide may not be recommended since the necessary irrigation that moves the systemic material where it needs to go would impact the effectiveness of the contact fungicide.

> **TURFTIP**
>
> Although most fungicide products are manufactured for sprayable functions, many fungicide formulations are also available to turf managers in granular or particle-based products. These products are excellent tools to have around because they are easy to apply for spot treatments, treatments during winter thaws, or during emergency situations.

on certain turf species and cultivars and can be even more detrimental if materials are applied when the plant is under stress. Awareness of a fungicide's potential for this effect and of the sensitivity of a particular turfgrass species can help you avoid unwanted difficulties. Fungicide formulations can also affect turfgrass health. Contact materials are designed to coat the surface of treated turfgrass leaves, invariably reducing the water-conducting abilities of the plants for a period of time. Applied under very hot conditions, these fungicides can impair turfgrass transpiration and cooling,

> **TURFTIP**
>
> Contact materials are designed to coat the surface of treated turfgrass leaves, invariably reducing the water-conducting abilities of the plants for a period of time. Applied under very hot conditions, these fungicides can impair turfgrass transpiration and cooling, resulting in unwanted environmental stress.

resulting in unwanted environmental stress. Although somewhat rare, this type of detrimental circumstance can be avoided by using contact materials with caution, particularly in extreme heat and on sensitive turf areas like putting greens.

CONTROL STRATEGIES FOR TURFGRASS DISEASES

For a turf manager, there are many fungicide options and alternatives when planning a disease management program. Accurate disease identification is the most important factor when selecting a fungicide to apply for correcting a disease problem on turfgrass. Furthermore, focus on establishing a threshold for disease presence before intervention and what is an acceptable level of appearance to customers, property owners, club members, and officials. Be sure to clearly and openly discuss your turf management program with the main goal being education and communication of these threshold factors. It is important to understand the concept of economic outcome and effects as it relates to threshold or how much disease we can live with. If disease pressure is present to the point of significant turf loss and damage, where replacement is above and beyond routine maintenance, this could be considered economic loss. When a turf manager must apply fungicides at curative rates to correct a situation, and then absorb the costs of turfgrass renovation, this can be considered economic loss.

If healthy, aesthetically pleasing turf is desired, a turf manager must implement some form of control strategy to elude pathogen infestation. However, cost-based approaches to fungicide use can vary. A golf course superintendent would take a different approach than a highway superintendent regarding turf management. Fine turf managers must place more emphasis on turf aesthetics, and disease control programs will mirror this objective. Aggressive cultural practices and a preventative fungicide program are common com-

> **TURFTIP**
>
> Accurate disease identification is the most important factor when selecting a fungicide to apply for correcting a disease problem on turfgrass. Furthermore, focus on establishing a threshold for disease presence before intervention and what is an acceptable level of appearance to customers, property owners, club members, and officials.

ponents of fine turf management, where little or no turf loss can be tolerated. Lower maintenance turf can usually tolerate lesser emphasis on disease control, saving these measures for extreme cases of need.

Understanding the three main components of the disease triangle is an important concept to grasp before implementing a disease management program. There are three segments that must each be in place for disease infection to occur:

- A susceptible host plant
- An active pathogenic entity
- A favorable environment for disease development

Understanding the triangular relationship of turfgrass diseases means first understanding the diseases and their life cycle patterns. Pathogenic spores naturally occur in the soil. If left alone and unchecked, these spores can perpetuate the disease cycle. Spores invade plant cells through multiple ports of entry, infecting the plants with detrimental pathogens. Symptoms appear on infected plants. The desirable host will essentially get sick and die from disease. If left unchecked and untreated, the pathogen will coalesce to involve larger areas of turf. Disease spores overwinter in the soil and fester in the thatch/soil interface, feeding on dead and dying organic matter. Understanding this phase of disease development will assist in developing a plan for well-timed control strategies.

The second key component of the disease triangle is the susceptible host. In this regard, positive identification of the turfgrass species you are dealing with is a must. It is important because some turfgrass species or cultivars are more susceptible to pathogens than others and some diseases are species specific. Annual bluegrass is a common turfgrass species that turf managers are in a constant struggle to maintain. This plant at any mowing height and almost any degree of maintenance is succulent, weak, and can be susceptible to many diseases such as anthracnose. Avoiding excessive maintenance to annual bluegrass can help reduce disease potential. Perennial ryegrass is also at particular risk from either gray leaf spot or pink snowmold. Dollar spot causes high levels of damage to bentgrasses on golf courses, while spring dead spot exclusively damages bermudagrass in southern regions. It is wise to design a fungicide program that suits the budget, aesthetics, and health requirements of the turfgrass stand being maintained. Choose the correct grass cultivar for the site. Seed producers, sod growers, and researchers are continually producing and introducing grass cultivars that are bred to withstand pests or are more tolerant to the sorts of pests that love to feed on turfgrass plants.

The last component of the disease triangle is environment. Weather significantly affects the presence of disease-causing organisms, as many of them are seasonal in nature. Humidity, soil temperature, sunlight, ambient temperatures, and air movement are important factors for disease development. Developing a weather-based approach to disease control is best based upon past experience. Historical documentation of growth habits and past weather conditions is also helpful, as is being able to react to weather changes that may induce disease development. To give you an idea of how the three components of the disease triangle work together, let's look at some example situations.

Example 1. With overnight temperatures of 68°F, 80 percent humidity, and no breeze, a practitioner should deem this representative of a prime climate for many diseases. Strains of pythium, dollar spot, brown patch, and anthracnose should be anticipated, control measures should be initiated, and sites that are prone to infection should be monitored.

Example 2. In the Great Lakes Region, it is not uncommon for pink snowmold to appear with air temperatures as high as 65°F under overcast skies and damp conditions. However, when air temperatures climb above 70°F with low humidity and adequate air movement, this disease activity will cease. By monitoring weather factors, there may be no need for the turf manager to intervene with a fungicide application. Conversely, gray snowmold is a disease where extended periods of snow cover are needed. If extended periods of snow cover are common to your area, the proper control strategy in this case would be a preventative fungicide application if economic loss of turf is to be avoided.

Example 3. On the Fourth of July, a thunder storm drops 2.5 inches of rain on already saturated soil. In this case, fine turf managers should be prepared to make fungicide applications immediately to avoid threats of diseases like brown patch, summer patch, or pythium.

Example 4. Overseeding of warm-season turfgrasses during periods of dormancy presents a different set of circumstances for the manager of fine turf. Germinating seedlings and immature turf are susceptible to many diseases. Newer varieties of perennial ryegrasses are constantly being introduced for disease resistance but pythium blight and gray leaf spot should never be taken lightly in these circumstances. Overseeded putting greens in warm-season turfgrass areas have the propensity to become infected with seedling diseases. *Poa trivialis* and bentgrasses have been known to contract take all patch, pythium, and *Helminthosporium* strains of pathogens. Carefully prescribed and timed applications of fungicides are necessary because the threshold on these turf areas is very low. There is a substantial financial investment at stake and plant protection from disease infection and potential loss could be potentially devastating. It is critical to keep this overseeded turf healthy.

Example 5. Many disease symptoms resemble one another. For example, pythium infection can resemble dollar spot incidence. The choice of control strategy in this case is critical because of the similarity of symptoms at the early stages. A turf manager could be applying fungicides for dollar spot when in reality, pythium infection is present. In this case the pythium could become more severe and, as a result, time and money could be wasted because of an improper diagnosis. Comparing observed symptoms to ambient weather conditions in this case can help determine the true identity of a disease.

Turfgrass cultural practices like mowing, irrigation, and fertilization can be contributing factors to disease infestations. Sufficiently irrigated home lawns that are fertilized adequately and mowed above two inches in height are less likely to develop disease problems. Lower mowing heights require higher degrees of turfgrass maintenance and are at greater risk for diseases. Raising the mowing height can be an effective disease control strategy that a turf manager can utilize but may be difficult or unacceptable in some circumstances. For example a three-inch mowing height may not be acceptable for a football field at the professional sports level but is satisfactory for recreational usage or a home lawn. Golf course putting greens require more intense mowing to achieve desired results, and are thus are more

---

TURFTIP

Turfgrass cultural practices like mowing, irrigation, and fertilization can be contributing factors to disease infestations.

likely to develop disease problems. However, raising mowing heights above ¼ inch on golf course greens would be not acceptable. Equipment care is as important a part of mowing as the mowing height, when it comes to diseases. Sharp and clean mowing equipment not only is a sign of sound and well maintained machinery, but is vital to curb spread of disease. A fungicide application should be scheduled in conjunction with invasive maintenance practices.

Excessive irrigation is usually a recipe for disease, especially during extreme hot, humid conditions. Moderate irrigation practices that do not result in moist turf for extended periods of time can help reduce this risk. Overfertilization can sometimes promote diseases by producing a succulent plant that may be less hardy than those receiving fewer nutrients. However, some diseases thrive in areas that receive low fertility (e.g. dollar spot) so provision of fertilizer at times of high disease pressure can help minimize the threat. Understanding the nature of the pests you encounter will help you make proper fertility decisions.

Other cultural issues to consider when designing a disease management program are core cultivation, verticutting, top dressing, and soil conditioning procedures that are invasive and injurious to turfgrass plants. The wounds produced during these aggressive practices could weaken the grass plants and provide points of entry for pathogens or other destructive pests. Turf experts often recommend

TURFTIP

Turf experts often recommend to golf course superintendents to avoid topdressing or other aggressive activities during periods of high stress because the sand granules and consequent brushing could damage leaf tissue and accelerate disease activity.

to golf course superintendents to avoid topdressing or other aggressive activities during periods of high stress because the sand granules and consequent brushing could damage leaf tissue and accelerate disease activity. Timing of these abrasive practices has to be judicious during periods of high disease pressure.

When selecting fungicides for a particular problem, always know what is registered for use in your region or state. A majority of fungicides are restricted use and should only be handled by certified pesticide applicators. Judicious care and handling by professionally trained applicators utilizing personal protection equipment is a wise and necessary practice (Figure 21.21). Reliable and properly calibrated spraying equipment is always important, whether it is a sprayer or a spreader, in order to make the task more efficient. Also pay attention to the specific spraying instructions for fungicides. Contact materials need to be applied in at least 40 gallons

FIGURE 21.21 A sprayer used for a fungicide application, being operated by an applicator with proper protective equipment.

of water per acre to properly coat the plant and to protect it from disease. Some systemic fungicides need to be irrigated into the root zone to maximize plant uptake. The newer forms of strobularin fungicides require tank-mixing with a fungicide that has a different mode of action because repeated use could promote other diseases like dollar spot. Pay attention to these sorts of details in order to get the most out of your applications.

When it comes time for a turf manager to implement a reliable disease management plan, the best defense is a good offense when it comes to fungicide selection and application in turf management. Table 21.2 lists some common fungicide products and the diseases they control. The primary control strategy is a good, healthy turf achieved by means of sound cultural practices, balanced nutrition,

TABLE 21.2 Common Turf Fungicides and Key Diseases They Control.

Fungicide Trade Name	Diseases Controlled
Heritage	Anthracnose, brown patch, fairy rings, gray leaf spot
Daconil, Echo, Concorde	Dollar spot, brown patch, fusarium blight
Koban	Pythium
Rubigan	Take all patch, spring dead spot
Forsetyl-Al	Pythium, summer stress
26GT	Brown patch, dollar spot, helminthosporium
Subdue	Pythium, seedling damping off
Turfcide 400	Snow mold, pink and gray
Endorse	Anthracnose, brown patch, zoysia patch
Banol	Root pythium
Banner MAXX	Brown patch, dollarspot, summer patch
Bayleton	Anthracnose, brown patch, dollar spot,
Compass	Anthracnose, brown patch, summer patch
Cleary's 3336	Dollarspot, fusarium blight, red thread
Curalan	Brown patch, dollar spot, pink snow mold

and consistent monitoring. When your acceptable disease threshold is threatened by pathogen activity or weather conditions, the control phase of disease management must be introduced.

In most cases, it is best not to delay disease control action until visual symptoms of disease are present. By that time, it may be too late to reverse the pathogenic damage. Early implementation of a fungicide application can halt or even reverse the trend of a destructive disease infection. Be sure to document and record the site conditions, the environmental conditions, and the materials applied, because concise record keeping, continued monitoring, and success evaluation will assist in better decision making for future practices, especially if a particular disease is a persistent problem. In summary, a manager of fine turf can use the following five-step plan when designing and implementing a fungicide program:

1. Early detection. Through scouting and monitoring, a turf manager can establish indicator areas to watch for pockets of disease development. Poor sites should be constantly observed.

2. Early diagnosis. Positive identification, by utilizing the information from signs and symptoms that are present to catch the pathogen as early in development and infection as possible, will make control more effective. A diagnostic lab, a plant pathologist, or a turf expert would be helpful.

3. Early intervention. Applying a corrective fungicide at label rates with reliable equipment at the earliest stage of disease development provides the best control.

4. Evaluate. Follow up monitoring of the infected site is important. Here a turf manager can observe if the desired results were achieved by the fungicide application and the pathogen activity has ceased.

5. Historical data. Record keeping is invaluable in designing and implementing future disease management practices. It will provide a map for future scouting and monitoring.

By following these key steps and committing to good comprehension of both diseases and their control methods, you will be able to make good judgements and optimize turfgrass disease control at your facility. Develop a sound disease management program through education. Be consistent, be proactive, and stick to what works. The next chapter takes this topic one step further to include disease problems and control strategies in landscape plants. Let's continue, shall we?

INDEX

2,4-D, 378
AVB see Atmospheric vacuum breaker
CEC see Cat ion exchange
EC see Electrical conductivity
EPA see Environmental Protection Agency
GCSAA, see Golf Course Superintendents Association of America
HOC see Height of Cut
IPM see Integrated Pest Management
LCO see Lawn care operator
LLC see Corporations, limited-liability
MCPP, 378
MSDS see Material Safety Data Sheet
NTEP see National Turfgrass Evaluation Program
PGR see Plant Growth Regulator
pH, 334
PPE see Personal protective equipment
PVB see Pressure vacuum breaker
RCV see Remote control valve
RP see Reduced pressure backflow device
SAD see St. Augustine decline
SAR see Sodium Absorption Ratio
TDS see Total dissolved solids

Aeration, 14 – 16, 199, 201, 287, 472 - 473
  aerator, 117 – 118
AerWay, 323
Agrostris Palustris, (Bentgrass), 49
Agrostris Stolonifera, (Bentgrass) 49
Anatomy of plants, 18 – 22
Annual Bluegrass (Poa Annua), 42 – 43, 291 – 303, 407 – 409
Annual ryegrass (Lolium multiflorum), 45
Annual weeds, 399 – 409
  reproductive phase, 399 - 400
  vegetative stage, 399 - 400
Anti-siphon protection, 219 - 221
Athletic field maintenance, 75 – 76
Atmospheric vacuum breaker (AVB), 220

Backpack sprayer, 361
Baits, 486
Banner Maxx (trademark), 381, 382
Beard, Dr. James, 272
Bentgrass, 289 – 290
Bermuda grass (Cynodon Dactylon) 55 – 57
Bids and contracts, 171 – 178
Biennial, 409 - 410
Biotechnology, 339 – 340, 342
Black ataenius, 385, 387
Blend, 39 – 40
Bluegrass, 290 – 291
Broadleaf weeds, 418 – 420, 421 – 443
Buffalo grass, 59
Buffer strips, 524 - 525
Bulbs, 416 - 417
Busch Stadium, 56

Cannestra, Joe, 317
Cat ion exchange (CEC), 117, 234
Cemeteries, 172 - 173
Centipede grass (Eremochloa ophiuroides), 58
Chemicals, 346 – 356, 475 - 478
  disposal, 366 – 369
  environmental impact, 521 - 535
  granular, 347 – 348
  safety, 504 – 516
  understanding labels, 516 – 521
Chinch bugs, 385
Chlorophyll, 26
Cities, 194 – 196
Coarse free and fines free, 320
Collars and fringes, 85
College maintenance, 73, 75
Commercial lawn care, 167 – 170
Contact pesticides, 356
Controllers, 211 - 213
Cool-season fertilization 95 – 100
Cool-season grasses, 34, 35, 36 – 51
Corms, 416 - 417
Corporations, 184 – 185
  limited-liability (LLC), 184
Crabgrass, 375 – 377, 401, 447

Creeping bentgrass, 49 – 51
  Agrostis Palustris, 49
  Agrostris Stolonifera, 49
  Penncross, 51
  Seaside, 50
Cultivar, 40
Cultivation, 267, 308 - 309
Cultural practices, 81, 269
  special residential, 163 – 165
Cutworms, 385
Cynodon Dactylon (Bermuda grass), 55 - 57

Daconil (trademark), 380, 382
Damage, turfgrass, 31 – 32, 198 – 206
  winter, 292 – 293
Dandelion, 441 - 442
Deep tine aerator, 201, 202
Dicamba, 378
Dicot weeds, 375 – 378, 419
Disease, 55, 56, 378 – 383, 531 - 533
Disposal of chemicals, 364 – 369
Downy mildew, 382 - 383
Drain valves, 218 - 219
Drip irrigation systems, 109 – 110, 221 – 222
Duich, Dr. Joseph, 323
Dursban, 387
Dylox (trademark), 384, 387

Ecosystem, role of plants, 27
Electric pump sprayer, 361 - 362
Electrical conductivity (EC), 335
Electrically actuated hydraulic valve, 212
Emission devices, 222 - 223
Endophyte, 47
Environmental Protection Agency (EPA), 339
Eremochloa ophiuroides (Centipede grass), 58
Erosion control, 192
Establishment, of turfgrass, 113
  of warm season grasses, 141 – 143
Expenses, 173 - 175

Fairways, 85, 279 – 280
  conversion, 280 – 288
  maintenance, 288 – 303
  Poa annua, 291 - 303
Fertility, 3 – 9

Fertilizer, 88
  application methods, 94 – 101, 471 - 472
  types, 88 – 94
Festuca arundinacea (Tall fescue), 46 – 48
Fine fescues, 48 – 49
Fines free, 320
Fixed spray, 221
Flail mower, 82, 83 – 84
Flood bubblers, 221
Floral bracts, 24
Flyking, 40
Fumigants, 360, 482 - 483
Fungicides, 360
Fungus see Endophyte

Gas-powered, 362 – 363
Golf course management, 243 – 248
  beautification, 253 – 254
  budgeting and finance, 256 - 261
  careers, 63 – 68, 69, 70
  irrigation, 104 – 105, 106, 108
  maintenance, 247 – 256
  mowing, 82 – 87
  raising money, 254 - 255
  water supply, 250
  winter play, 277 - 278
Golf Course Superintendents Association of America (GCSAA), 65 – 66
Granular, 347 - 348
Grass bunker, 249
Grasses, 418, 420 – 421, 443 - 467
Gray leaf spot, 382
Green speed, 269 - 276
Greens, 84
Ground squirrels, 392 - 393
Grubs, 384 - 385

Handheld equipment, 361
Height of cut (HOC), 271
Herbicides, 357 – 387
Heritage (trademark), 382
High yielding cropping systems, 530
Homeowners, 148 – 153 see also Lawn care operator
Humans, effect of on turfgrass, 31 – 32
Hydraulic clock, 211
Hydrozoning, 232, 233

Inflorescence, 22, 24, 25
Insect control, 344 – 348
Insect damage, 530 - 531
Insecticides, 357 – 359, 383 - 384
Integrated Pest Management (IPM), 110 - 111
Irrigation, 101 – 110, 209, 265 – 266, 307, 471 - 472
  commercial systems, 223 – 229
  drive-by maintenance, 224 – 226
  conservation, 226 - 229
  controllers, 211 – 213
  effects of water, 336 - 337
  efficiency, 237 – 241
  management, 229 – 241
  scheduling, 236 – 237
  system hardware, 209 – 223
  water quality, 328 - 336

Jamestown, 37

Kentucky, 31, 48
Kentucky Bluegrass, 36 - 42

Lakeshore dune sand, 320
Landscape design, 193 – 194
Latham, Jim, 317
Lawn care operator (LCO), 68 – 73
  appearance, 161 - 162
  billing and invoicing, 156 – 159
  charges and finance, 159 – 161
  commercial, 167 - 170
  maintenance schedule, 153 - 155
  residential, 145 – 148
  special cultural practices, 163 - 165
Lawn scalping, 18
Leaching, 526 - 527
Lemma, 24
Licenses and taxes, 183 – 186, 332 - 334
Light, 28, 31
Liquid fertilizer, 99
Lolium multiflorum, 45
Lolium perenne (Perennial ryegrass), 43 - 44

Madison, Dr. John, 315, 318
Material Safety Data Sheet (MSDS), 504 - 516
Meyer, 52
Mixture, 40 – 41

Moguls, 249 - 250
Moisture, 29 – 30
Mole, 388 – 389, 390 - 392
Mole cricket, 387
Monocot weeds, 375
Mowing, 82 – 87, 178 – 180, 196, 264 – 265, 269 – 271, 306, 470 - 471
  areas, 84 - 87

National Turfgrass Evaluation Program (NTEP), 276
Nitrogen, 8 – 9, 94, 99, 100, 101
Non-native species, 372 - 373
Nozzles, 105
Nutlets, 416 - 417
Nutrition, 266 – 267, 273 – 275, 307 - 308

Overseeding, 120 – 121
Ozaukee Country Club, Mequon, Wisconsin, 273

Palea, 24
Panicle inflorescence, 25
Pedicels, 24
Penncross, 51
Pennstar, 40
Per-time contracts, 178
Perennial ryegrass (Lolium perenne), 43 – 44
Perennial weeds, 411 – 418, 452 - 467
Personal protective equipment (PPE), 348 - 350
Pest control, 267, 310 – 311
Pest management, 371 – 373
Pest thresholds, 528 - 535
Pesticide application
  equipment, 360 - 364
Pesticide treatments, 186 – 189
  application, 339
  granular, 347 – 348
  label, 350 - 356
  liquid, 348
Pesticides, 475 – 478
  alternative fates, 524 - 526
  contact, 356
  dusts, 481, 483 – 486
  environmental impact, 521 - 535
  formulations, 478 - 516
  fumigants, 360

fungicides, 360
granules, 481, 486 – 490, 502 - 503
herbicides, 357 – 387
insecticides, 357 – 359
liquids, 492 – 498, 499, 502, 503 – 504
mobility, 526 - 528
powders, 481, 490 - 491
restricted-use, 361
rodenticides, 360 – 361
safety, 504 - 516
systemic, 356
types, 356 – 360
understanding labels, 516 – 521
Phloem, 26
Phosphorous, 94
Photosynthesis, 26
Plant Growth Regulator (PGR), 276
Plant science, p.63
Plants, anatomy, 18 - 22
  responses, 17 – 18
  tissue, 26
Plugging, 141 - 142
Poa Annua *see* Annual Bluegrass
Pocket gopher, 390
Pollution, reduction of, 196, 251
Post-emergence chemicals, 375, 377
Potassium, 94
Power rake, 118 – 120
Practice tee, 311 - 312
Prairies, 58
Pre-emergent chemicals, 376, 377
Pressure vacuum breaker (PVB), 220
Pumps, 105

Quick coupling irrigation system, 210
Quick coupling valves, 216

Raceme, 24
Raleigh, 58
Reduced pressure backflow device (RP), 221
Reel mower, 82, 83
Reestablishment *see* Seeding
Remote Control Valve (RCV), 213 – 215, 220
Renovation, 114 – 124
  seeding, 120 - 124
Research, 76 – 77
Restricted-use pesticides, 360

Rhizomes, 37, 414 - 416
Rhizoctonia (Brown patch), 279 – 380
Rodenticides, 360 - 361
Rolling the greens, 271 – 273
Roots, 414
Rotary mower, 82, 83
Rotary spreaders, 362
Rotor spray, 221
Rough-stalked Bluegrass, 42
Roughs, 85
Roundup (trademark), 121, 340, 342
Ryegrass, 40

St. Augustine decline, 58
St. Augustine grass (Stenotphrum Secundatum), 57 – 58
  Raleigh, 58
Safety, 180 – 183, 345 - 356
Salt-related problems, 331 - 336
Sand areas, 248 – 249
Scalping, 83
Seaside, 51
Seasonal contracts, 177 – 178
Sedges, 418, 420 – 421, 443 - 467
Seeded German Bentgrass, *see* Creeping bentgrass
Seeding, 113 – 138, 286 – 287
  new installations, 125 - 136
Seeding structures, 22 – 26
Sessile, 24
Sevin (trademark), 385
Sickle mower, 83
Silt dressing, 318
Skid-mounted sprayer, 363 - 364
Soaker hoses, 109 – 110
Sod cutters, 138
Sod production, 77
Sodding, 138 – 141, 201
Sodium Absorption Ratio (SAR), 335
  adjusted SAR, 335
Soil pH, 3 – 6
Soil textural triangle, 12 - 13
Soils, 1 – 3, 233 – 237, 325 - 328
  aeration, 14 – 16
  moisture, 273
  salt-affected, 332 – 334
    saline, 332, 334
    saline sodic, 332 - 333
    sodic, 332 - 333
  sampling, 327 – 328

source of plant food, 326 - 327
types, 9 – 13
Solid set irrigation system, 210
Spikelets, 24, 25
Sprigging, 142 - 143
Spring dead spot, 55, 56
Sprinklers, 105
Stenotphrum Secundatum (St.
    Augustine grass), 57 – 58
Stolons, 58, 414 - 416
Straw, 130 – 132
Subdue Maxx (trademark), 383
Summer annual weed, 400 – 404,
    445 - 448
Surface irrigation system, 109
Surface runoff, 526
Systemic pesticides, 356

Take-all patch, 380 - 381
Tall fescue (Festuca arundinacea),
    46 – 48
Taproots, 414
Tees, 85, 305
    maintenance and cultural practice,
        305 - 311
Temperature, 28 – 29
Testing, 342 - 344
Thatch, 53
Tiff series, 57
    Tiffgreen, 57
    Tiflawn, 57
    Tifway, 57
Tillers, 37, 407
Tissue, plant, 26
    phloem, 26
    xylem, 26
Topdressing, 268 – 269, 307 – 308,
    315 – 316
    benefits, 318 – 320
    incorporating sand, 321 - 324
    sand, 317 - 318
Total dissolved solids, (TDS), 335
Transition zone, 34 – 35, 52
Trimec (trademark), 378
Tupersan, 133
Turf management, 208
    science of, 264
Turfgrass,
    careers, 61 - 63
    damage to, 31 – 32, 198 - 206

environment, 28 – 32
establishment, 113
in cities, 194 - 196
in landscape, 191
seeding, 113 - 138
zones, 33 - 35

Vacuum relief valves, 216, 217
Valves, 213 – 219
Vertical mower, 322
Volatility, 524

Warm-season fertilization, 100 – 101
Warm-season grasses, 33 – 34, 52 – 59,
    141 – 143
Water, ion toxicity, 336
Water pressure, 105
Weed control, 373 – 378
    cultural practices, 470 - 474
Weed treatments, 346
Weeds, 395 – 398, 533 - 535
    annual, 399 – 409
    biology, 418 – 421
    ecology, 467 - 474
    life cycles, 398 – 414
Whorf, Dr. Gayle, 276
Wilting, 16
Wind, 30 – 31
Winter annual weeds, 404 – 410,
    445 – 446, 450 - 452

Xeric irrigation system see Drip irrigation
Xylem, 26

Zones, turfgrass, 33 – 35, 52
Zoysia grass, 52 – 55, 382 – 383,
    387 - 388
    Meyer, 52
    Zoysia japonica, 52
    Zoysia matrella, 52
    Zoysia tenuifolia, 52